Mathematical Card Magic

Fifty-Two New Effects

Mathematical Card Magic

Fifty-Two New Effects

Colm Mulcahy

CRC Press
Taylor & Francis Group
Boca Raton   London   New York

CRC Press is an imprint of the
Taylor & Francis Group, an **informa** business

AN A K PETERS BOOK

All photographs in the text taken by Dan Bascelli of Spelman College.

CRC Press
Taylor & Francis Group
6000 Broken Sound Parkway NW, Suite 300
Boca Raton, FL 33487-2742

© 2013 by Taylor & Francis Group, LLC
CRC Press is an imprint of Taylor & Francis Group, an Informa business

No claim to original U.S. Government works

Printed on acid-free paper
Version Date: 20130507

Printed in India by Replika Press Pvt. Ltd.

International Standard Book Number-13: 978-1-4665-0976-4 (Hardback)

Library of Congress Cataloging-in-Publication Data

Mulcahy, Colm Kevin, 1958-
   Mathematical card magic : fifty-two new effects / Colm Mulcahy.
      pages cm
   "An A K Peters book."
   Includes bibliographical references and index.
   ISBN 978-1-4665-0976-4 (hardback : acid-free paper)
   1. Game theory. 2. Card tricks. I. Title.

QA269.M85 2013
510--dc23
                                        2013010461

Visit the Taylor & Francis Web site at
http://www.taylorandfrancis.com

and the CRC Press Web site at
http://www.crcpress.com

Dedicated with great fondness to the memory of departed friends
Ron Keith, Steve Sigur, Monika Decker,
Chris Orrall, Tom Rodgers, and John Beechinor,
who displayed such grace and dignity playing the cards
they were dealt over the past six or seven years,

and to
Vicki, Ann, and Molly
for being my loveliest assistants down through the decades.

Contents

Colm: Cool and Collected

Among magicians, "mathemagic" is most often the object of ridicule. Ask the average magician for his or her thoughts about this type of conjuring, and you'll get an answer that is virtually guaranteed to include the words "long" and "boring." And there's a great likelihood that there will also be a disdainful mention of dealing piles of cards.

As if that's a bad thing. Curiously, if you ask those same magicians to tell you the greatest card trick of all time, a solid majority will name "Out of This World," a wonderful piece created by Paul Curry in 1942 that involves the dealing of 48 cards into piles—dealt by the spectator, no less. But magicians are easily distracted, and because the method for the Curry trick is not overtly mathematical, they forget that it negates their criticism.

Among mathematicians, the field has fared somewhat better, although there is a sort of intellectual pecking order, wherein math-derived magic is frequently dismissed as trivial unless it is based on some sort of exceedingly complex methodology, which in turn often leads to the very type of material that magicians deride as "long" and "boring" . . . .

None of this, of course, has disrupted the constant flow of good mathematical magic over the past couple of millennia. And, happily enough, we currently find ourselves in a quite healthy phase of that history.

For a variety of reasons, most of them sociological and not particularly germane to this discussion, in the early 1900s there began a new blossoming of mathematical magic, via such brilliant inventors as Charles T. Jordan, Stewart James, Robert Hummer, and Martin Gardner. The last named, starting around the midpoint of the twentieth century, became the modern era's greatest champion of recreational math. Through his hundreds of articles and books, Martin introduced several generations to the glorious intersection of math and magic.

You'll meet the above-mentioned people, plus many other worthwhile thinkers, in the book that you are about to read. My friend Colm Mulcahy has joined those ranks. His bi-monthly *Card Colm* has appeared on the website of the Mathematical Association of America since 2004, and via those entries he has provided a splendid stock of math-based card magic. Some of it has been raised to impressive heights by standing on the shoulders of his predecessors, and some has managed to reach those heights without external support—which is to say, he's come up with both clever variations on existing ideas and ideas that are brand new and very exciting.

If you have been a *Card Colm* reader, you already know the quality of his imaginative output. You will find much of the best material from that column here in this collection, revised and improved. (Colm is one of those fellows who simply never stops tinkering.) You will also encounter a wide variety of previously unreleased material, with some lovely surprises.

It's delivered in the author's customary style: When Colm's words dance across the page, it's a true reflection of the twinkle in his eye. Colm may have relocated to the United States over three decades ago, but the sense of whimsy he brought along from Ireland continues unabated.

All of which goes to say that the book you're about to read is a lot of fun. Indeed, more fun than this introduction. So, let's get our priorities in order. I'll stop here; you grab a deck of cards and jump in.

Max Maven
Hollywood, California
June 2013

Hit the Deck Running

Mathematical card effects have been around for hundreds of years, and they offer both the novice and experienced magician an opportunity to entertain with a minimum of props: *a simple deck of cards does the trick every time!*[1] We offer thirteen chapters, named after the card values from Ace to King. The last three chapters focus on two-person mathemagic, and are the true royalty of mathematical card magic. Each chapter contains four card effects, generally starting with simple applications of a particular mathematical principle and building up to more complex ones.

Many of the principles are new and have not appeared in print before. For a list of all key principles, in the order in which they appear, see page xxiii. Those believed to be original principles are flagged with a ▶◀ symbol. For an alphabetical list, see the index, under *principle*.

The effects have names—the logic of which should make sense once you have read through the descriptions—but please note that in many cases you may not want to share those titles with your audience, for various reasons. Each effect is also tagged with a specific card name determined by its position within its chapter, using the CHaSeD suit order convention: Clubs, Hearts, Spades, and Diamonds. For instance, the effects in the first chapter are assigned the tags A♣, A♥, A♠, and A♦, those in the second chapter are tagged 2♣, 2♥, 2♠, and 2♦, and so on. The end of each effect—and the end of some discussions between effects—is marked with a centered, set-off display as follows:

There is one further use of suit symbols, discussed in "The Ratings Game" to follow.

---

[1] Okay, on rare occasions you'll need two or three decks.

Over half of the chapters are independent of each other, and may be read in any order. Note, however, that Chapters 5 and 9 are extensions of Chapter A, and Chapters 7 and 8 both assume familiarity with Chapter 6.

Practice a handful of the chapter-opening effects, and in no time at all you'll establish your reputation as a bona fide mathemagician, amazing and amusing friends, colleagues, and family alike. Delve a little deeper, and the mathematics gets interesting.

Here are four surefire crowd pleasers. As their placements within their respective chapters reveal, two can be done more or less off-the-bat, and two require significant practice to get right. The first effect below—indeed the first one in the book—invokes a new dealing and dropping principle that can be explained in several ways: pictorially, by using permutations, or in terms of a fundamental decomposition of a packet into three parts.

Three Scoop Miracle—Done Magic Before? (A♣)

Hand about a quarter of a deck of cards to a spectator, and ask her to shuffle freely. Take those cards back, and mix them further in your hands as you ask the spectator what her favorite ice cream flavor is. Let's suppose she says, "Chocolate."

Deal from the packet to the table, one card for each letter of "chocolate," then scoop those up with one hand, commenting that this represents one scoop of ice cream, and with the other hand, drop the remainder on top ("as a topping"). Repeat this spelling (and scooping and topping) routine twice more, for a total of three times.

Emphasize how random the dealing was: since the cards were shuffled repeatedly and the spectator named the ice cream flavor. Ask her if she has done magic before. Regardless of the answer, now ask her to press down hard on the top card of the packet on the table, requesting that it be miraculously turned a specific card, say, the Four of Diamonds. When the card is turned over, it is seen by all to be the desired card. Congratulate the spectator on a job well done.

The above effect is one of the few here that can be repeated, under some circumstances.[2] The same cannot be said for the next effect, which showcases your prowess in the arena of *unaddition* but requires advance deck preparation.[3] (If you insist on doing something like this twice, see the "Presentational options" listed on page 92.)

---

[2]Note that just because you can repeat something doesn't mean you should, as we caution in "Tips of the Trade."

[3]It does, however, make a nice followup to either "Alphabetical Triple Addition" (2♣) or "Subtler Bracelet" (2♥).

Little Fibs (4♣)

Give the deck several shuffles, then deal six cards face down to the table, setting the rest aside. Turn away, requesting that those six cards be thoroughly mixed up. Have any two cards selected by two spectators, who then compute and report the total of the two card values. From that information alone, you promptly name each card.

Here's one that will baffle any gamblers in the audience. As is the case for the first effect above, it can be done with any deck you are handed, and it can be repeated. In our experience, audiences find this one more intriguing upon repetition.

Poker with Any Ten Cards (3♦)

A spectator shuffles the deck, and deals ten cards face down. You glance at the card faces and pause to write down a prediction. Then the spectator is given numerous choices for distributing the cards between the two of you. Yet, without fail, not only do you end up with the winning poker hand, but your specific prediction, made in writing in advance—which may say something like, "You will have a pair of Jacks, but I'll have three Fives"—also comes true.

Finally, we have another repeatable effect that seems to defy belief, even if you are upfront about the fact that the so-called volunteer is actually your mathematically trained accomplice. It's our book closer and requires more mental gymnastics than most of the other effects.

Fitch Four Glory (K♦)

Select a volunteer and give her the deck of cards. After you have stepped out of the room, an audience member shuffles the cards and gives the volunteer any four of them. The volunteer then hides one of them and arranges the other three in a row, with some of them face down. You reenter the room, glance at the display, and soon name the hidden fourth card—even in the case where all of the cards are face down!

Those are just four of the fifty-two mathematical card effects highlighted in these pages. They—and a few others buried in here—are possible because of surprising mathematical principles; to pull them off, there is no need for you to learn fancy sleight-of-hand moves (or, for that matter, advanced mathematics). To perform most of them, you don't even have to fully grasp the underlying logic—those effects are truly self-working.

However, it's always fun and instructive when you also understand the logical underpinnings. The bulk of the mathematics encountered is easy to grasp and is explained as needed. For interested readers, additional details, background, and suggested further explorations are given.

Along the way, we slip in standard card magic themes and concepts, from card control and forcing to prediction, spelling, and mind reading, not to mention the not-so-innocent use of "lucky cards." We also suggest some routines geared specifically toward those who enjoy mathematics for its own sake.

A special feature involves three chapters on effects that require two mathemagicians who communicate—strictly via the cards, of course!—entirely mathematically. (One such effect was highlighted above.)

Each chapter closes with some parting thoughts. These are often in the form of questions—sometimes of a more mathematical nature—and we don't know all of the answers ourselves. Feel free to ignore these and press on to the following chapters, perhaps returning to them later when the material has had a chance to sink in.

Feel free to drop into the "Convention Center" to familiarize yourself with the assumptions we make about cards and mathematics. Then, whenever you're ready, head for page 25 where the real fun begins.

Tips of the Trade

For entertainment purposes, we recommend dressing up the routines in these pages in a magical cloak. Remember the first rule of magic: *Engage your audience!*

A vital ingredient in many magic effects is misdirection: distracting onlookers with words or gestures, so that people either fail to notice or fail to remember some key action on the performer's part. For us, it's the subtle use of mathematics that makes the magic possible, but we also accept occasional help from the world of magic. Ironically, we find that when performing many of these effects, the average audience won't spot any connection to mathematics. Sleight-of-hand or other magical powers not actually possessed are sometimes attributed to the mathemagician. If this happens to you, take the credit and run!

A good card magic demonstration consists of three, or perhaps four, well-chosen and diverse effects. Be sure to include some audience participation, never do two effects that are likely to resemble each other from the spectators' point of view, and save your best effect until last. Leave the crowd wanting more, but resist the temptation to show them another two or three you just happen to be keen on that day: for most people a total of five or six effects is definitely overkill. Maybe do one encore, another stunner you know never fails to impress, then quit while you're ahead—if the audience leaves the room before you do, you know you've overstayed your welcome!

- In general, do not tell your audience in advance what (you hope) is going to happen, because you don't always know how things will turn out. For example, on occasion, some effects have a kicker, but announcing that at the outset will make you look less than magical if the second punchline doesn't pan out.

- In spite of the impression we may have given in "Hit the Deck Running," you should usually resist all pleas to repeat an effect.

Adapted from "Mathematical Card Tricks" [Mulcahy 00] at AMS.org.

- Don't underestimate people's ability to get simple directions wrong, and save effects that require non-trivial participation for audiences whose cerebral abilities you can personally vouch for.

- Be crystal-clear when speaking, and demonstrate anything (e.g., cutting the deck) that you expect others to do. Sometimes this also gives you a chance to pull a fast one such as peeking at a certain card, so it can serve two purposes!

- Have chosen cards shown around to others in the audience so that there can be no disagreement later as to their identities. Yes, on occasion, participants have been known to forget their own special cards—not to mention lie.

- And always remember: Never reveal your secrets!

If you are performing for a "lay audience" (i.e., nonmathematical and nonmagician) you really should think twice before divulging how an effect works. *As soon as you explain it, people cease to be entertained.* They are inclined to think—and say!—"Oh, is that all it is?" The (magic) spell is permanently broken, and your stature as an entertainer is diminished in their eyes.

If you are performing for intellectually curious spectators, such as colleagues or motivated students, it's different, but you still need to be careful. We strongly believe that people should, in some sense, earn the right to this kind of insider information—merely seeing you do an effect or two, being impressed, and saying "How did you do that?" in an endearing fashion just isn't enough!

On the other hand, being mathematically curious is a good start. Give people hints and let them work out some of it themselves; it's far more rewarding than being handed the whole effect on a plate. This applies to you, too. Try to work out the effects as you go along; you'll be glad you did.

Listen to a pro on the subject:

> The hardest thing is to convince people that the value of a secret is in keeping it secret ... The secret of a card trick (mathematical or not) is like the punchline of a joke. Both are secrets and both are valuable in front of the current audience until you tell them. Nobody wants to hear the same punchline twice. (Steve Beam)

The Ratings Game

Each of the fifty-two effects highlighted in these pages is rated to provide information on the level of mathematics involved, the likely impact on a general audience, the amount of setup required, and difficulty of performance. These ratings are indicated in the margins, as two examples below illustrate.

The number of ♣s displayed reflects the relative sophistication of the underlying mathematics, compared to other effects in this book. Think of this as insider information, available only to members of the (math-emagical) *club*. If the mathematics is as invisible to the performer as it is to the audience, then a high rating here is nothing to fear.

Having performed these effects for several years for many lay (i.e., nonmathematical and nonmagician) audiences, we use ♥s to draw attention to those effects that seem to entertain the most. There is one (maybe two) ♥♥♥♥ rating per chapter, reserved for what seem to be the best *loved* effects for general audiences. Conversely, items that tend to go over better with more mathematically curious audiences get lower ♥ ratings.

How much setup is required is indicated with ♠s (think "how much groundwork or *spade*work is needed?"). It can range from none at all (a single ♠), meaning you are ready to perform with any deck, say, one handed to you by a suspicious spectator, to needing to know the identity of one or two cards at the top and/or bottom of the deck (♠♠), to having a desired stack, perhaps of up to a quarter-deck (♠♠♠), to having a completely prearranged deck (♠♠♠♠).

Finally, ♦ denotes how *hard* an effect is to perform (diamonds, of course, being hard). This takes into account both card handling (e.g., false shuffles, peeking) and desired memory work or on-the-fly mental calculations. So, ♦ effects are very easy to do, ♦♦ require a little concentration to get right, and the hardest effects to pull off are the ♦♦♦♦s.

For instance, the first effect in Chapter A, "Three Scoop Miracle—Done Magic Before?" (A♣), gets the rating in the margin because it is based on a nontrivial mathematical principle, goes over very well with all audiences, can be done with any shuffled deck, and requires just a little "hanky panky" (and some care in handling).

By contrast, the final effect in Chapter K, "Fitch Four Glory" (K♦), which also goes down very well indeed with the masses and can be done with any deck handed to you, takes advantage of deeper mathematics and requires quite a bit of practice to pull off.

Given Any Deck

For easy reference, here are twenty-one effects with rating ♠ that can be performed with any unprepared deck, full or not.[4] Some of them require a lot more practice than others. For the ones marked ⊕, however, you will also need the services of a mathematically savvy assistant.

---

[4]At no point in these pages is it assumed that the reader is playing with a full deck.

Principles

Here we list the main mathematical principles upon which the effects are based, in the order in which they appear in the pages to follow. They are also listed in the index, under *principles*, in alphabetical order. Those believed to be original—as applied to card magic—are indicated with the ►◄ symbol. Many of the names given to longstanding principles are new and purely for our own amusement.

Convention Center

Here we gather, in one central place, basic card-handling information, several conventions that it will be convenient to observe, and a summary of most of the background mathematics that is needed. We assume that readers have basic familiarity with a standard deck of fifty-two cards and its four suits of thirteen values. Occasionally we also add in a Joker.

Feel free to jump ahead to Chapter A after a quick glance and refer back here as required.

Handling Matters

Descriptions and depictions of card-handling moves are generally presented from the perspective of right-handed persons. Obvious adjustments should be made by left-handed readers.

As depicted in the bulk of the images to follow, these moves are also presented from the performer's perspective, the assumption being that many may wish to try out the moves while learning about the fun applications of the mathematics. It is, however, very important to be mindful of what the audience sees: from angles of view (perhaps you can see something you don't want others to spot) to the nature of displays (e.g., a row of overlapping cards looks different from the "other" side). Like a dentist, the mathemagician must be extra careful when using the words "right" and "left."

Putting Our Cards on the Table

We recommend using only high-quality cards in very good condition. We never suggest any kind of card marking, crimping, or even slight bending, even though these methods are used in other forms of magic. Worn cards have natural imperfections that can be incorporated into magic effects, but that's far from the spirit of mathemagic.

Having said that, for the sake of entertainment, we often recommend bending the truth a little in the presentation of the effects. As magician Eugene Burger has wisely observed, "Magic teaches us how to lie without guilt." Hence, when we say, "Shuffle the cards," there are occasions on which that means "Make it look like you shuffle the cards." Likewise, it's not unheard of to say, "The deck is totally mixed up, I couldn't possibly know where anything is," when in fact you do know a little (or a lot) about the card distribution, due to either an unadvertised setup or some secret mathemagical principle. Deal with it.

A central theme of card magic, which we see over and over in the effects that follow, is creating illusions of various types. As just mentioned, the shuffled-deck illusion is especially helpful in the entertainment context. We show how to use mathematics in many different ways to create the illusion that numbers or cards are freely chosen by spectators, or were the result of some random processes. Such *force* cards can also be predicted in advance (in writing if desired) or used in so-called mind-reading stunts.

A similar word of caution applies to claims such as "Given any six cards, we can ...." Generally we really do mean that the stated result holds for any six cards, but there are other times when we make such a claim in performance—having just pretended to shuffle a deck—before inviting a spectator to take off the top six seemingly random cards that were planted there all along.

Before discussing such nonmathematical matters as pretending to shuffle, we need to cover some basics.

Cutting (and Completing the Cut)

The phrase "cutting a packet of cards" can refer to different but equivalent activities, carried out to one of two possible stages of completion. To cut a packet is simply to separate it into two parts, via a break that is not necessarily near the middle. This is often done with the packet on a table, as shown in the first two images of Figure 1. We simply lift off some of the top cards as a unit and set them down beside the rest, to yield two tabled piles. Reassembling the packet by picking up the pile on the dealer's left and placing it on top of the pile on the right, as shown in the third and fourth images in Figure 1, is called "completing the cut." The original top card is now lost within the packet.

While we seem to have cut from top to bottom, it can also be argued that we've cut from bottom to top, as the original bottom card is now lost within the packet. Such cutting merely cycles all of the cards in the packet around, and can be viewed as going in either direction.

In most cases we simply say "cut the cards" when we really mean both "cut" and "complete the cut." On the rare occasion where we really do mean cut, but don't complete the cut, we'll make that clear.

Figure 1. The completed table cut.

There are also various in-hand options, for which a table is dispensed with. In-hand cutting is also usually understood to mean completing the cut. One common in-hand cut is depicted in Figure 7 later, under the guise of overhand shuffling.

Riffle Shuffling

Riffle shuffling refers to dividing a packet into two piles and dovetailing those together on a table with no particular regularity, as depicted in Figure 2. The thumbs are used to release cards from each pile. Finally, the cards are pushed together and the resulting packet squared up.

Figure 2. Riffle shuffle.

A key point is that the cards within each pile maintain the same order relative to each other after the shuffle. Imagine shuffling a Red Ace–King pile into a Black Ace–King pile: if the Red cards are then extracted, they are still in Ace–King order, as are the Black cards left behind.

Rosette Shuffling

For our purposes, riffle shuffling is equivalent to the signature rosette shuffling of Swedish magic maestro Lennart Green, as depicted in Figure 3:

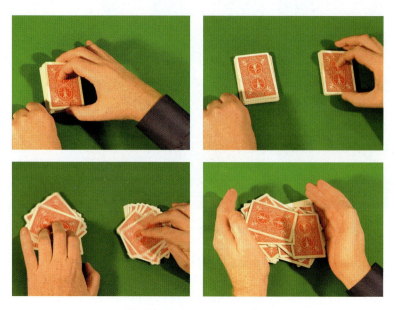

Figure 3. Rosette shuffle

Take a packet of cards and cut it into two piles, side by side on a table. Use your fingers to "twirl" the piles into rosettes, and push those together, squaring up the resulting packet.

Rosette shuffling works particularly well with small packets, where riffling can present physical challenges. It also works for larger packets, even whole decks, provided that a lot of twirling is done to ensure that this action reaches the cards at the bottom of each pile. Otherwise, those cards may remain too well aligned in blocks due to slippage, causing problems when pushed together later. This is often an issue if the piles are on a smooth surface. We recommend rosetting on rougher surfaces, as then the bottom card of each pile is more likely to resist motion, facilitating more widely distributed twirling of the cards above them.

Rosette shuffling is ideal for spectators large and small who are not comfortable with riffle shuffling. As a move that isn't well known outside certain magic circles, it also looks intriguing to the average audience.

In-Hand Options

We have been assuming that you have access to a reasonable surface to riffle or rosette the cards on. But what if these options are not on the table? Perhaps you need to riffle while sailing in rough seas or while perched atop a cliff. The best mathemagician is prepared for any eventuality. In what follows we ignore air resistance due to high winds.

Some people can riffle shuffle "in the air"—but it requires a high level of dexterity. A much easier alternative is in-hand rosetting. Hold one pile in each hand, between your fingers and thumbs, anchoring the top cards with your thumbs. Now twirl the piles underneath with your fingers, and push the piles together in due course, as shown in Figure 4. Practice a little and you'll soon wow audiences with this exotic-looking flourish, with or without matching flamenco dance moves.

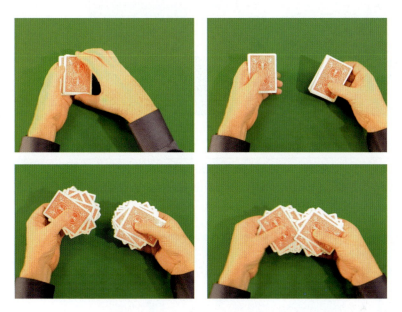

Figure 4. In-hand rosette shuffle.

There's another more practical option, known as a butt shuffle. Cut the deck and square up the resulting two packets by butting them against each other at an angle, then align them and push them together. This sequence is shown in Figure 5.

It's important to attempt this only with high-quality cards, such as Bicycle decks. Using cheap materials here usually leads to total failure because the cards will strenuously resist intermingling, and the very act of trying to push them together damages their edges for all eternity.

We're not at all concerned here with the nature or precision of the interweaving of either a riffle or butt shuffle, so give both a try. The butt shuffle is fast and has the advantage of being easier to see for large audiences, as you can hold the cards aloft while doing it.

Faro (or perfect) shuffling can be thought of as in-hand butt shuffling done with great precision: the cards interweave in a perfectly alternating fashion, as shown in the images in Figure 6. While not nearly as difficult

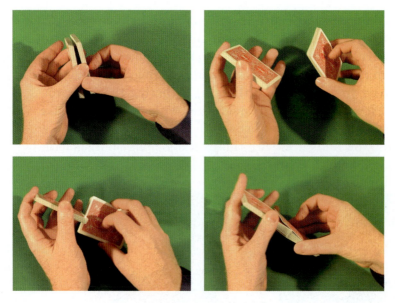

Figure 5. Butt shuffle.

to do as the tabled Faro—which achieves the same goal but looks like a riffle shuffle—learning in-hand perfect Faro shuffling requires much careful hands-on personal instruction and weeks of practice.

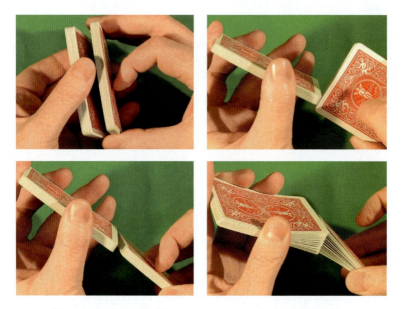

Figure 6. Faro shuffle—one way to achieve perfection.

Nothing in these pages assumes an ability to Faro shuffle perfectly, but if you rise to the challenge and master it, a whole new world of mathemagical possibilities opens up [Minch 91, Minch 94, Morris 98, Diaconis and Graham 11]. (Who knows—perhaps, over time, you may find yourself doing a perfect Faro if you can get your butt to a high state of perfection.)

Different people Faro differently. Figure 6 shows the method we favor, mastered years ago, after (1) an hour of intensive personal attention from innovative toy designer and magician Mark Setteducati, followed by (2) a day or two of hopeless, soul-destroying failure, (3) a month of on/off, frustrating and seemingly unproductive practice, and finally (4) a totally unexpected "Aha! moment" late one night. It looks effortless when somebody else does it, but is completely impossible when you first try it yourself. Yet, once you get the hang of it, like bicycle riding, it's never forgotten.

The first step is particularly difficult: splitting a packet in half exactly, as shown in the first image in Figure 6. The second image presupposes that the resulting halves have been squared up first, as in the third image in Figure 5. Then—and this is the part that seems impossible to beginners—the two halves are "engaged" with the assistance of pressure exerted by the left index finger and right little finger, as depicted in the third image in Figure 6, the other fingers and thumbs keeping everything on track.

If you are lucky, the cards will suddenly spring together as shown in the last image in Figure 6; perfectly interwoven to boot. It is customary to facilitate this by applying a little more pressure on either the bottoms or tops of the two halves, using your thumbs. In the images shown, this pressure was applied to the touching corners nearest the viewer, and as a result, the springing together propagated from front to back. Some people in similar situations find it easier to have the action develop from back to front. The final step—not shown here—is pushing the two piles into each other and squaring up the reconstructed packet.

We're glossing over one important consideration here: there are actually two distinct types of Faro shuffling with rather different mathematical properties: *out-Faro shuffling* and *in-Faro shuffling*. The distinction is briefly discussed at the start of Chapter 8.

There's also reverse Faro shuffling: deal a Faro-shuffled packet into two piles on the table, either left to right or right to left (depending on which Faro was done), then deal each of those piles to the table again to reverse the card order, and finally place one pile on top of the other. The Faro-shuffled packet effectively is restored to its initial state.

There is an equivalent move known as up-jogging (and stripping out), which is also not considered in these pages. Such alternatives are used to good effect to take advantage of many fascinating Faro properties in both *Magical Mathematics* [Diaconis and Graham 11, Chapter 6] and *RedivideR* [Goldstein 02].

Overhand Methods

Overhand shuffling refers to a common type of in-hand shuffling, as shown in Figure 7. Typically, the deck is held between the thumb and middle finger of the right hand, with the left hand waiting underneath, and cards from the middle or bottom of the deck are then lifted up by the right hand, allowing the rest of the deck to fall back into the cradled left hand, before some or all of the raised cards are dribbled or dropped onto what is already in the left hand. This can be repeated speedily, sometimes using the left thumb to pull down some cards from the lifted packet.

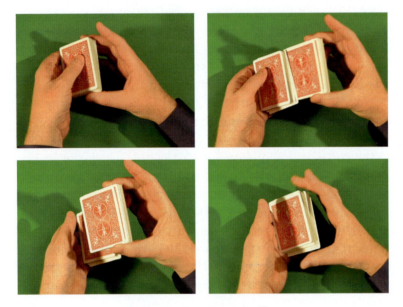

Figure 7. Overhand shuffling

Figure 7 depicts a special case where all the lifted cards come from the bottom of the packet and are then dropped on top of the rest as a single unit. This is really a simple in-hand cut. In general, the lifting up and dropping may be less regular, if not quite sloppy.

The way we've shown it, you never see any card faces, though your audience may, because the original deck faces the right palm at the outset. It can also be done with the cards facing the other way, so that you see card faces as they drop into the left hand. Done discreetly, this allows you to shuffle in a quite irregular and seemingly honest fashion, with your eyes engaging your audience as you chat, yet one peek at the very last minute tells you the identity of the bottom card.

Peeking

Although completely nonmathematical, the ability to surreptitiously gain knowledge of one or two cards in a shuffled deck, usually right under the noses of your audience, greatly enhances your ability to pull off seeming miracles. For instance, upon completing a shuffle, it's perfectly natural to square up the deck by tapping it against the table, thus providing an opportunity to peek at the bottom card, as shown in Figure 8.

Figure 8. Peeking.

Paying attention to audience angle is critical. If you hold the cards as shown in Figure 8, then, provided that the audience is to your right, people assume that your view is as depicted in the left image, whereas in reality, it's as shown in the right one. Also, you can shuffle the peeked-at bottom card to the top, if that's where you want it. It's an easy matter now to peek at the new bottom card, if you need to, so that you'll then have knowledge of both the top and bottom cards.

Actually, you can get away with something like this, suitably modified, as long as there is no one too far to your left, assuming you are right-handed. This is worth remembering if performing for people seated around a dinner table.

Like many magic moves, actions such as peeking are best done under the cloak of misdirection. Distract your audience with a hand gesture—using the hand that is not holding the cards—or an irrelevant question or a good, bad, or indifferent joke, and you probably have an opportunity to do what you need to do without anyone being any the wiser.

Running (Off)

Running or *running off* refers to an in-hand way of rearranging some cards—either to genuinely shuffle in a haphazard way, or for a very specific purpose—and hopefully not to the reaction of audience members to your efforts to entertain them. In the case of interest to us, the goal is generally to transfer a fixed number of cards from the top of a packet to the bottom, in reverse order, while giving the impression of sloppy shuffling.

Hold the packet firmly in one hand, with the card backs toward your audience, who should be to your right if you hold the cards at the angle suggested in Figure 9, so that only you can see the face of the packet. The extreme angle depicted here is for illustrative purposes—in practice, you usually don't need to be able to see the bottom face this clearly.

The thumb and fingers grasp the short ends of the cards, and the other hand is held cupped open underneath, and moved up and down in a slow steady rhythm while that thumb pulls off cards from the first hand one by one, so that they fall into the waiting palm.

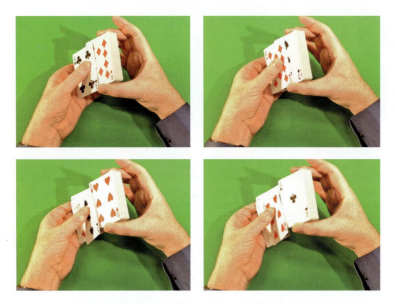

Figure 9. Running off cards.

The cards run off—in Figure 9 it's 4♣, 10♦, 3♠, and 6♥—are thus reversed in order. When the desired number (or more) have been moved from the face of the packet to the waiting hand, the remainder are dropped on top of them as a block. In the image, that results in the A♣ being at the face. Running cards in this way, casually and with no obvious attention being drawn to the card faces, also has the advantage that it allows you to peek at card faces right before the end of the process, so that you gain knowledge of the resulting top card or two.

Underhand Methods

There will be occasions when you plant from one to a dozen cards at the top or bottom of the deck before showtime, possibly in a specific order, and you want them to stay there despite some fairly convincing (not to mention convincingly fair) shuffling. Such groupings are called top or bottom stocks. There might also be a need to know what the middle card of a packet is, or the middle two cards (those in positions 26 and 27) of a full deck, or a middle clump of other sizes; we refer to these as middle stocks.

Stocks can be kept in position throughout some careful and casual false shuffling done right in front of your audience. There is additional flexibility if you merely need to keep, say, the four Aces at the bottom of a packet, but their exact order is not important.

For instance, it's not hard to riffle while preserving top and bottom stocks: simply ensure that several cards from one hand fall first, and several from the other hand fall last. In the images in Figure 10, the pile on the left was the original bottom part of the packet shuffled; its cards fall first, and the original top cards in the right-hand pile fall last. In practice, do all of this fast and most people will have no idea that you are up to anything.

Figure 10. Maintaining top and bottom stocks

If only the top or bottom stock, but not both, needs to be preserved, then this can be done more casually, and hopefully without raising any suspicions. Likewise, either top or bottom stocks (but not both) can be maintained via judicious use of overhand shuffling; the same applies to middle stocks.

The *pinky break* is the simple sleight-of-hand move depicted in Figure 11. The first image shows the tip of the little finger of the left hand discreetly holding a break between two parts of the deck. We assume that

Figure 11. The pinky break.

the identity of one of the two cards this finger touches is known, either the top card of the bottom part or the bottom card of the top part. To the audience, hopefully watching from the right here, all seems fair, but you can now cut the cards as shown in the second image, thus gaining access to the known card.

How the pinky break is achieved in the first place is another question. Options include (1) planting a key card on one side or other of that little finger at the outset, perhaps allowing the left hand to hang casually to your side before bringing it and the deck into view, or (2) having a selected card placed on top of the deck before cutting the cards in hand, maintaining the cut surreptitiously with that little finger.

Just so you know what you're dealing with.

Cards Matters

Asymmetry of Cards

Consider a face-down packet running A–K♦ from the top down. Most commercial cards have identifying marks (pips) in only two of their four corners, specifically the top left and bottom right corners. As a result, whether the handler is right- or left-handed, when fanning this packet face up, it's natural to have the King on the right at the face of the packet, as seen in Figure 12, with the other (mostly hidden) cards spread out to the left underneath the King, identifiable from the pips visible in their top left corners.

Figure 12. Fanning face-up cards reverses their perceived order.

The upshot is that (1) fanning a face-down packet face-up, as shown in Figure 12, displays the cards in the expected order, viewed from left to right, but (2) fanning a packet that is considered to be face-up in the first place actually displays the cards in a counter-intuitive order. In the latter case, the display should be read from right to left to reflect the true order of the cards.

CHaSeD and Other Suit Orders

A sequence of cards—anywhere in size from four to fifty-two—cycling Clubs, Hearts, Spades and Diamonds over and over as necessary, is said to be in a CHaSeD order. It's not strictly required that the packet size be a multiple of four. The colors alternate here, as they do for the near-reversed SHoCkeD order.

Deck Orders

Sometimes it's convenient to consider the entire deck ordered by suit, and then by value. Common orders include

$$A\clubsuit, 2\clubsuit, \ldots, K\clubsuit, A\heartsuit, \ldots, K\heartsuit, A\spadesuit, \ldots, K\spadesuit, A\diamondsuit, \ldots, K\diamondsuit,$$

with entire suits in CHaSeD order, or

$$A\clubsuit, 2\clubsuit, \ldots, K\clubsuit, A\diamondsuit, \ldots, K\diamondsuit, A\heartsuit, \ldots, K\heartsuit, A\spadesuit, \ldots, K\spadesuit,$$

with entire suits in alphabetical order. An order that intertwines both numerical and CHaSeD cycling has advantages for some purposes:

$$A\clubsuit, 2\heartsuit, 3\spadesuit, 4\diamondsuit, 5\clubsuit, 6\heartsuit, \ldots, 8\spadesuit, 9\diamondsuit, 10\clubsuit, J\heartsuit, Q\spadesuit, K\diamondsuit.$$

Here's an order that intertwines both alphabetical ordered values—namely, Ace, 8, 5, 4, Jack, King, 9, Queen, 7, 6, 10, 3, 2—and CHaSeD cycling:

$$A\clubsuit, 8\heartsuit, 5\spadesuit, 4\diamondsuit, J\clubsuit, K\heartsuit, 9\spadesuit, Q\diamondsuit, 7\clubsuit, 6\heartsuit, 10\spadesuit, 3\diamondsuit,$$
$$2\clubsuit, A\heartsuit, 8\spadesuit, 5\diamondsuit, \ldots, 6\clubsuit, 10\heartsuit, 3\spadesuit, 2\diamondsuit.$$

Sometimes it's handy to have at our disposal a particular value order that is easy to remember but unlikely to be thought of by a general audience as having any significance. The above is one such value order, and so is the one below favored by mathematician John H. Conway:

$$3, 5, 10, A, J, 9, \text{Joker}, 2, 8, 7, Q, 6, 4, K.$$

It is associated with the mnemonic "The Five Tenacious Boys Nicely Joke To Hated Servant Girls Sick For Absent Kings." ("The" is pronounced *thee* here.) This too can be used in conjunction with some kind of suit cycling for a full deck setup.

A popular order of a full deck, although we don't use it in this volume, is the one known as *Si Stebbens*, where the values increase by 3 each time, with wraparound as needed, and the suits cycle CHaSeD:

$$A\clubsuit, 4\heartsuit, 7\spadesuit, 10\diamondsuit, K\clubsuit, 3\heartsuit, 6\spadesuit, 9\diamondsuit, Q\clubsuit, \ldots, 2\clubsuit, 5\heartsuit, 8\spadesuit, J\diamondsuit.$$

The above foreshadows some natural numerical associations discussed next, and a thirteen-hour clock discussed on page 15.

Natural and Integral Assumptions

We assume that Ace, 2, 3, 4, ..., 10, Jack, Queen, King, have the values 1, 2, 3, 4, ..., 10, 11, 12, 13, respectively, regardless of suits. (Hence, except in poker considerations, Aces are always considered low.)

Sometimes, however, it's handy to also have at our disposal cards that correspond to negative values and zero, to give us the full range of integers from -13 to 13. It's easy enough to assign the value zero to Jokers, and we sometimes adopt the convention that Black cards have positive values (think credit), whereas Red cards have negative values (think debit). That is to say, Diamonds and Hearts may be viewed in an unromantic negative light when it suits us.

Binary Identifiers

It's helpful to consider different ways to split the cards into two sets, not necessarily of the same size. Sometimes you don't want that division to be obvious to casual observers.

Among the possibilities are

- Red (Diamonds and Hearts) or Black (Clubs and Spades),

- Roundies (Clubs and Hearts) or sharpies (Diamonds and Spades),

- Evens (2, 4, ..., Q) or odds (Ace, 3, ..., Jack, King), or

- Primes (2, 3, 5, 7, J, K) or composites (4, 6, 8, 9, 10, Q), with the Aces either all lumped in with one of those sets or split equally between them in some way you can remember.

Mathematics Matters

Modular Arithmetic

It's about time. We are all familiar with the fact that five hours after nine o'clock (or nine hours after five o'clock), a twelve-hour clock will show two o'clock: it's called *clock arithmetic*. To a mathematician, this reads $9+5 = 5+9 = 2 \mod 12$. Meanwhile, five hours before nine o'clock is four o'clock, but nine hours before five o'clock is eight o'clock, which can be written $9 - 5 = 4 \mod 12$ and $5 - 9 = 8 \mod 12$, respectively. When using a twenty-four-hour clock, these considerations need to be modified, assuming we know whether the times are a.m. or p.m. (in the latter case, they first need to have 12 added to them). Note that $12 = 0 \mod 12$ and $24 = 0 \mod 24$. Here mod is short for the word modulo—it can be read (or said) either way. Similarly, if you drive 50 miles in a car whose odometer reads 99,990 miles, then if the odometer tops out at

99,999 miles, it will read 40 at the end of the trip, since $99{,}990 + 50 = 40$ mod 100,000 (here, $100{,}000 = 0 \mod 100{,}000$).

These are examples of *modular arithmetic*, and such additions and subtractions can be performed modulo any positive whole number. For example, $3 + 5 = 0 \mod 4$, and $3 - 5 = 2 \mod 4$. (Those can be interpreted in terms of quarters of an hour: an hour and a quarter after three quarters past an hour is something o'clock, and an hour and a quarter before three quarters past an hour is half past something o'clock.)

Likewise, $5 + 8 = 3 \mod 10$ and $5 - 8 = 7 \mod 10$. Mod 10 is all about the last digit of a number, subject to the convention that the last digit of a negative number, like that of a positive number, refers to its distance from the highest power of 10 to its left. With this convention, the last digit of 52 is 2 but the last digit of -52 is 8, since $-52 = -60 + 8$.

The cards of one suit, say A–K♦, can be thought of as a thirteen-hour clock, as shown in Figure 13.

Figure 13. Thirteen-hour clock of Diamonds.

The card seven past the 3 is the 10, and the one seven before the 3 is the 9: for the latter we must think of wrapping four cards back around from the Ace in reverse (counterclockwise) order. Similarly the card five past the 9 is the Ace. We can express these facts succinctly as $3 + 7 = 10$ mod 13, $3 - 7 = 9$ mod 13, and $9 + 5 = 1$ mod 13. It's the same as observing that if such a thirteen-card packet is cut so that the 3 is on top, and seven cards are then cut one by one from the top to the bottom, then the 10 is on top, whereas if seven cards are instead cut one by one from the bottom to the top, then the 9 is on top. There are many such *mod*-ified arguments in this volume.

Mod 2 is particularly interesting because it gives rise to binary arithmetic, which simply encodes evenness and oddness. All even numbers (e..g., 2, 6, 14) are equal to 0 mod 2, and all odds (e.g., 1, 3, 19) are equal to 1 mod 2. This is known as (even versus odd) *parity*. Now the $1 + 1 = 0$ mod 2 rule of binary arithmetic seems perfectly reasonable, as it just says that the sum of two odd numbers is even. Two numbers are said to have the *same parity* when both are even or both are odd, and *opposite parity* when one is odd and the other is even.

We'll also use mod 3 (ternary arithmetic): here we have $0 = 3 = 6$ mod 3, $1 = 4 = 7$ mod 3, and $2 = 5 = 8$ mod 3, along with the rules of ternary arithmetic such as $1 + 2 = 0$ mod 3 and $2 + 2 = 1$ mod 3. Sometimes it's convenient to use -1 in place of 2 here, in which case we refer to this as balanced ternary arithmetic.

When $a = b$ mod n, we say a and b are *equal* mod n (or *the same* mod n) or *congruent* mod n. This holds precisely when n divides into $a - b$ exactly, or equivalently, when a and b have the same remainder upon division by n. For instance, 17 and 52 are equal mod 5 because 5 goes into $17 - 52 = -35$ exactly, or equivalently 17 and 52 both have remainder 2 upon division by 5.

Formulas That Count

The *multiplication rule* says that if there are a ways to do one thing and b ways to do a second thing, then there are ab ways to do both. For instance, if there are four different restaurants to dine in and five possible films to see later, then there are twenty different dinner–film experiences available.[5] This extends in the obvious way to more than two things done sequentially.

Arranging several things in a row can be thought of as filling in slots from left to right, which lends itself to application of the extended multiplication rule. In how many ways can we arrange the three letters in the word ADO? Since there are three choices for the first letter or slot,

---

[5]In contrast, if you either dine out or go to a film, but not both, then there are only nine available experiences, since $4 + 5 = 9$. That's the *addition rule* in action.

then two choices for the second, and only one[6] remaining for the third and final slot, there are six ways. These are as ADO itself and as AOD, DAO, DOA, OAD, and ODA. We have $6 = 3 \times 2 \times 1$, which is written as 3! and is known as *three factorial.*

Similarly, there are five factorial, or $5! = 5 \times 4 \times \cdots \times 1 = 120$, ways to arrange the five letters in the word MAGIC. Listed alphabetically, the 120 possible arrangements are ACGIM, ACGMI, ACMIG, ..., MIGCA. (Mathematicians refer to such ordered arrangements as *permutations.*)

So there are $13! = 13 \times 12 \times \cdots \times 2 \times 1 = 6{,}227{,}020{,}800$ possible ways to arrange all of the Hearts—that's about what the world population was in 2002. This last observation and the next several paragraphs appeared in the online *Aperiodical* article [Mulcahy 12_05].

The number of ways to order all the red cards in a deck is 26!, which is about 4×10^{26}, or more than the number of grains of sand on all of the beaches on earth. Similarly, there are $52! = 52 \times 51 \times \cdots \times 1$, which is about 8×10^{67}, ways to arrange all of the cards in a deck. For comparison, the number of elementary particles believed to be in the universe is about 10^{80} [Mulcahy 12_05], so we don't recommend trying to list all possible arrangements of a deck of cards, as we suspect you'll run out of time *and* paper.

For a deck of cards, it means that there are far more shuffled states than have ever been written down. Hence, the totality of all deck orders that have ever been achieved with actual decks in the history of the world is a very thin set within the set of all possible deck orders. In other words, "A well-shuffled deck, which is far from being in any obvious or recognizable order, is probably unique in the sense that nobody else has ever come up with it before." You can shuffle till the cows come home, but you'll still miss the vast majority of the possibilities.

There are only $52 \times 51 \times 50 = 132{,}600$ ways to arrange three cards in a row, using any cards from a full deck. That number can also be written somewhat impractically as $\frac{52!}{49!}$ (the numerator and denominator are both well beyond the reach of a calculator display). This is "the number of permutations (arrangements) of fifty-two objects, three at a time."

Another useful application of factorials is to count selections, rather than arrangements. For selections, where order is not important, factorials are combined in what appear to be fractional expressions, but these always turn out to be whole numbers.

For instance, if we wanted to know how many different ways there are to decide which three cards should be turned over in a display of seven face-up cards, the answer is $\frac{7!}{3!4!}$. This is known as "the number of selections of seven objects, three at a time" or "seven choose three." It

---

[6]Sometimes confusingly expressed as "none."

comes out to be 35. We're selecting or choosing which three cards will be turned over, but here the order in which they are flipped is irrelevant.

That's the same as counting how many ways there are to decide which four of the seven cards should *not* be turned over. That can be done in $\frac{7!}{4!3!}$ ("seven choose four") ways, unsurprisingly also yielding 35. (Yes, that's a special case of a general equivalence principle.)

It is not, however, the same as deciding on one of seven cards for the first one to be flipped over, then one of the remaining six for the second one, then one of the remaining five for the third one. That yields $7 \times 6 \times 5 = 210$ ways, and is the answer to a subtly different question of a type already considered above, namely, "how many arrangements of three cards are there if seven are available overall?" Here, it's a serious overcount, as each selection possibility (say $\{3, 4, 6\}$) has been accounted for $3! = 6$ times (as $3, 4, 6$; but also $3, 6, 4$; $4, 3, 6$; $4, 6, 3$; $6, 3, 4$; and $6, 4, 3$). That's why the correct answer is $\frac{7!}{4!}$ divided by $3!$, or $\frac{7!}{4!3!}$. Mathematicians refer to such (unordered) selections as *combinations*.

Loco-Motions

There is a well-organized way to describe each of the $n!$ particular ways to rearrange a list numbered 1 to n. A particular rearrangement might correspond to a very specific shuffle, such as a perfect out-Faro, or to COATing a certain number of cards, as explored in Chapter A. Any such arrangement is a *permutation*. There are $n!$ different permutations of 1 to n, or indeed of any n different objects.

Suppose 1 to n is rearranged in some prescribed fashion. We can think of this as a shuffle of a packet that starts out running Ace, 2, 3, ..., n, from the top down (there may be more than thirteen cards). Issues to consider include which card ends up on top (or on the bottom); which cards— if any—are invariant (i.e., end up where they started); which pairs are interchanged, which sets of three, four, or five cards cycle around among themselves, and so on. Also of interest, especially for small packets, is the *period*, or *order*, namely, the minimum number of times the rearrangement must be done to restore the packet to its original order.

Fixed points are numbers f such that the card that starts in position f ends up in the same position.

A 2-*cycle* (more commonly known as a *transposition*) refers to switching pairs, namely numbers (s, t) such that the cards that start in positions s and t end up in positions t and s, respectively. Transpositions are especially important because it's well known that all possible arrangements can be attained using only these. By that, we mean that, for instance, any arrangement of n people in a row can be attained by having people sit in any order to start with, and then asking appropriate pairs of people just switch seats until the group is in the desired order.

A 3-*cycle* refers to numbers (s, t, u) such that the cards that started in positions s, t, and u end up in positions t, u, and s, respectively. More generally, a *k-cycle* refers to numbers $(s_1, s_2, s_3, \ldots, s_k)$ such that the cards that start in positions $s_1, s_2, s_3, \ldots, s_k$ end up in positions $s_2, s_2, s_4, \ldots, s_1$, respectively. Hence, 2-cycles are just switching pairs (transpositions), and 1-cycles are fixed points.

For instance, consider rearranging 1–8 to end up with $6, 7, 8, 5, 4, 3, 2, 1$. (In Chapter A, where we study many such situations, this is known as "COATing five cards.") Let's give this permutation a name, say, σ. So σ sends card 1 to position 8, card 2 to position 7, card 3 to position 6, card 4 to position 5, card 5 to position 4, card 6 to position 1, card 7 to position 2, and card 8 to position 3.

Note that σ switches the positions of cards 4 and 5, a fact that can be expressed succinctly by the transposition (4 5). Also, σ sends 1 to 8, then 8 to 3, then 3 to 6, and 6 to 1. So if the COAT was applied four times, 1 would be back to 1 (and 8, 3, and 6 would also be back to their respective positions). All of that information can be encapsulated in the 4-cycle (1 8 3 6). Meanwhile, σ sends 2 to 7 and 7 back to 2, which is another transposition.

It is conventional to write

$$\sigma = (1\ 8\ 3\ 6)(2\ 7)(4\ 5),$$

which is called the (disjoint) cycle decomposition of the permutation.

This conveys the same information we started with, but in another form. There is a lot of flexibility here, as we can also write

$$\sigma = (4\ 5)(1\ 8\ 3\ 6)(2\ 7),$$
$$\sigma = (2\ 7)(4\ 5)(1\ 8\ 3\ 6),$$

or even

$$\sigma = (5\ 4)(6\ 1\ 8\ 3\ )(2\ 7),$$

among others. In other words, it doesn't matter in what order we list the constituent cycles, and we are free to cycle around within each one. (For transpositions, this looks like reversing the order, but don't try that for longer cycles!)

Now let's rearrange 1–9, instead, to end up with $6, 7, 8, 9, 5, 4, 3, 2, 1$. (This is another example of "COATing five cards" in the language of Chapter A.) This time, we obtain (1 9 4 6)(2 8 3 7)(5), which is usually written as (1 9 4 6)(2 8 3 7). Any omitted symbols are assumed to be 1-cycles, namely, fixed points, such as the middle card in position 5 here.

Try this for the rearrangement $4, 5, 6, 7, 8, 3, 2, 1$ of the numbers 1–8. This time 3 and 6 are switched, and the other six cycle around, yielding (1 8 5 2 7 4)(3 6).

Notice that cycles of length one make no real contribution to the period, they merely correspond to fixed cards, that don't move at all. Ignoring those, we are really interested in the cycles of length two or more: note that the sum of their lengths cannot exceed n (the sum would be n if we included the cycles of length one).

What is important above is the lengths of the cycles that show up in the permutation decomposition. They encode key information:

> Each permutation σ of the packet 1–n corresponds to a list of cycles of lengths $c_1, c_2, \ldots, c_t$, and the period (or order) of σ is the least common multiple of the cycle lengths, namely, $LCM\{c_1, c_2, \ldots, c_t\}$.

Of course, some (or all) of the c_i can be the same. Returning to the three examples above, can you see why the periods are 4, 4, and 6 respectively?

You Takes Your Chances

Bearing in mind the unavoidable reality that a card effect known to be watertight can go wrong—either you the mathemagician or a well-meaning audience member may make a mistake and cause the whole thing to blow up in your face—it's prudent to be prepared for such emergencies and to keep in reserve a snappy crowd pleaser that you know cannot fail.

Several of the effects in Chapters 3, 4, and J merely have a high probability of working, but we highly recommend considering some of these. Mathematics may not be able to guarantee 100% success every time, but the payoffs when things do work out arguably outweigh the red faces on the rare occasions when they don't.

If the dice seem to be loaded against you on the day that you try one of these probabilistic effects in public, simply act as you should when one of your old reliable effects lets you down. Say nothing to hint at your disappointment, calmly continue with words such as, "Let's try something different," and quickly switch to something sure-fired that works with a totally unprepared deck. For such eventualities, we particularly favor "Three Scoop Miracle" (A♣) or "Poker with Any Ten Cards" (3♦).

Probability is a measure of the likelihood of the occurrence of something specific, considered in the context of all possible occurrences. For us, it will usually be estimated by a fraction (or its approximate decimal equivalent), or as a percentage (or "1 in ... chance"). That fraction is the ratio of two counts, the numerator being the number of possible occurrences of A, and the denominator being the total number of equally likely possible outcomes.

For instance, if a card is selected at random from a shuffled deck, and A is the outcome that this card is Red, then the probability of A is the number $P(A) = \frac{26}{52} = 0.50$, or 50% (representing a 1 in 2 chance), because 26 of the 52 equally likely possible outcomes correspond to Red cards. Note that this is to be interpreted as meaning that if a single card is randomly selected from a deck over and over, the deck always being full and shuffled, then about half of the time, on average, the card will be Red. We are certainly not implying that if this is done ten times, we'll get a Red card exactly fives times.

Similarly if A is the outcome that this card is a Diamond, then $P(A) = \frac{13}{52} = 0.25$, or 25% (a 1 in 4 chance), since 13 of the cards are Diamonds. The probability that the card is an Ace is only $\frac{4}{52} = 0.0769$, or about 7.69% (a 1 in 13 chance). From here on we round these numbers a lot, and the equal sign really means approximately equal. The probability that the card is a Red Ace drops to $\frac{2}{52} = 0.0385$, or roughly 3.85% (a 1 in 26 chance).

When several cards are selected at random from a shuffled deck, it's helpful to use the tools discussed in "Formulas That Count." For instance, for two cards, there are $\frac{52!}{2!50!} = 1{,}326$ equally likely possible outcomes, of which $\frac{26!}{2!26!} = 325$ correspond to both being Red. Hence, the probability that both cards are Red is $\frac{325}{1{,}326} = 0.2451$, or approximately 24.51% (a 1 in 4.08 chance). For five cards, representing a poker hand, there are $\frac{52!}{5!47!} = 2{,}598{,}960$ (over two and a half million) equally likely possible outcomes, of which $\frac{26!}{5!21!} = 65{,}780$ correspond to all five cards being Red. Hence, the probability that all five are Red is $\frac{65{,}780}{2{,}598{,}960} = 0.0253$, or roughly 2.53% (a 1 in 39.5 chance).

In poker, getting five cards of one color is of no interest, but getting five of one suit is, such a hand being known as a flush. Arguing as above, we find that the probability of getting a flush in Spades is about $\frac{1{,}287}{2{,}598{,}960} = 0.000495$, and the same is true for getting a flush in any other specific suit. Hence, the probability of getting a flush in some suit is that tiny number added to itself four times—one for each possible suit—since there are $4 \times 1{,}287$ ways out of 2,598,960 to get five cards of one suit. The probability of getting a flush comes out to be roughly 0.00198, or about a fifth of 1% (a 1 in 505 chance).

The argument just given can be viewed as a repeated application of the addition formula, which states that $P(A \text{ or } B) = P(A) + P(B)$ strictly on the condition that A and B cannot both happen at the same time (compare to the addition rule on page 16). For instance, when selecting just one card at random from a full deck, the probability that it's an Ace or a King is $\frac{4}{52} + \frac{4}{52} = 0.1538$, or about 15.38%, since the card can't be an Ace and a King. However, in general, the correct formula is $P(A \text{ or } B) = P(A) + P(B) - P(A \text{ and } B)$, whether A and B can happen at the same time or not. A good example of the need for the adjustment subtraction term—which is zero when it's impossible for A and B to occur together—is finding the probability that a single card selected at random from a deck is an Ace or a Diamond. It's $\frac{4}{52} + \frac{13}{52} - \frac{1}{52} = \frac{16}{52} = 0.3077$, or roughly 30.77%. The point is that there are sixteen (not seventeen) cards in a deck that are Aces or Diamonds; we must not double count the Ace of Diamonds!

Throughout Chapter 3, the relative values of various desirable poker hands are of great importance. These are ranked based on their scarcity, the rarer (hence, less likely to occur) hands being prized the most. Table 3.1 lists in order the values of poker hands resulting from the number of ways they can occur among all possible 2,598,960 hands, as well as the corresponding chances ("1 in ...").

Another fundamental idea we need is based on the simple observation that if there is a 30% chance of rain, there is a 70% chance of no rain. In probability, we write

$$P(\text{not } A) = 1 - P(A)$$

or, equivalently,

$$P(A) = 1 - P(\text{not } A).$$

There are times when $P(\text{not } A)$ is much easier to work out than $P(A)$. For instance, when selecting four cards from a deck, what is the probability that at least two of them share a value (ignoring suits)? It could be two or three of them, or all four, and it's not easy to track the possibilities and count them correctly. However, *not* having at least two of the same value is the same as having four different values: that we can count without too much difficulty. Temporarily considering the selected cards as being in some definite order, there are 52 ways to select the first one, 48 ways to select the second (avoiding the value of the first), 44 ways to select the third (avoiding the values of the first two), and 40 ways to select the fourth (avoiding the values of the first three). The product of these, namely, $52 \times 48 \times 44 \times 40$, overcounts the number of selections by a factor of $4!$:[7] the actual number of selections of four cards of different values is $\frac{52 \times 48 \times 44 \times 40}{4!} = 183{,}040$. Since there are $\frac{52!}{4!48!} = 270{,}725$ possible selections of four cards from a deck, the probability that four randomly selected cards do not have any shared values is $\frac{183{,}040}{270{,}725} = 0.6761$ and so the probability of getting at least one pair (i.e., two cards sharing a value) is $1 - 0.6761 = 0.3259$. A key pattern can be discerned here by noting that this is also equal to

$$1 - \frac{52 \times 48 \times 44 \times 40}{52 \times 51 \times 50 \times 49}.$$

---

[7]Because it counts as being different all of the ways to arrange each specific set of four differently-valued cards, such as the $4! = 24$ ways to order 4♠, A♦, 9♣, and 2♠.

For five cards randomly selected from a deck, the corresponding probability comes out to be

$$1 - \frac{52 \times 48 \times 44 \times 40 \times 36}{52 \times 51 \times 50 \times 49 \times 48} = 0.4929,$$

which is to say, almost half of all possible poker hands contain at least a pair. That fact is reflected in the final row of Table 3.1.

A Certain Certainty

Above we hinted that impossible events happen with probability 0 (or 0%). That's why the general formula $P(A \text{ or } B) = P(A) + P(B) - P(A \text{ and } B)$ reduces to the simpler addition formula when A and B cannot happen at the same time. Likewise, certain events—yes, we mean events that are absolutely certain to occur—happen with probability 1 (or 100%). A good example of the latter is that if any fourteen cards are selected from a deck (shuffled or not), then it's totally guaranteed that at least one value is repeated—as only thirteen values are available—in other words, there must be at least one matching pair. That's a consequence of the following fairly obvious fact that proves to be tremendously useful later.

> **Pigeonhole Principle**
> *If we have several boxes occupied by pigeons, and there are more pigeons than boxes, then some boxes must have more than one pigeon in them.*

Strictly speaking, "some boxes" here may just mean one box. For instance, with fourteen cards, only one value has to be repeated.

An extension of the Pigeonhole Principle guarantees that given any seventeen cards, there must be at least two values repeated, and given twenty-seven cards, there must be at least one example of a value occurring three times. Can you see why those statements are true?

Low-Down Triple Dealing

We've all met some low-down double dealing types in our lives. Have you ever considered the advantages of being a low-down *triple* dealer? You may warm to the idea once you've tried some of the items below.[1]

The recurring theme here is a reversed transfer of cards from the top to the bottom of a packet (see Figure A.1), often done under the guise of spelling out words, and generally repeated several times to interesting effect. We revisit this topic and some variations on it in Chapters 5 and 9. Throughout all dealing, we suggest that the cards be held low, so that the audience doesn't see any flashed card faces; hence, our chapter title.

Our opener is based on nontrivial mathematics (♣♣)—even though, in our experience, that's rarely suspected by onlookers. It never fails to please (♥♥♥♥), requires no setup (♠), and is easy enough to perform (♦♦) for people who can deal and count at the same time.

A♣ Three Scoop Miracle—Done Magic Before?

How it looks: *Hand about a quarter of a deck of cards to a spectator, and ask her to shuffle freely. Take those cards back, and mix them further in your hands as you ask the spectator what her favorite ice cream flavor is. Let's suppose she says, "Chocolate."*

Deal from the packet to the table, one card for each letter of "chocolate," then scoop those up with one hand, commenting that this represents one scoop of ice cream, and with the other hand, drop the remainder on top ("as a topping"). Repeat this spelling (and scooping and topping) routine twice more, for a total of three times.

---

[1] In Chapters 5 and 7 we'll even extol the virtues of single and double dealing.

Figure A.1. Spelling "chocolate" reverse transfers nine cards.

Emphasize how random the dealing was: since the cards were shuffled repeatedly and the spectator named the ice cream flavor. Ask her if she has done magic before. Regardless of the answer, now ask her to press down hard on the top card of the packet on the table, requesting that it be miraculously turned a specific card, say, the Four of Diamonds. When the card is turned over, it is seen by all to be the desired card. Congratulate the spectator on a job well done.

How it works: There are two secrets here: (1) a key relationship between the number of letters in the word being spelled out and the size of the "quarter"-deck being used, and (2) the fact that you must know the identity of the bottom card at the start of the spelling and dropping.

There are many ways to address the second point, such as peeking at the bottom card after you get the shuffled packet back (see page 9, perhaps while tapping the cards on the table to square them up. A more sophisticated handling is suggested later.

As to the first point, the size of the "quarter"-deck used must be at least as big as the number of letters in the word being spelled, yet no larger than twice that number. For instance, using *chocolate*, you need to work with between nine and eighteen cards; we recommend a number not too close to those extremes, such as eleven to fifteen. About a quarter of a deck works well for the flavors most commonly mentioned, which seem to be *chocolate*, *vanilla*, or *strawberry*. If *rum* is desired, try to force *rum raisin*, and for selections with long names like *mint chocolate chip*, use about half of the deck.[2]

---

[2]That particular flavor suggests another presentation, as we'll see on page 112.

Why it works: The mathematical analysis of what is going on here is deferred to page 30.

Source: Original. The principle involved was stumbled on in the spring of 2003 while living in Las Rosas, on the northwestern outskirts of Madrid, playing around with Jim Steinmeyer's "The Nine Card Problem" [Steinmeyer 93, Steinmeyer 02] (perhaps more well known today as "Nine Card Speller"), Bob Hummer's CATO Principle [Diaconis and Graham 11], and the George Sands Prime Principle [Fulves 75] (we have more to say about these three in Chapter 9).

This was published online at MAA.org in October 2004 as the inaugural *Card Colm* "Low Down Triple Dealing" [Mulcahy 04_10], dedicated to Martin Gardner, the best friend mathematics ever had, on the occasion of his ninetieth birthday. It also appeared in print at that time [Mulcahy 04_11], and a few years later [Mulcahy 08] in slightly expanded form, with mention of a generalization that we discuss in Chapter 5.

Presentational options: You could have the spectator handle the cards throughout if you think you can get a peek at the bottom card after the first round of shuffling is done—it's surprising how careless and "revealing" some people's handling of cards can be! In such cases, the bottom quarter of the deck is used—something you must make seem natural—and the additional in-hand mixing possibility is forgone. Most importantly, you will need to direct the three rounds of spelling and dropping carefully.

One way to ensure you learn the bottom card is to first take the deck back after the initial shuffling, then talk about the coming dealing, sneaking a peak when squaring up the deck as suggested earlier, while talking about the upcoming dealing. Finally, say, "Why don't you do all the work?," handing the spectator about a quarter of the deck, taken from the bottom.

The following presentation is popular, and baffling for most audiences: ask for the ice cream flavor to be called out as you shuffle freely, peeking at the bottom card at the last minute before you set the deck on the table. Then, having explained that there are three low-calorie scoops and topping to follow, pick up the deck and set aside the top three-quarters or so. Start mixing the remaining cards in hand, being careful to keep the bottom card in place. Next, discreetly move this key card to the top, and shuffle the rest with abandon, keeping the peeked-at card on top. You can even flash many of the card faces, including the ever-changing bottom ones, saying, "Note that these cards are all different and hopelessly jumbled." Add, disingenuously, "I haven't a clue what any of them are." As you lower the cards, boldly shuffle the top card back to the bottom. You are now ready to proceed, and the audience has the impression that the cards are totally randomized, which makes the conclusion positively perplexing.

Other options are discussed in the *Card Colm* cited above [Mulcahy 04_10]. Magician and mentalist Max Maven utilized the Low-Down Triple Deal principle in the "Final Destination" routine from his "One Man Parade" in the November 2006 issue of *The Linking Ring* [Maven 06].

An extension of the principle, which was first explored by others [Sirén 08, Miller 10], is discussed in detail in Chapter 5.

Our second example of mathemagic also happens in a spectator's hands, and uses the same mathematical principle as our opener. It could be performed instead of that effect, but we advise against doing both for the same audience. This one requires a little sure-footed bravado, and should not be attempted by anyone of a nervous disposition; hence, the ♦♦♦♦ rating. You need to start with a completely rigged deck, as indicated by the ♠♠♠♠ designation, so it's not repeatable. If you can pull it off, the crowd will love you (♥♥♥).

A♥ Any Card (and Any Magician)

How it looks overall: *Shuffle the deck over and over, as you ask a spectator to shout out the name of any card. Next ask for the name of any magician. You deal cards from the deck to the table, one for each letter in the magician's name, pick up those cards and do additional in-hand mixing. You then demonstrate a spelling and dropping routine twice, and hand the packet to the spectator, who now spells and drops. At the conclusion of this, the named card is miraculously found to be at the top of the packet.*

How it looks in detail: Suppose the spectator responds with "Eight of Clubs" and "Martin Gardner." Shuffle the cards one more time, and then deal cards to the table, face down, of course, one for each letter in the magician's name. Set the rest of the deck aside. Pick up the (thirteen) cards on the table and say, "Random cards, in a random order," as you mix them some more in your hands. "Let's see if Martin Gardner can help us to locate the desired card. I want to show you how to use his name to further randomize the cards."

Deal out cards to the table again, one for each letter in "Gardner," as you spell that word out loud. Drop the rest (there should be six) on top. Now peek at the top card and say, "No." Pick up the cards and repeat the spelling and dropping, again peeking at the top card and expressing mild disappointment at what you see. "Something tells me that it won't work until the third try, or maybe I just don't have the touch," you say, as you give the cards to the spectator and request that the experiment be repeated one more time.

Upon completion of this last round of spelling and dropping, pause and remind the audience that you had absolutely no control over what card or magician was named. Have the top card turned over, as you tentatively say, "Third time lucky?" It is indeed the Eight of Hearts. "Congratulations! Maybe I should have let you do all the work earlier on, too!"

How it works: At the outset, the deck should be arranged in such a way that as you peek at the ever-changing bottom card while doing simple overhand shuffles, it's easy for you to get the named card to the bottom, and then pause. For instance, new deck order, in which the cards of each suit are together and in either ascending or descending order, works. The idea is that you cycle the cards around as you wait to be told the name of the desired card, and then keep "shuffling" until that card is on the bottom. Yes, it takes a little nerve to pull this off while being watched, but try it; it's a skill you can acquire.

For instance, to get the Eight of Clubs on the bottom, you could cycle the deck with overhand shuffles until some Club is on the bottom, and then either shuffle off a few more cards as necessary or move about twenty cards to the bottom twice in succession, and then perform that maneuver. This is the only tricky part, and yes, it requires practice and some nerve, but it is doable! It's also vital that nobody other than you sees this force card at the bottom.

The rest is easy enough. Shuffle the desired card to the top and do several more sloppy shuffles that keep it there. You may now safely let the audience glimpse the new bottom card, as it plays no role in what follows. Whatever magician's name is now suggested, deal out a packet as you spell out that name, one card for each letter. (The named card is now on the bottom of this packet, and care must be taken not to flash it.) Set the rest of the deck aside.

In the case of *Martin Gardner*, deal again to the table, from the thirteen-card packet, while spelling out loud the letters of Gardner. Drop the rest on top. In the case of *Ricky Jay*, deal from the resulting eight-card packet, while spelling the letters of Ricky, and then drop the rest on top. If one name is shorter than the other, it's important to use the longer of the two words when spelling and dealing before dropping.

When you peek at the new top card, as suggested, it will not be the desired one. Repeat with a second round of dealing and dropping, with similar results. Then hand the packet to the spectator, who by now should be familiar with the dealing and dropping routine. At the conclusion of the final dealing and dropping, the named card will indeed be on top. Like the previous effect, it's a straightforward application of the Low-Down Triple Dealing principle which we are about to explain.

Source: Original. The use of the magician name spelling in conjunction with Low-Down Triple Dealing appears in the October 2004 *Card Colm* [Mulcahy 04_10], but in the context of the next highlighted effect.

A First Look at Low-Down Triple Dealing

The mechanism behind our first two effects is a *reversed transfer* of some fixed number of cards in a packet—at least half—from top to bottom, done three times in total.

The dealing out (and hence reversing) of k cards from a packet that runs $\{1, 2, \ldots, k-1, k, k+1, k+2, \ldots, n-1, n\}$ from the top down, and then dropping the rest on top as a unit, yields the rearranged packet $\{k+1, k+2, \ldots, n-1, n, k, k-1, \ldots, 2, 1\}$.

For instance, reverse transferring eight cards from the packet A♥–K♥ yields: 9♥, 10♥, J♥, Q♥, K♥, 8♥, 7♥, 6♥, 5♥, 4♥, 3♥, 2♥, A♥.

Figure A.2 shows how this would appear if the cards were face up. Of course, in actual performance, all cards would be face down.

Figure A.2. Reverse transferring eight cards from A–K♥.

Figure A.3 shows how such a face-down packet would look before and after, if fanned face up. Note that the Ace moves from top to bottom.

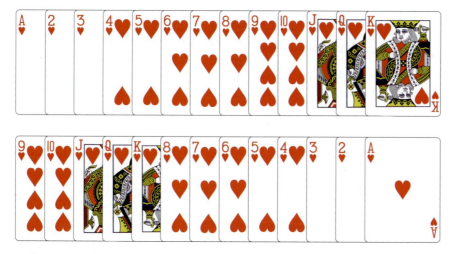

Figure A.3. Before and after reverse transferring eight cards from A–K♥.

If we reverse transfer another eight cards from this packet, face down, and then fan it face up, we obtain the first image in Figure A.4.

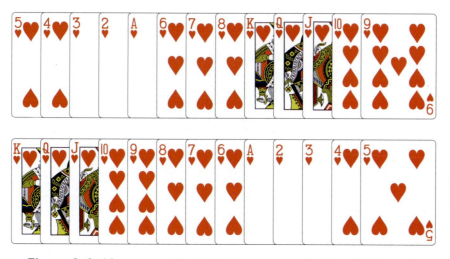

Figure A.4. After a second and third reverse transferring of eight cards.

A third such reverse transferral of eight cards yields the second image in Figure A.4.

Note that the original bottom card of our packet K♥ is on top after three such reverse transferrings of eight cards. This holds in general, and is the real secret behind our first two effects above:

Bottom to Top Principle ▶◀

The original bottom card of the packet ends up on top after three such reverse transferrings of k cards from n, provided that $k \geq \frac{n}{2}$.

It is certainly easy to see how this works if $k = n$ or $k = n - 1$, since we are actually reversing all of the cards in both cases. It's almost as easy to see if $k = \frac{n}{2}$, when we're reversing exactly half of the cards.

As we later demonstrate visually, it's not so hard to *see* in all cases. Before we explain why this magic property holds, we point out something that is more obvious and has its own applications.

Low-Down Deal Deck Separation

Reverse transferring at least half of a packet preserves top and bottom halves in a certain sense. For instance, if the packet starts with five Red cards on top of five Black cards, then no matter how many ($k \geq 5$) are reverse transferred, the Red cards will end up on the bottom (reversed, but that's not our focus here), and the Black cards will end up on top of those (rearranged a little).

Specifically, reverse transferring seven cards from the packet running 2♦, 10♥, 5♥, 6♦, 9♥, A♣, K♠, 7♠, Q♣, 4♣, from the top yields: 7♠, Q♣, 4♣, K♠, A♣, 9♥, 6♦, 5♥, 10♥, 2♦. Figure A.5 shows such a face-down packet before and after, fanned face up.

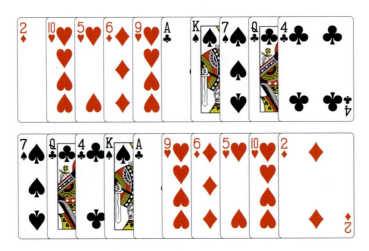

Figure A.5. Color separation if reverse transferring seven from ten.

This generalizes as follows.

> ### Low-Down Deal Packet Separation Principle ▶◀
> *If k cards are reverse transferred from the top to the bottom of a packet of size n, and $k \geq \frac{n}{2}$, then the top and bottom halves switch places, subject to some internal reordering.*

If the packet size is an odd number, then the middle card is effectively fixed, and the sets of cards initially above and below this fulcrum are the ones that exchange places, getting rearranged in the process.

Odd Location Method

If you're feeling cheated because our first two effects are so similar, here's an extra offering that is easy to master. It's based on the fact that the middle card in an odd-sized packet is doomed to return to that location over and over when at least half of the cards are reverse transferred.

A spectator picks a card at random from a shuffled deck, noting and remembering its face. The selected card is sandwiched between two groups of cards that you offer, following which the spectator mixes the cards repeatedly using words chosen at random from a list you have left in full view. These are words that might describe you, the mathemagician, such as "affable," "engaging," "persistent," "bewildering," and so on.

Remind the audience that you have never touched the selected card. Announce that the spectator can now find it herself, without seeing any card face, "Using an odd method of location I have devised."

Have the cards dealt into two new piles, left to right. The one with an odd number of cards in it is picked up, the other being set aside. Repeat. In short order, the spectator's packet of cards will reduce to just one. Ask what her selection was, and have the surviving card turned over to confirm that she found it successfully despite all her mixing.

The only thing required for this to work is that the two groups of cards you offer be of the same size. We suggest six of one and half a dozen of the other, resulting in the selected card being in the middle of a thirteen card packet. You may offer to have the selected card sandwiched between groups of four and eight cards if you don't mind handling them yourself, later casually shuffling two cards from front to back while distracting the audience with the word list.

An arbitrary number of reverse transfers of at least half of the cards can now be done without altering the fact that the selected card is in position 7, provided that each word spelled out has between seven and thirteen letters. We suggest providing humorously immodest words of no more than eleven letters.

The elimination deal is well known: starting with thirteen cards we then get piles of size seven, then three, then a lonesome card, and that final card is the one that started out as the middle card of the thirteen.

It can be modified to work for other-sized packets. You can hide the selected card between two groups of size seven if the words on the list are adjusted appropriately (e.g., omit "affable" and throw in "infuriating"). This could be the basis of a repeat performance.

Two months after this was conceived, a more romantic take on the idea appeared as the *Huffington Post* blog "In My Heart of Hearts: Valentine's Day Special" [Mulcahy 13_02b].

The Top, Middle, Bottom Decomposition

Let's do some bookkeeping. It turns out that when reverse transferring the same number of cards over and over, you have to keep track of three portions of the packet. These three portions move around intact, subject at most to some internal reversals.

Note that since $k \geq \frac{n}{2}$, we have $k - (n - k) = 2k - n \geq 0$.

Gallia est omnis divisa in partes tres[3]

Writing $n = k + (n - k) = [(n - k) + (k - (n - k))] + (n - k) = (n - k) + (2k - n) + (n - k)$, *we see that a packet of n cards naturally breaks symmetrically into three pieces T, M, B, (top, middle, and bottom) of sizes $n - k$, $2k - n$, $n - k$, respectively.*

Starting with the packet $\{1, 2, \ldots, n\}$, we thus get

$$T = \{1, 2, \ldots, n - k\},$$
$$M = \{n - k + 1, n - k + 2, \ldots, k\},$$
$$B = \{k + 1, k + 2, \ldots, n\}.$$

Needless to say, if $n = 2k$, i.e., exactly half of the packet is dealt each time, then M is nonexistent (that is not a problem).

For instance, if $n = 13$, and $k = 8$, and we start with $\{1, 2, \ldots, 12, 13\}$, we have $T = \{1, 2, 3, 4, 5\}$, $M = \{6, 7, 8\}$, and $B = \{9, 10, 11, 12, 13\}$, as seen in Figure A.6. Counting out eight cards (i.e., T and M together) and dropping the rest on top yields $\{9, 10, 11, 12, 13, 8, 7, 6, 5, 4, 3, 2, 1\}$, that is B followed by M reversed followed by T reversed. Note that the middle card, in position 7 in this case, remains fixed throughout.

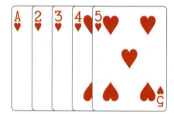 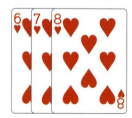 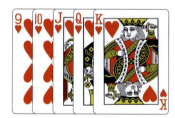

Figure A.6. *T, M, B when reverse transferring eight cards from thirteen.*

In general, counting out (hence reversing) the first k cards (i.e., T and M together) and dropping the rest on top leads to B followed by M reversed followed by T reversed. We denote this basic count out and transfer operation by using the notation $X, Y, Z \rightarrow Z, \overline{Y}, \overline{X}$, where the bar indicates a complete subpacket reversal.

---

[3]Classical Roman Empire references are divided into three parts.

So under the first count out and drop, we find that $T, M, B \rightarrow B, \overline{M}, \overline{T}$. Note that the middle card in the packet is fixed if n is odd, and the middle two cards trade places if n is even. In fact, cards equidistant from the center trade places, as long as they are not too close to either end.

The second round of counting and dropping yields $B, \overline{M}, \overline{T} \rightarrow \overline{T}, M, \overline{B}$, since two reversals of a subpacket restore it to its initial order. The third round results in $\overline{T}, M, \overline{B} \rightarrow B, \overline{M}, T$, and hence the original bottom card is now on top, just as we wished to show.

We can actually say a lot more, and we will shortly.

A COAT by Any Other Name

Before returning to the analysis of Low-Down Triple Dealing, let's give the move involved a new name. Note that reverse transferring is different from simply *cutting* cards from top to bottom, without altering their order. (Cutting k cards—either individually or as a group—merely cycles everything around, changing $\{1, 2, \ldots, k-1, k, k+1, k+2, \ldots, n-1, n\}$ into $\{k+1, k+2, \ldots, n-1, n, 1, 2, \ldots, k-1, k\}$.)

The key order reversal in a reverse transfer amounts to **C**ounting **O**ut **A**nd **T**ransferring cards from top to bottom, which suggests an easy-to-remember acronym, COAT. It is vital to note that the first two letters stand for Count Out, not Cut Off! (In Chapter 9, we'll examine other situations in which we *do* cut off and transfer.)

> ### COAT (Count Out And Transfer)
> *Given a packet of n cards, COATing k cards refers to counting out that many from the top into a pile, thus reversing their order, and transferring those as a unit to the bottom.*

Up until now, we have implied that the counting out is done to a table, with the rest of the packet then being dropped on top. An alternative handling is to **C**ount **O**ut (or push off with a thumb) cards from one hand to a second, one by one, to form an ever-growing new pile, **A**nd then **T**uck that behind the remainder of the initial packet, as shown in Figure A.7. Any way you COAT it, the result is the same.

Since the focus in this chapter is on the case where $k \geq \frac{n}{2}$, we often refer to that as *overCOATing*. If $k = n - 1$, we end up reversing the whole packet, just as when $k = n$, and doing this twice puts everything back where it started. Also, if $k = 1$, we are just cutting a single card to the bottom. These borderline cases are mostly ignored in what follows, as they are a bit confusing when handling cards in the reverse transfer context.

Figure A.7. In-hand COATing.

A Second Look at Low-Down Triple Dealing

As hinted above, overCOATing the same number of cards three times yields more than we have highlighted so far. Not only is the bottom card moved to the top, the entire bottom half of the packet (and then some) is moved intact to the top, in reverse order.

> ### Save at Least 50% Principle ▶◀
> *If k cards from n cards are dealt out into a pile, reversing their order, and the remaining $n - k$ are dropped on top, and this process is repeated twice more, then provided that $k \geq \frac{n}{2}$, the original k bottom cards become the top k cards, in exact reverse order. That is to say, three overCOATs preserve at least half the packet—the bottom half—only in reversed order, at the top.*

A close examination of our earlier analysis reveals why this holds. We know that three rounds of overCOATing transforms T, M, B to $\overline{B}, \overline{M}, T$, and since the subpacket M, B is the original k bottom cards, they end up on top here, reversed, as the subpacket $\overline{B}, \overline{M}$.

For example, overCOATing $\{1, 2, 3, 4, 5, 6, 7, 8, 9, 10\}$ three times, seven cards at a time, transforms it into $\{10, 9, 8, 7, 6, 5, 4, 1, 2, 3\}$, yielding what one might call a (reversed) saving of 70%, on top of everything else.

To put it another way, if you want to get the bottom five cards of a thirteen-card packet to the top, in reverse order, there is an alternative to the obvious method of counting out all of the cards into a pile to reverse their order. Simply select a number between seven and eleven, and COAT that many cards, three times over. This can be done casually in-hand, to give the illusion of mixing the cards (see Figure A.7).

Shortly, we present a visual proof of all of the above. It is instructive to experiment with a face-up packet of about a dozen cards in a known order (e.g., numerical), and COAT them, say, eight at a time.

The next effect requires no setup, and is particularly impressive if you know some sleight-of-hand card control. If that sounds daunting, don't worry, we also suggest a more open handling that is perfectly acceptable.

A♠ Triple Revelation

How it looks: *Have three spectators each pick one card at random, look at it, and memorize it. Return those three cards to the deck, and shuffle.*

Ask a fourth person to name their favorite magician; let's suppose Bill Simon is selected. Hold the deck in one hand and peel cards off the bottom into the other hand without altering their order, one for each letter, as you spell out the whole name.

Hand the stack of cards (here, nine) to the first spectator and ask that "Simon" be spelled out while dealing five cards to the table, then dropping the other four on top. Give the cards to the second spectator, with the same directions, and finally to the third spectator for one last deal and drop. Take the cards behind your back and immediately produce three cards, handing one to each spectator, face down.

Have the chosen cards named, as they are turned over, to reveal that you have correctly located each one.

How it works: The selected cards must first be returned to the bottom of the deck, in a known order. You can either do this openly, or, if you have some magic chops, "control cards to the bottom"; this means that you seemingly return each card to random parts of the deck but secretly get them all to the bottom in due course. The pinky break (see page 11) may help. However you pull it off, these three key cards will then remain in position throughout some riffle shuffling, if you are careful first to drop at least three cards from the bottom of the deck each time.

Let's assume that, as a result, the third spectator's card is at the bottom of the deck, the second spectator's card is one up from the bottom, and the first spectator's card is two up from the bottom. Peel cards off the bottom of the deck without altering their order, one for each letter of the name of the magician called out, as you spell out both words in full. Hand the resulting packet of cards to the first spectator and ask that the longer of the two names (Simon in our example) be spelled out, as cards are dealt into a pile, before dropping the remainder on top. Now give the cards to the second spectator and finally to the third spectator for two more deals. The three chosen cards are now on the top of the packet of cards, with the order reversed, and you are all set to conclude in triumph.

Why it works: This works because of the Save at Least 50% Principle.

Source: Original. It appeared online as the October 2004 *Card Colm* [Mulcahy 04_10].

Low-Down Quadruple Dealing

We have one further confession to make. The Low-Down Triple Deal explored above is actually 75% of a special quadruple deal. Note that since three overCOATs takes the bottom part of a packet to the top, reversed, one more overCOAT will restore those cards to their original positions at the bottom. Hence at least half of the packet is restored to its original order. Here's the full scoop (or real low down):

> ### Four OverCOATs Principle   ►◄
> *If four reversed transfers (overCOATs) of k cards are done to a packet of size n, where $k \geq \frac{n}{2}$, then every card in the packet is returned to its original position.*

In other words, you'll never notice the effect of four overCOATs. In a sense, they cancel each other out. Thus, we have a false shuffle worth adding to one's portfolio.

Casually running off a number of cards gives the appearance of shuffling (see page 10). If done with purpose, it does the opposite.

> ### Quad False Shuffle Principle   ►◄
> *Four applications of the following move restores a packet of size n to its original order, assuming $k \geq \frac{n}{2}$. Run off k cards from the top of the packet into a waiting hand, thus reversing their order, and then tuck them as a unit behind the remainder (or just drop the rest on top).*

Done casually four times in-hand (as shown on page 10) to a small packet (e.g., COATing six cards out of ten each time) gives the illusion of mixing the cards. As noted earlier, three such COATs moves the original bottom stock to the top, in reversed order. *Top* or *bottom stock* simply refers to a clump of adjacent cards of interest at one end of the deck (see page 10).

One way to see why this quad property holds is to go back to our earlier analysis: we already saw that the symmetric breakdown of the packet into subpackets T, M, B (top, middle, bottom) of respective sizes $n - k$, $2k - n$, $n - k$, results in $\overline{B}, \overline{M}, T$ after three overCOATs. After a fourth overCOAT, $\overline{B}, \overline{M}, T \to T, M, B$, which is indeed back to where we started, as desired.

A Visual Approach to Low-Down Dealing

Perhaps the best way to *see* what really is going on here is with pictures. Suppose, for the sake of concreteness, that $n = 13$ and $k = 8$. Let's represent a packet of thirteen cards by a vertical strip of gray-scale panels in decreasing order of brightness, from white for the top card to black for the bottom card, as depicted in the leftmost strip of Figure A.8.

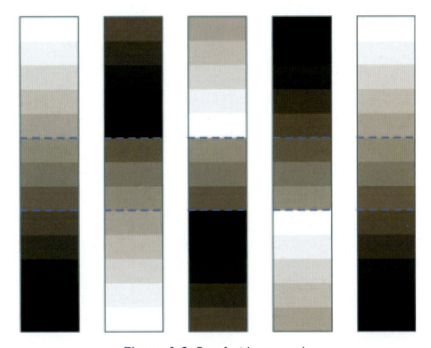

Figure A.8. Proof without words.

The results of the four overCOATs of eight cards are given by the successive vertical strips. The last strip shows a fully restored packet, so this overCOAT sequence has a period of four: after four reverse transfers, we are back to where we started.

It is also clear from these images why the original bottom card (represented as a black panel here) has risen to the top after three overCOATs, in preparation for its final journey back to the bottom under one more overCOAT.

Moreover, the eight bottom cards become the eight top cards, reversed, after three overCOATs. We have to keep track of three portions of this packet with sizes of five, three, and five—and they move around intact, subject at most to some predictable internal reversals.

The only relationship between 13 and 8 that is needed to make this sequence of images totally generalizable is $8 \geq \frac{13}{2}$, so a similar succession of strips can be constructed to illustrate any case of interest to us here.

The period is two, not four, if $k = n - 1$ or $k = n$, and for very small packets the period is only one or two.

Interested readers can find other effects that take direct advantage of this principle in the October 2004 *Card Colm* [Mulcahy 04_10].

The Special 4-Cycle

Our final effect depends on a deeper analysis of overCOATing. Recall that if $n = 13$ and $k = 9$, then $\{1, 2, \ldots, 12, 13\}$ can be thought of in terms of these three subpackets: $T = \{1, 2, 3, 4\}$; $M = \{5, 6, 7, 8, 9\}$; and $B = \{10, 11, 12, 13\}$. This facilitates the tracking of COATing nine cards at a time, and a little exploration reveals that the top position is occupied by 1, 10, 4, and 13, and then 1 again, in that order, as we go through four COATs.

Note that those four top visitors are the bookends of T and B. These cards are also the bottom visitors, and the inhabitants of the middle M are kept away from the top and the bottom of the packet by the more experienced travelers in T and B.

More generally, when repeatedly COATing nine of thirteen cards, the cards that start in positions 1, 10, 4, and 13 cycle around those four key slots in a well-determined fashion, and can be relied upon to be in one of those slots no matter how many COATs are applied.

Also, it turns out that the starting top card moves to positions 13, 4, and 10, in that order, before returning to position 1 again. So the orbit of any of those four positions is the same as the top visitors.

We can explain that in the language of permutation decompositions, as presented in summary form in the "Convention Center" (see page 19). The good news is that we already have explained the origins of the resulting special 4-cycle in the example above.

The disjoint cycle decomposition of the permutation associated with any COAT varies somewhat, depending on the specific values of k and n. In the cases in this chapter, however, where at least half of the cards are COATed, the associated permutations must be products of disjoint 2-cycles and 4-cycles, since their period (namely, four) is the least common multiple of the lengths of the cycles they comprise (see page 20).

For example, let's assume that the cards are numbered $1, 2, \ldots, n$. Then if $n = 10$ and $k = 6$, we find that the associated permutation, let's call it τ, is $(1\ 10\ 4\ 7)(2\ 9\ 3\ 8)(5\ 6)$, whereas if $n = 11$ and $k = 9$, we obtain $\tau = (1\ 11\ 2\ 10)(3\ 9)(4\ 8)(5\ 7)$ with card 6 being fixed.

As suggested by those examples, four key card positions are closely linked here. We saw them before if $n = 13$, and $k = 9$, and for the record, in that case, the cycle decomposition of τ is $(1\ 13\ 4\ 10)(2\ 12\ 3\ 11)(5\ 9)(6\ 8)$ with the card in position 7 being fixed.

In general, considering four applications of $X, Y, Z \to Z, \overline{Y}, \overline{X}$, we get the following result.

Special 4-Cycle Principle ▶◀

If $k \geq \frac{n}{2}$, then under a sequence of four COATs, the initial top card orbits through positions n (the last card in B), $n - k$ (the last card in T), and $k + 1$ (the first card in B), in turn, before returning to the top (the first card in T).

Equivalently, the top position is visited by the cards originally in positions $k + 1$ (the first card in B), $n - k$ (the last card in T), and n (the last card in B), in turn, before the original top card returns there.

Hence, τ includes the 4-cycle $(1\ \ n\ \ n-k\ \ k+1)$ in its decomposition. In other words, the cards in positions 1, $k + 1$, $n - k$, and n—namely, the top and bottom cards and those k places from each end—have a special relationship to each other. Algebraically, there is a factorization $\tau = (1\ \ n\ \ n-k\ \ k+1)\ \sigma$, where σ fixes each of $1, n, n - k$, and $k + 1$.

As we saw above, if $n = 10$ and $k = 6$, the special 4-cycle is $(1\ 10\ 4\ 7)$, and if $n = 11$ and $k = 9$, it is $(1\ 11\ 2\ 10)$. The cards that start in the positions 1, 10, 4, and 7 cycle around those key slots under four overCOATs. Similarly, if $n = 13$ and $k = 8$, the special 4-cycle is $(1\ 13\ 5\ 9)$, and the cards that start in positions 1, 13, 5, and 9 cycle around those key slots in a well-determined fashion.

Let's now utilize all of this for a four Ace effect.

A♦ Ace Combination

How it looks: *A spectator is invited to select and write down three digits a, b, c at random. Tell the audience, "It's for a combination lock, so it's best to avoid obvious ones such as $1, 2, 3$ or $3, 2, 1$." The first two digits, a and b, are used to determine the sizes of three packets that are removed from a shuffled deck before being assembled and handed to the spectator for c rounds of dealing and dropping.*

Take the combined packet back and review the randomness of the proceedings so far. Say that you suspect the cards have a secret locked in them that you will try to discover by two more rounds of dealing and dropping, with your hands under the table (or behind your back). Upon completion of this, spread the cards face up to reveal that four cards now face the other way. Turn the spread packet over to reveal the four Aces, as you conclude, "I guess that was an Ace combination you picked!"

How it works: The deck is set up with two Aces at the top and the other two at the bottom. This condition can be maintained through some convincing-looking shuffles. Have a spectator select a, b, and c and write them down; it's actually best if both a and c are at least 3. For instance, $4, 5, 4$ or $9, 1, 3$ are good combinations for this effect.

Deal out a pile of a cards, so that two Aces will be at the bottom, and then pick it up and hold it face down in your hands. Hand the rest of the deck to the spectator to deal out a second pile of b cards. While this is taking place, casually shuffle the bottom Ace in your pile to the top, perhaps with your hands lowered or behind your back.

Now place your pile on the table beside the spectator's, and request that one of the piles be placed on top of the other. This will, of course, form a packet of size $a + b$. While this is taking place, discreetly move the other two Aces from the bottom of the remainder of the deck to the top. Next, deal another pile of size a; pick this up, and remind the spectator that you had no input into the size of the earlier piles, or the order in which they were assembled. While you are saying this, slip to the top the bottom Ace of the pile in your hand. Finally, casually assemble the two packets in such a way that the pile of size b that the spectator dealt is now sandwiched between the two piles of size a that you dealt.

This elaborate setup ensures that, unknown to all but you, the four Aces are in positions $1, a, a + b + 1$, and $2a + b$ in a packet of size $2a + b$. In other words, the Aces are in positions 1 and a from the top and 1 and a from the bottom. No matter what value c is, that number of dealings and droppings (overCOATs) automatically cycles the Aces around within these four key positions.

When you take back the apparently totally randomized packet, you know where the Aces are. Hold the cards out of view, and silently turn over the top and bottom cards. Next, noisily transfer $a + b$ cards from top to bottom, reversing their order as you do so, and again silently turn over the new top card. Repeat. Overhand shuffle just before you bring the cards back into view, to conceal the regularity of the key positions. As all four Aces have been flipped over, you are ready for the dramatic finale.

Why it works: This is a direct application of the Special 4-Cycle principle. The numbers 1, a, $a + b + 1$, and $2a + b$ play the roles of 1, $n - k$, $k+1$, and n, respectively. For all positive values of a and b, corresponding values of k and n are uniquely determined, and, most important, $n \leq 2k$ since $2k - n = 2(a + b) - (2a + b) = b$.

For instance, if $a, b, c = 4, 7, 5$, then $k = a+b = 11$ and $n = 2a+b = 15$. In practice, you start off by dealing four cards to the table, the first two of which are Aces, then pick up this packet. While the spectator deals seven cards to the table, surreptitiously move the bottom Ace in your pile to the top. Let's assume that the spectator opts to reassemble the piles with yours on top; while that is being done, slip the last two Aces on the bottom to the top of the rest of the deck. Deal out four cards again, thus putting those Aces on the bottom of that pile. Pick it up and distract the spectator with claims of randomness while you slip the bottom card to the top. In this case, the final reassembly is done by placing those four cards underneath the packet that the spectator assembled.

Unsuspected by anyone, the Aces are now in positions 1, 4, 12, and 15, and they'll cycle around those slots when the spectator indulges in all of those overCOATs—the five rounds of dealing eleven of the fifteen cards, and dropping the remaining four on top each time.

Source: Original. From June 2003.

Parting Thoughts

- Show that COATing k cards from a face-down packet of n cards can be undone as follows:

    ### UnCOATing
    Assume k cards have been COATed from a packet of size n, that puts the original top card on the bottom. Turn the whole packet face up, and COAT k cards again, before finally flipping the whole packet face down once more.

    This restores any packet to its initial order, regardless of the value of k—for once we don't need to assume that k is at least half of n. Try it for three cards COATed from eight.

- Show that an overCOAT sends any card in the first half of a packet of size n, say, the one in position i ($\leq \frac{n}{2}$), to position $(n + 1) - i$ in the second half. Where does it send the rest of the cards?

- Having drawn attention to a special 4-cycle in the cycle decomposition of the permutation corresponding to an overCOAT, it's natural to ask if there are other 4-cycles of interest from a magic perspective? Or 2-cycles (i.e., transpositions or card switches) not involving simple position trades in the middle part?

- All that's necessary to pull off an effect such as "Three Scoop Miracle" (A♣) is to get the bottom card to the top. We did it with three overCOATs of k from n cards. In a sense, only the middle of those had to be a real COAT. Take any packet of about a dozen cards and peek at the bottom one. Fix a number such as eight. Push off eight cards from the left hand to the right, collecting them in any random order, some on top of the growing pile, some underneath. Tuck the new pile of eight cards underneath the rest. Now COAT eight cards carefully, i.e., reverse their order by counting out and then tucking them behind the others. Finally do another sloppy transfer of eight cards from the left to the right hand, in any order, and tuck those behind what's left. The resulting top card is still the one you first peeked at, as if you'd done three COATs of eight cards.

  ### COAT Sandwich Principle   ▷◁
  If $k \geq \frac{n}{2}$, the original bottom card of the packet ends up on top after three "transfers" of k cards from n as above, as long the middle one is a COAT, i.e., a proper count out and transfer.

  There's lots of flexibility here: the first and third transfers can be straight cuts,[4] the favored reversed transfers (cutting cards that are first counted out precisely), or any jumble of the k cards involved.

- Our focus has largely been on getting the bottom card to the top with three overCOATs. Assuming that $k \geq \frac{n}{2}$, discover a dual move to COATing k cards from n that also has period four but takes the original top card to the bottom after three applications. Hint: COATing refers to **C**ounting **O**ut (hence reversing) the top k cards **A**nd **T**ransferring (cutting) those to the bottom. The desired dual move corresponds naturally to the acronym TACO, however, different numbers of cards are moved in its two key parts.

  Adapt "Three Scoop Miracle" (A♣) to this COAT dual setting. Simply have the spectator peek at and remember the top card after mixing a given packet, and take the cards back at the end of the three requisite moves. Under the guise of shuffling the packet

---

[4] As if cutting bread to make a sandwich.

further, you can peek at the bottom card and use the information to good effect, perhaps dressing up your performance as a mind-reading stunt. There is a downside, however, in that the person handling the cards essentially needs to know how big each topping is (or, equivalently, how many cards are in the packet in total). We say more on this topic at the end of this section.

- Returning to COATs, how many are needed to restore a packet to its original order if we deal out fewer than half of the cards each time? In other words, what is the period of an underCOAT, if the same number k of cards is always dealt out?

  Suppose that for some k and n (the latter more than twice the size of the former), you find that twelve COATs restores the packet to its original order. What happens after three, four, or six COATs? Are there interesting small sets of cards that get interchanged only among themselves each time? (This is the way in which the cards in positions 8 and 10 are exchanged repeatedly with any overCOAT of fifteen cards.)

- If $k = 2$ and n is even, what is the period of the corresponding COAT?

- If $k = 2$ and $n = 2s + 1$ is odd, see whether the original top card returns to the top after $s + 1$ rounds. If so, must the period of this COAT also be $s + 1$?

- If $k < \frac{n}{2}$ and n is a multiple of k, say, $n = dk$, see what happens after $2d$ COATs.

- If $k < \frac{n}{2}$ and $n - 1$ is a multiple of k, say, $n - 1 = dk$, see what happens after $2d + 1$ COATs.

- In September 2012, Jay Cummings, a doctoral student of Ron Graham's at the University of California at San Diego, shared the following result with proof. Suppose we COAT k cards from n. Then to find the period, we first divide k into n to get $n = ak + b$ with remainder $0 \leq b \leq k - 1$. Next peel off as many powers of 2 as possible from a, so that $a = 2^\alpha \gamma$ with γ odd, and $n = (2^\alpha \gamma)k + b$.

  Case 0: As hinted above, if k divides n exactly (so that $b = 0$), then it is not difficult to determine the period. It is n if $k = 1$, and $\frac{2n}{k}$ if $k > 1$.

  To understand what is going on if $b > 0$ (i.e., when $\frac{n}{k}$ is not a whole number), we must first consider the architectural product $\lfloor \frac{n}{k} \rfloor \lceil \frac{n}{k} \rceil$, built from the floor and ceiling of the fraction $\frac{n}{k}$.

For definitions of these terms, the *floor* of 5.2, $\lfloor (5.2) \rfloor$, is 5, and its *ceiling*, $\lceil (5.2) \rceil$, is 6. For instance,

$$\left\lfloor \frac{11}{8} \right\rfloor \left\lceil \frac{11}{8} \right\rceil = 1 \times 2 = 2, \qquad \left\lfloor \frac{11}{5} \right\rfloor \left\lceil \frac{11}{5} \right\rceil = 2 \times 3 = 6,$$

$$\left\lfloor \frac{11}{4} \right\rfloor \left\lceil \frac{11}{4} \right\rceil = 2 \times 3 = 6, \qquad \left\lfloor \frac{11}{3} \right\rfloor \left\lceil \frac{11}{3} \right\rceil = 3 \times 4 = 12,$$

$$\left\lfloor \frac{11}{2} \right\rfloor \left\lceil \frac{11}{2} \right\rceil = 5 \times 6 = 30, \qquad \left\lfloor \frac{14}{4} \right\rfloor \left\lceil \frac{14}{4} \right\rceil = 3 \times 4 = 12,$$

$$\left\lfloor \frac{14}{3} \right\rfloor \left\lceil \frac{14}{3} \right\rceil = 4 \times 5 = 20, \qquad \left\lfloor \frac{20}{3} \right\rfloor \left\lceil \frac{20}{3} \right\rceil = 6 \times 7 = 42.$$

(For the record, $\lfloor \frac{14}{2} \rfloor \lceil \frac{14}{2} \rceil = 7 \times 7 = 49$, although we won't need this.)

Here is the less obvious result in the cases not explicitly covered above. The Four OverCOATs Principle, which we have explored at length, falls out of this in due course, as you should check.

Case 1: If $k \geq 3$, then the period is $\lfloor \frac{n}{k} \rfloor \lceil \frac{n}{k} \rceil$ if either $b = 1$ and $\alpha = 0$, or $b = k - 1$ and $\alpha > 0$. The last condition says that n is within 1 of an odd multiple of k.

Case 2: Otherwise, the period is $2 \lfloor \frac{n}{k} \rfloor \lceil \frac{n}{k} \rceil$.

- On page 44 we suggested using TACO for a natural dual of COATing: a kind of **T**ransferring **A**nd **C**ounting **O**ut of cards from top to bottom of a packet. It's faster and easier to do all of this in-hand, dispensing with a table, and implementing the counting out by passing cards from one hand to the other as necessary.

What we didn't mention, but readers may have discovered, is that in order for three applications of that dual move to bring the top card of a packet of size n to the bottom, the transferring actually refers to first cutting $n - k$ cards from top to bottom, followed by counting out (to reverse their order) the new top k cards *and replacing them on top again*. As usual, we assume $k \geq \frac{n}{2}$, so that $n - k \leq k$. Hence we are transferring no more cards than we are reversing, and we emphasize that by using TACO(LM) for **T**ransferring **A**nd **C**ounting **O**ut (**L**ess and **M**ore) cards (respectively).

The replacing-on-top part makes the whole thing feel like a three stage move and hence more complicated than COATing. Also, it seems inevitable that the card handler knows the exact value of n in order to know how many cards to cut. However, there is a way around both of these issues.

Note that in practice, TACO(LM)ing is generally done several times over. Next, cutting $n - k$ cards from n is the same as leaving k of them uncut. Hence, to cut $n - k$ cards without consciously knowing n, simply fan the cards from the left hand to the right hand, using the left thumb and fingers to pull off k cards from underneath, then cut what remains in the right hand under those (without needing to count them, or even fan them fully). Now use something like a pinky break (see page 11) to maintain a gap between the $n - k$ cards at the back and the k that have just come to the front. This makes it especially easy to use the right hand to take those top k cards from the left, one at a time, reversing their order. They can effectively be counted out even if you don't consciously enumerate them. Drop them on top of the others and repeat: cut $n - k$ (which is less than or equal to half of the packet size) from top to bottom, and reverse the rest before replacing them on top.

Under a single TACO(LM), assuming that $k \geq \frac{n}{2}$, show that in the notation used on page 34, we have $T, M, B \rightarrow \overline{B}, \overline{M}, T$. Deduce that, as claimed earlier, such a overTACO(LM) move has period four, whereas three applications of it take the original top card to the bottom.[5]

In addition to obtaining a Four OverTACO(LM)s Principle and a Top to Bottom Principle, there are corresponding Low-Down Deal Packet Separation, Save at Least 50%, and Special 4-Cycle principles. Not to mention a TACO(LM) Sandwich Principle! Is the corresponding Quad False Shuffle Principle of any practical value?

Let's switch attention from k to $j = n - k$. Given any j and a packet of size at least $2j$, overTACO(LM) by cutting j cards from top to bottom, maintaining a break between the resulting two parts with any finger—it doesn't have to be at all subtle—then reversing what's on top by passing them from the left to the right hand one by one, before dropping them back on top and letting go of the break. Repeat as often as is required. Admittedly, if the person doing this is paying attention, the size of the packet can be deduced, but the point is that given a desired value of j, it can be done by a spectator using a packet whose size has only been crudely estimated. For instance, if $j = 4$, it can be done with any packet of eight or more cards: three times cutting four to the bottom and then reversing the rest, it being understood that the cards thus reversed are dropped back on top each time, takes the bottom card to the top.

---

[5]In fact, with the third application, we can be lazy and skip the reversal that is supposed to follow the cutting. Is there a corresponding shortcut we can take when triple overCOATing, which still puts the original bottom card on top?

A family of off-centered generalizations of overCOATs is explored
in Chapter 5, and COATs and TACO(LM)s (and surprising generalizations of both) resurface near the end of Chapter 9.

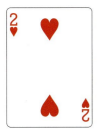

For Richer or Poorer

In his final years, Martin Gardner liked to share the observation that the number 8,549,176,320 had a curious property that could be discovered by dividing it by 5, then by 5 again. It has a more obvious special feature of a type that will play a role in this and some later chapters.

We're about to stumble upon several coincidental and fortuitous interactions involving numbers and alphabetical order, in English, that are fun to ponder. The magic effects they give rise to are more interesting, and totally independent of the language in which they are performed.

We'll consider several kinds of magic bracelets, namely, circles of numbers with characteristics that we can put to work for entertainment purposes.

Additional certainties are effects in which knowledge of the sum of a selection of numbers allows one to identify with certainty each of the constituent parts. This chapter starts with two of them, and Chapter 4 has two more.

The most impressive item here might well be our opener.

2♣ Alphabetical Triple Addition

How it looks: *Shuffle a deck, and deal out eight cards in a face-down circle. Address three spectators, saying, "While I turn away, please select three cards in a row next to each other in the circle, flipping those face up." Turn away as promised, and continue, "Now I want each of you to touch one of those three cards with your forefinger. Have you done that? Now please remember your card, and flip it face down again." When all*

of this has been done, turn back. Have the three card values summed. Ask what the total is, or find out surreptitiously.

Shuffle all eight cards back into the rest of the deck, and announce the identities of the three selected cards.

How it works: Of the $8! = 40{,}320$ ways that $1, 2, \ldots, 8$ can be arranged (see page 16), only $\frac{7!}{2} = 2{,}520$ count if we think of the numbers as arranged in a *bracelet*, namely, as a reversible and revolvable circle.

We focus on what we call the *g4g8 bracelet*, which is determined by connecting the ends of [8 5 4 1 7 6 3 2], as shown in card form in Figure 2.1. Note that the suits cycle in CHaSeD order (see page 13).

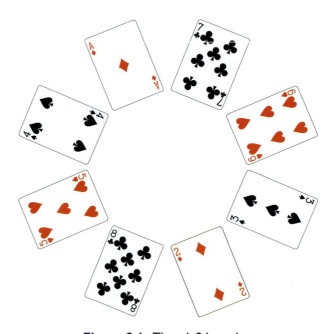

Figure 2.1. The g4g8 bracelet.

Assume we have such a face-down card bracelet, with values and suits memorized. Counting out in the order $1, 2, \ldots, 8$ reveals a pleasing geometric pattern.

Disguise the fact that you know anything by starting with these cards in order on the top of a deck, maintaining that stock during some apparently fair shuffling (see page 10). Then deal eight cards in a face-down circle following the outline of a capital omega, Ω, starting at the bottom left with the 8♣, and working around clockwise.

Surprisingly, the eight possible sums of three adjacent *beads* (i.e., card values) are distinct.

g4g8 Bracelet Principle ▶◀

If any three adjacent numbers in the g4g8 bracelet obtained by connecting the ends of [8 5 4 1 7 6 3 2] are summed, then it's possible to deduce which numbers were used, because we obtain the distinct triple sums 17, 10, 12, 14, 16, 11, 13, 15.

Here we have listed the triple sums in the order in which they turn up, as we go around the circle, starting with $17 = 8 + 5 + 4$, followed by $10 = 5+4+1$, and so on, ending with $15 = 2+8+5$. Ordering them, we get the consecutive numbers $10, 11, 12, \ldots, 17$. We have a *sum-rich* bracelet, meaning that as many different (triple) sums as possible are generated, for a circle of its size. The same can be said for the bracelet obtained from the numerical [1 2 3 4 5 6 7 8] that generates the eight triple sums 6, 9, 12, 15, 18, 21, 16, and 11. Compare those to [1 8 2 3 6 4 5 7], which yields *sum-poor* results if looped, because it generates only four distinct triple sums, namely, 11, 13, 15, and 16.

It would be nice, given a triple total arising from [8 5 4 1 7 6 3 2], to know which numbers it arose from. First look at what happens if we subtract 8 from each of the possible totals: we obtain 9, 2, 4, 6, 8, 3, 5, and 7. The order in which these occur can be deduced from the phrase:

Mnemonics to very easily remember one–eight desired.

(The first word in that phrase contains nine letters, the next has two, the third four, and so on.)

So if the total is 14, first subtract 8 to get 6, then *easily* determine that this corresponds to the fifth bead in the original bracelet. Provided you remember the suits and values of the beads, you deduce that the 7♣ and its two neighbors the A♣ and 6♥ were the selected cards, which you identify to thunderous applause.

Remembering the all-important [8 5 4 1 7 6 3 2] is not a problem. These numbers enjoy a very special property:

αΩ Principle ▶◀

Read in the shape of a capital Omega, the numbers 8, 5, 4, 1, 7, 6, 3, 2 are in alphabetical order: eight, five, four, one, seven, six, three, two.

It may be all Greek to the casual observer, but it's actually as easy as A, B, C.

Source: Original. Published online at MAA.org in June 2008's *Card Colm* as "Sum-Rich Circulants" [Mulcahy 08_06], having been unveiled that March as the "g4g8 bracelet" at the Gathering 4 Gardner 8 conference in Atlanta.

Presentational options: Those of an octal bent may wish to work with the equally alphabetic [5 4 1 7 6 3 2 0] instead, making modifications as required, and perhaps using a Joker in place of the zero.

A book page force implementation is suggested in the *Card Colm* cited [Mulcahy 08_06].

Sum-Rich Bracelets

There are sum-rich arrangements other than [8 5 4 1 7 6 3 2] that have distinct triple sums. Straight numerical order $[1, 2, 3, \ldots, 8]$ yields these triple totals: 11, 6, 9, 12, 15, 18, 21, 16, for instance, but this linear array has an obvious disadvantage when it comes to the kind of application considered in the last effect.

Other arrangements, for instance, [8 2 3 5 1 6 7 4], yield distinct double sums instead of distinct triple sums, and can be used to identify just two cards based on the sum of their values. Our favored [8 5 4 1 7 6 3 2] doesn't yield distinct double sums, however, since $5 + 4 = 6 + 3$.

Numerous arrangements, such as [2 3 5 7 8 4 6 1], yield distinct quadruple sums as well as distinct triple sums. For such arrangements, the sums of four adjacent card values could be asked (or fished) for, and all four cards successfully named.

Consider the general problem of identifying n numbers from their sum. It's trivial when $n = 1$ or $n = 13$ (i.e., using all possible card values, with none repeated). One could ask, for what number between one and thirteen would pulling off such a stunt be the most impressive?

Also of interest is the more vague query posed next, worth pondering here in the context of the above sum-rich bracelets.

> ### Conundrum (More or Less?)
> *Which is more impressive, being able to perform a certain effect when there are just a few cards involved, or when there are more?*

This nagging question should pop into readers' minds many other times in the pages and chapters to follow, in a variety of situations.

Product-Rich Bracelets

The g4g8 bracelet [8 5 4 1 7 6 3 2] is also product-rich for pairs, triples, and quadruples of adjacent beads. This means that when all possible products of adjacent sets of beads of those sizes are computed, we get different numbers. It follows that it is product-rich for any fixed number of adjacent beads.

For adjacent pairs, we get the distinct products 40, 20, 4, 7, 42, 18, 6, 16. For triples, we get 80, 160, 20, 28, 42, 126, 36, 48, and for four adjacent beads, we get 160, 140, 168, 126, 252, 288, 240, 320.

Those large numbers can be used in conjunction with a lengthy book, such as *Magical Mathematics* [Diaconis and Graham 11], for a kind of book force. Have that volume handy, and a helper on standby elsewhere. Have the cards chosen, the triple product reported, and the first word on the corresponding page of the book noted. Ask a spectator to share that word with your helper by phone or some electronic means (such as email or instant message).

Your helper will be able to respond promptly by naming the three selected cards with the help of a cheat sheet linking words to products to factors (and suits). Here are the required word/product matches for that particular book: one/20, to/28, and/36, subjective/42, forward/48, Stanford/80, the/126, cards/160.

The triple sums and triple products considered are also distinct from each other, which allows for this variation: have three adjacent cards chosen and say, "I'd like you to add or multiply the values you got—I don't care which—just tell me your overall answer."

Open-Ended Arguments

If we snip off the first bead of the g4g8 bracelet and open up the result to form [5 4 1 7 6 3 2] (which is 1–7 in alphabetical order), we have a doubly rich row: it is both sum- and product-rich for adjacent triples.

The open [8 5 4 9 1 7 6 3 2] (with the inclusion of 8 and 9, still in alphabetical order) is sum-rich for four adjacent beads, yielding the sums 26, 19, 21, 23, 17, 18.

Alphabetizing all ten digits—with cards, a Joker can be used to represent zero—we obtain the row [8 5 4 9 1 7 6 3 2 0], which is also sum-rich for four adjacent beads, the sums being 26, 19, 21, 23, 17, 18, 11.

Perhaps readers can come up with a good mnemonic to help out in the last two cases: given the total, first subtract 10, to get numbers that are within word-length range. With an appropriate mnemonic, the payoff is the ability to name any four cards in a row, knowing only the total value, from a display that can be passed off as random.[1]

Curiously, the modified row [0 8 5 4 9 1 7 6 3 2] (with the zero moved to the start) has adjacent sums of five beads alternating between 26 and 27, so it's sum-poor; we consider such observations later in this chapter.

♣   ♥   ♠   ♦

---

[1]Magician Chris Morgan suggests the following for 26, 19, 21, 23, 17, 18, 11 (after subtracting 10 from each): "Electromagnetics, chemistry, archaeology, ophthalmology, delimit syllabus I."

We have seen that sometimes we can succeed without knowing whether we have a sum or a product. What if we stick to sums, but don't know how many values were added up? We can still deal with it, so to speak, if we use the right bracelet.

2♥ Subtler Bracelet

How it looks: *Shuffle a deck, and deal out ten cards in a face-down circle. Turn away, and request that two or three spectators peek at and remember adjacent cards in the circle, one each. Ask one of the spectators to sum the values and tell you only the total. You shouldn't know how many spectators or cards are involved. Next, have all ten cards from the circle shuffled back into the rest of the deck.*

Turn back, and say, "I bet I can find your cards by going through the deck once, face up. Since I don't know how many cards are involved, it won't be easy, there are so many possibilities." Sure enough, you soon extract the two or three cards in question.

How it works: We focus on the ten-bead bracelet obtained by connecting the ends of [1 3 4 5 9 10 6 7 8 2], shown in Figure 2.2 in card form. It is one very special way to arrange $1, 2, 3, \ldots, 10$ in a reversible circle, among all $\frac{9!}{2} = 181{,}440$ possible ways to do so. We call it the *g4g10 bracelet*. Note that the suits cycle in CHaSeD order as much as possible.

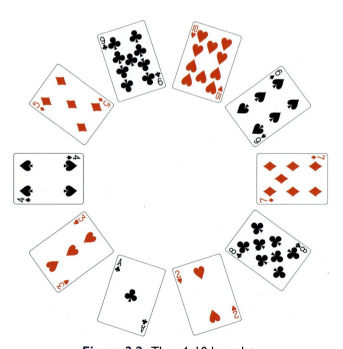

Figure 2.2. The g4g10 bracelet.

Assume you have such a face-down card bracelet, where both the values and suits are memorized. Disguise the fact that you know anything by starting with these cards in reverse order on the top of a deck, maintaining that stock during some seemingly fair shuffling, and then dealing ten cards in a (face-down) circle, following the outline of a capital omega, Ω (i.e., starting at the bottom left and working around clockwise).

The surprising thing is that the possible sums of any two *or* three adjacent card values are all distinct. Not only is this bracelet sum-rich, but knowing the sum here also tells you whether two or three numbers were summed.

g4g10 Bracelet Principle

If any two or three adjacent numbers in the g4g10 bracelet obtained by connecting the ends of [1 3 4 5 9 10 6 7 8 2] *are summed, then it's possible to deduce which numbers were used, because we obtain the distinct sums* 4, 7, 9, 14, 19, 16, 13, 15, 10, 3 *(for double sums) and* 8, 12, 18, 24, 25, 23, 21, 17, 11, 6 *(for triple sums).*

Coming up with a strategy for converting a total into its summands, without at first knowing whether two or three numbers are involved, is an exercise that we leave to the reader.

Remembering the all-important [1 3 4 5 9 10 6 7 8 2] is no problem here, if you simply think of the number of letters in the words of the mnemonic phrase:

A big card-based summation-prediction circle (ordered bracelet OK?).

Source: The delightful magic bracelet used here was shared by mathematician Steve Butler, in March 2012, and is used with his permission. (The effect title given is an anagram of "Butler's Bracelet.") He included it in his paper contribution to the Gathering 4 Gardner 10 conference in Atlanta that month [Butler to appear], so it seems appropriate to dub it the g4g10 bracelet.

Presentational options: Since you start with a circle of ten cards, you may wish to come up with a "digital clock" storyline.

The additional certainties in the last two effects use sum-rich bracelets—for which all sums that can be formed turn out to be distinct—permitting the deduction of what numbers are involved knowing only their totals.

Let's next switch our attention to sum-poor bracelets, for which the sums that show up are so restricted that we can perform miraculous prediction or forcing effects. We ignore the bracelets in which all beads have the same value.

2♠ From Alpha to Omega

How it looks: *A half-dozen cards from a shuffled deck are dealt into a face-down circle, and a volunteer points to any one of them. Its value is added to that of the cards on either side of it, and this triple total is used to determine another card in the remainder of the deck. It turns out that the card arrived at was predicted in advance, in writing.*

How it works: The top half-dozen cards of the deck are in a special order, namely 5, 4, Ace, 6, 3, 2. Suits are irrelevant, but we recommend a healthy mix so that no suspicions are aroused if anybody later inspects the cards. Such a top stock can be maintained despite some shuffling (see page 10), for example, by casual overhand shuffling that ultimately leaves the top part of the deck undisturbed. Deal the six key cards into a face-down counterpart of what is shown in Figure 2.3. (Tracing out 1, 2, 3, 4, 5, 6 in order reveals a geometric pattern that we explain in due course.)

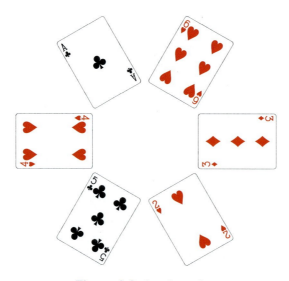

Figure 2.3. Sixy bracelet.

Here is the real gem of note:

n, n + 1 Principle
If we sum any three adjacent numbers in the bracelet obtained by joining the ends of [5 4 1 6 3 2], *then we get* 10 *or* 11. *Specifically, triple sums centered on even numbers are always* 10, *and those centered on odd numbers are always* 11.

It's all very predictable. Even better: 1, 2, 3, 4, 5, 6 here are in alphabetical order. In other words, like the g4g8 bracelet encountered earlier in this chapter, this one has the $\alpha\Omega$ property:

αΩ Property
Read in the shape of an Ω, the numbers 5, 4, 1, 6, 3, 2 are in alphabetical order (five, four, one, six, three, two)—a most fortuitous coincidence in the English language.

You get to decide which card from the rest of the deck is forced, its identity perhaps written on a prediction slip ahead of time. It must start in position 17 in the deck. That's over and above—or should we say eleven positions under and below—the six special cards planted at the top.

As already mentioned, some careful shuffling can be done early on that does not alter this setup. After the circle of six is dealt, more convincing shuffling can be done, as long as the top eleven cards of the rest of the deck are not disturbed.

Recall that triple sums are either 10 or 11, numbers close enough to permit the same magical conclusion—the forcing of the eleventh card in the remainder of the deck—via the following casual con:

Whatever Principle
We can work with whatever number is determined. If a total of 10 *is reported, simply have the volunteer count out that many cards from the remainder of the deck, and* look at the next card, *whereas if* 11 *is the total arrived at, have that many cards counted out and* look at the last one. *Think to yourself (but don't say out loud), "10, 11, whatever!"*

We defer examination of the general mathematical principle behind such bracelets until a little later.

Source: Original, having been stumbled on during a very productive flight to Hawaii in December 2006. Published online as the August 2007 *Card Colm* "Sixy Alpha Omegas" [Mulcahy 07_08].

Presentational options: Unbeknownst to your audience, this effect works because you begin with six key cards arranged in an alphabetized circle. You can openly start from such a configuration if you wish, perhaps following it up with this *alphabetical counting* kicker.

Announce, "I'm going to take any Ace, two, three, four, five and six from the deck, and arrange them alphabetically. The Ace we think of as one. Let's see, I think the first card then would be the *five*, right? Then the *four, one, six, three* and finally the *two*." Arrange them in a face-down packet in the appropriate order, with the Five on top. Then say, "Now I'm going to demonstrate how to 'count it' out, from one to six. For instance, if I count out seven, which I wouldn't here, of course, I'd move seven cards to the bottom, one by one, like this." Count out,

"One, two, three, four, five, six, seven," as you transfer a total of seven cards to the bottom, one at a time.

Now hand the face-down packet to a spectator, and ask that this counting and transferring be done for the numbers one to six in turn, each time having the new top card set aside face down in a growing row from left to right.

That is to say, after one card is transferred, the new top card is set on the table. After two more cards are transferred, the new top card is set on the table to the right of the existing one, and so on. Once the row has grown to five cards, have the last card in hand "air transferred" six times for some comic effect, then placed on the right to complete the row. Turn over the cards, one by one, starting on the left: they are in numerical order. "Good job!" you say to the spectator. "Now that's what we magicians call 'alphabetically counting.'"

Sum-Poor Bracelets

Max Maven points out that the use of circular stacks (i.e., bracelets) designed to force alternating totals for sums of just two adjacent values was published by H. S. Paine in 1922, totals of 14 or 15 being inevitable using a fifty-card deck consisting of two cycles of Ace, King, 2, Queen, 3, Jack, 4, 10, 5, 9, 6, 8, 7, 7, 8, 6, 9, 5, 10, 4, Jack, 3, Queen, 2, King, in mixed suits (note that two Aces are simply omitted). A book force version appears as "The Printed Word" in [Fulves 79, page 48].

What can be said in general about sum-poor bracelets for triple sums? It's easy to show that any magic bracelet that alternately forces triple sums of n or $n + 1$ must have length six, or be repeats of such bracelets, e.g., with twelve or eighteen (or some other multiple of six) beads. Indeed, a little experimentation leads one to conclude that they must basically be of the form $[a \quad b \quad n - a - b \quad a + 1 \quad b - 1 \quad n + 1 - a - b]$, as depicted in Figure 2.4, for some integers a, b, yielding a two-parameter family of possibilities. This just means that given a target n (and $n + 1$), we can select any values we wish for a and b and figure out what the other four beads must be. If we wish to avoid zero or negative numbers, or numbers larger than 13, then there are some restrictions on a and b.

We're not assuming that all of the beads are distinct. The smallest positive example $[1 \ 2 \ 1 \ 2 \ 1 \ 2]$ (with $n = 4$) consists of two beads repeated, but it is hardly impressive. To get an example with distinct positive beads the smallest possible value for n is 10, since $1+2+3+4+5+6 = 21$ would also have to equal $n + (n + 1)$. The alternating triple sums of any such bracelet are therefore 10 and 11, and it's easily verified that the perfect

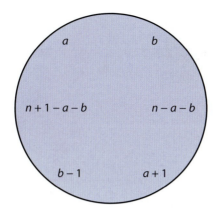

Figure 2.4. Sixy circle.

[5 4 1 6 3 2] seen earlier is basically unique (up to cycling and flipping direction; for instance, we view [1 4 5 2 3 6] as being equivalent).

If we drop the distinctness restriction, which is reasonable with cards, then there are five other bracelets with triple sums 10 and 11, such as those given by [5 4 2 4 5 1] or [2 4 4 3 3 5].

Three-One Modular Mantra
The perfect [1 4 5 2 3 6] may be obtained by starting with 1, and three times repeating the modular mantra "First add 3 and then add 1." Addition is mod 6.

This yields $1 \to 4 \to 5 \to 8 = 2 \to 3 \to 6 \to 7 = 1 \to \ldots$ whereupon it repeats, which—as just noted above—is merely another manifestation of [5 4 1 6 3 2].

Crazy Clocks

It's time to think bigger and generalize. Consider any Ace, 5, 9, 10, 2, 6, 7, Jack, 3, 4, 8, and Queen dealt into a face-down circle like the crazy clock in Figure 2.5.

It doesn't really matter where you start dealing, or in which direction you deal, even though we have followed the usual clockwise convention above.

The values can be obtained as follows:

Four-One-One Modular Mantra
The crazy clock may be obtained by starting with 1, and four times repeating the modular mantra "First add 4, next add 4 again, and then add 1." Addition is mod 12.

This yields $1 \to 5 \to 9 \to 10 \to 14 = 2 \to 6 \to 7 \to 11 \to 15 = 3 \to 4 \to 8 \to 12 \to 13 = 1 \to \ldots$ whereupon it repeats, duplicating the values [1 5 9 10 2 6 7 11 3 4 8 12] in the crazy clock depicted in Figure 2.5.

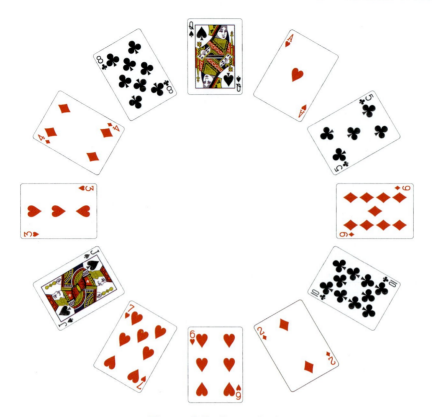

Figure 2.5. Crazy clock.

Now, if any four adjacent cards are selected, and their values are summed, then something special happens:

n − 1, n, n + 1 Principle   ▶◀

If we sum any four adjacent beads in the bracelet obtained by joining the ends [1 5 9 10 2 6 7 11 3 4 8 12], *then we always get 25, 26, or 27. In fact, quadruple sums of adjacent beads cycle through those possibilities in that order, starting by summing 1, 5, 9, and 10 and ending by summing 12, 1, 5, and 9.*

There are other examples of bracelets for which quadruple sums of adjacent beads cycle through $n-1, n, n+1$, in some order, for some n (above, we had $n = 26$). There is no loss of generality in assuming that $n-1, n, n+1$ cycle in that particular order. In principle, the values $n-1, n, n+1$ may repeat, in the same fixed order, in any of $3! = 6$ possible ways.

In Figure 2.7, it's 26, 25, 27 read clockwise, starting with the 1, 7, 7, 11, then moving on to 7, 7, 11, 0, and so on. Totals of 25, 26, 27 in that order may be obtained by reading counter-clockwise instead, starting with 7, 1, 12, 5, for instance. By rotating and/or reflecting as needed, the three totals may be obtained in any desired order.

It's not difficult to show that magic bracelets that force consecutive quadruple sums $n-1, n, n+1$ have length 12 (or a multiple of 12). Moreover, upon reflection (or rotation, or both), they must be of the form $[a \ b \ c \ n-1-a-b-c \ a+1 \ b+1 \ c-2 \ n-a-b-c \ 2+a \ b-1 \ c-1 \ n+1-a-b-c]$, as depicted in Figure 2.6.

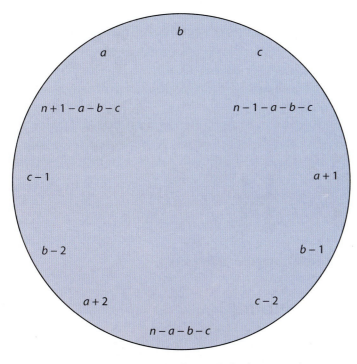

Figure 2.6. General clock.

Other examples, also with $n = 26$, are obtained by connecting the end beads of $[1 \ 7 \ 8 \ 9 \ 3 \ 6 \ 7 \ 11 \ 2 \ 5 \ 9 \ 10]$ or $[1 \ 7 \ 7 \ 11 \ 0 \ 9 \ 6 \ 10 \ 2 \ 8 \ 5 \ 12]$. Figure 2.7 depicts a card version of the second one, with a Joker representing the 0.

In these examples, no beads are negative or exceed 12; however, 7 is repeated. We have already seen that positive-beaded bracelets avoiding repeats exist, and even better, one exists that uses precisely $1, 2, \ldots, 12$. In any such bracelet, the total of its beads is $1+2+\ldots+12 = \frac{12 \times 13}{2} = 78$, and it can be checked that this total also equals $(n-1) + n + (n+1)$, so that n must be 26, as above.

It turns out that there are six essentially different ways to arrange 1, 2, ..., 12 as desired; another is given by $[1 \ 5 \ 12 \ 7 \ 2 \ 6 \ 10 \ 8 \ 3 \ 4 \ 11 \ 9]$.

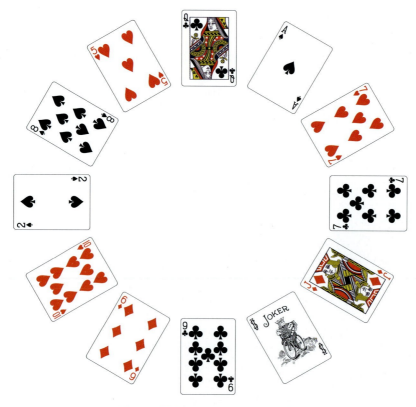

Figure 2.7. Another crazy clock.

2◆   Volunteer Four Hours

How it looks overall: *Twelve cards from a shuffled blue-backed deck are dealt face down in a circle on the table, like the hours of a clock. Four adjacent cards (hours) are selected randomly by a spectator, and the resulting card values are looked at and summed.*

The total is used to determine a card in a shuffled red-backed deck. It turns out that this card was predicted in advance, in writing.

How it looks in detail: Shuffle a blue-backed deck of cards, and deal twelve cards into a face-down circle, saying, "These represent the hours of a clock." Remove a red-backed deck from its case, and casually shuffle it, announcing, "I need a volunteer." Once somebody steps forward, say, "Actually, I need you to volunteer four hours." Pause. "And I do mean *four* hours. Is that okay?"

As a look of alarm spreads across the volunteer's face, add, "In theory, you are supposed to do four hours of volunteer work, but don't worry, this will only take a few minutes, and I promise there's no heavy lifting involved."

Continue, "Please mentally select a four-hour shift: decide on four adjacent cards from the clock of cards displayed here. These will determine the four hours you're volunteering for. Don't give me any hint as to which cards you've selected. Each card has a number on it that will reveal how many heavy boxes you're supposed to lift during that hour. Actually, the only lifting you'll have to do involves a corresponding number of cards from this red-backed deck." As you say those words, put the red-backed deck back in its case. Take a sealed prediction envelope out of your pocket, and say, "Earlier on, I made a prediction about the outcome of your volunteer work." Place the envelope on the table and put the red card case on top of it.

Ask the volunteer to indicate which four adjacent blue-backed cards they have mentally selected. Gather up the other eight cards and set them aside. Have the selected cards turned over and the values summed, while you turn away briefly, with the usual reminder that Aces count as 1, Jacks as 11, and so on.

Once the total is reported—let's suppose it's 26—pick up the red card case, and take out the cards. Hand them to the volunteer and say, "Here comes that heavy lifting you've been worrying about. I want you to lift off twenty-six cards. Do them one at a time to make it easier; that way the count will be perfect, too."

The volunteer lifts off one card at a time, setting them all aside until the twenty-sixth card is lifted off. "What is the last card you lifted?" you ask. The volunteer reveals the identity of that card. Have the sealed envelope opened and your prediction read out; it matches the selected red-backed card. The cards may all be inspected.

How it works: It should come as no surprise that the top dozen cards of the blue-backed deck are in a special order, try any Ace, 5, 9, 10, 2, 6, 7, Jack, 3, 4, 8, Queen. These can be maintained in place despite some shuffling. As a result, the circle dealt can be made to look like a face-down version of the first crazy clock shown earlier (Figure 2.5).

As we saw, if any four adjacent cards are selected, and their values summed, then we obtain 25, 26, or 27 as the total. Despite this seeming ambiguity, there is one particular card in the red-backed deck that is forcible and hence can be predicted in advance.

Extended Whatever Principle ▶◀
We can work with whatever number is determined.

1. *If the announced total is 25, have that number of cards counted off, and declare that the next card is the chosen one.*

2. *If the total is 26, then, of course, the twenty-sixth card is deemed to be the one.*

3. *If the total is 27, flip the deck over and have the cards run through face-up. Since the twenty-seventh card from the bottom is the twenty-sixth one from the top, all is well in this case too.*

(Think to yourself, "25, 26, 27, whatever!")

The strategy then is to plant your prediction card in position 26 in the red-backed deck and to make sure it remains there throughout some shuffles (see page 11); the rest works automatically.

It's not easy to remember [1 5 9 10 2 6 7 11 3 4 8 12]; a catchy mnemonic would really help. There is one on this page for this very purpose. Can you see it? Keep reading!

Source: Original. Developed on the banks of the Ohio River over Labor Day weekend, 2007. This effect first appeared online at MAA.org as the October 2007 *Card Colm* "A Magic Timepiece Influenced By Martin Gardner (Celebrating His 93rd Birthday Incidentally!)" [Mulcahy 07_10]. (A minimally modified version of that title worked up to his 95th birthday—the last one he celebrated in person.)

Presentational options: You can arrange the predicted card to be "something significant," one option being the 4♠ (representing four hours of spade work).

Parting Thoughts

- Show that all bracelets forcing alternating totals of n and $n + 1$ for two adjacent beads are in the same spirit as the 14/15 forcing example of H. S. Paine on page 58.

- Is the triple sum-rich g4g8 bracelet [8 5 4 1 7 6 3 2] using 1–8 unique? If not, how many are there in total?

- Is Steve Butler's double and triple sum-rich g4g10 bracelet [1 3 4 5 9 10 6 7 8 2], using 1–10, unique? If not, how many are there in total?

- For exactly which values of n do there exist triple sum-rich bracelets of n beads using 1–n? How many are there for a given n?

- What about bracelets that are rich for sums of j adjacent beads, for $j > 3$?

- Show why a sum-poor bracelet that alternately forces triple sums of n or $n+1$ must be of the form $[a \ \ b \ \ n-a-b \ \ a+1 \ \ b-1 \ \ n+1-a-b]$, for some a and b, or consist of multiple copies of such a bracelet.

- Show why a sum-poor bracelet that forces consecutive quadruple sums of $n-1$, n, or $n+1$ must be of the form

$$[a \quad b \quad c \quad n-1-a-b-c \quad a+1 \quad b+1 \quad c-2$$
$$n-a-b-c \quad 2+a \quad b-1 \quad c-1 \quad n+1-a-b-c]$$

for some a, b and c, or consist of multiple copies of such a bracelet. Why are there essentially six different ones using the numbers 1–12?

- Extend to sum-poor bracelets for which quintuple sums of five adjacent beads can be only one of four consecutive numbers, such as $n-1$, n, $n+1$, or $n+2$ (in some order) for some n. How many beads would such a bracelet have? Does it matter in what order the four possible sums cycle? (For the earlier smaller bracelets, it doesn't.) Note that although the Extended Whatever Principle can be modified to work here, the numbers obtained are beyond the scope of a deck of cards.

- Fix m, and seek bracelets for which the m sums of $m+1$ consecutive beads cycle through values $t, t+1, \ldots, t+(m-1)$ (in this, or some other order) for some t. Show that the number of beads must be $m(m+1)$ (or some multiple of this number).

Let's explore the relationship between t and m for such a bracelet with $m(m+1)$ beads. Because m and $m+1$ are relatively prime, in other words, they share no common factors other than 1, it can be shown that if we sum the m consecutive values $t, t+1, \ldots, t+(m-1)$, we get the sum of 1, 2, $\ldots$, $m(m+1)$. Hence,

$$mt + \sum_{i=0}^{m-1} i = mt + \frac{(m-1)m}{2} = \sum_{j=1}^{k} j = \frac{k(k+1)}{2},$$

where $k = m(m+1)$. Solving for t, we get,

$$t = \frac{(m+1)[m(m+1)+1]}{2} - \frac{m-1}{2} = \frac{m^3 + 2m^2 + m + 2}{2}.$$

These numbers are known: t is the $(m+3)$th term of sequence A162607 in *The On-Line Encyclopedia of Integer Sequences*, which is in turn related to sequence A159798. The latter gives the "triangle read by rows in which row n lists n terms, starting with 1, such that the difference between successive terms is equal to $n-3$." The terms in sequence A162607 are the row sums of the triangle of sequence A159798. See http://www.oeis.org/A162607 and http://www.oeis.org/A159798.

- It is easy enough to arrange 1, 2, ..., $m(m+1)$ in a circle so that the sum of any $m+1$ consecutive beads is $t, t+1, \ldots,$ or $t+(m-1)$ for some t. For example, we can start with 1, and $m+1$ times repeat the modular mantra: "Add $m+1$, add $m+1$ exactly $(m-1)$ more times, and then add 1."

  This generalizes the special cases seen earlier, and it can be explained using the language of group theory. Each round of adding "$m+1$ exactly m times" steps us through the elements of a coset of the additive group $Z_{m(m+1)}$ modulo the subgroup $\langle m + Z_{m(m+1)} \rangle$. Every time 1 is added, we switch to a new coset and repeat, until the group (or the mathemagician) is exhausted.

  Verify that the claimed summing property holds here.

- How many desirable bracelets with beads 1 to $m(m+1)$ are there in general? Is there an algorithm for finding them all?

Poker Powers

The effects in this chapter all concern seemingly impossible control on your part over how ten cards are distributed between you and a spectator, while giving the latter the illusion of free choice. Since you know the outcome beforehand in each case, you may wish to make verbal or written predictions to enhance your magic reputation.[1]

The principles explored naturally lend themselves to two-hand poker effects. Generally, we suggest that you award yourself the better cards, but you may reverse the roles and allow the spectator to come out ahead, at least some of the time. Please note, however, that none of this will be of any help in real games of poker, in which different rules of card distribution apply. We merely present surprising and entertaining material dressed up as being poker-related.

"Poker with Any Ten Cards" (3♦) is as good as it gets: it never fails to get a reaction and engages audiences even more upon repetition.

> *A spectator shuffles the deck, and deals ten cards face down. You glance at the card faces and pause to write down a prediction. Then the spectator is given numerous choices for distributing the cards between the two of you. Yet, without fail, not only do you end up with the winning poker hand, but your specific prediction, made in writing in advance—which may say something like, "You will have a pair of Jacks, but I'll have three 5s'—also comes true.*

---

[1]Opinions do vary here. Some see such predictions as overkill.

Poker Hands Worth

Before we embark, we need to be clear as to the ranking of poker hands, namely selections of five cards from a full deck. There are $\frac{52!}{5!47!}$, or roughly 2.6 million, possible poker hands (see page 16). The relative scarcity (and hence value) of some of the commonly desired hands is reflected in Table 3.1 (in which most of the chances are rounded). For instance, about half of all possible poker hands contain a pair or better.

Type of poker hand	Number of such poker hands	Chance of such a poker hand
Royal flush	4	1 in 649,740
Straight flush	36	1 in 72,193
Four of a kind	624	1 in 4,165
Full house	3,744	1 in 694
Flush	5,108	1 in 509
Straight	10,200	1 in 255
Three of a kind	54,912	1 in 46
Two pairs	123,552	1 in 21
One pair	1,098,240	1 in 2.36
No pair	1,302,540	1 in 1.995

Each row excludes any better hands listed in the rows above it.

Table 3.1. Values of key poker hands among the 2.6 million possibilities.

Three of a kind refers to exactly three cards of one value, accompanied by two cards with different values, such as 7♠, 7♥, 7♣, 5♠, K♥. Similarly, *four* of a kind means all four cards of a particular value, and one other, and *a pair* is used in place of "two of a kind" (accompanied by three boring cards). *Two pairs* means two different pairs, such as 7♠, 7♥, 9♠, 9♣, J♣.

A *full house* is a pair and three of a kind, such as 7♠, 7♥, 9♠, 9♣, 9♦ (this one is sometimes referred to as "nines over sevens").

A *flush* is five cards in one suit, such as 3♣, 5♣, 8♣, 9♣, K♣, or 7♥, 8♥, 9♥, 10♥, Q♥. A *straight flush* is five consecutive cards in one suit, such as 7♥, 8♥, 9♥, 10♥, J♥. Aces can be considered high as well as low, so that both A♠, 2♠, 3♠, 4♠, 5♠ and 10♦, J♦, Q♦, K♦, A♦ are straight flushes. The latter one has a higher value and is known as a *royal flush*. Note that wraparound past an Ace is not currently deemed to have special status among flushes, so that J♥, Q♥, K♥, A♥, 2♥ is a flush, but not a straight flush.[2]

A *straight* is any five consecutive cards in more than one suit, such as 7♠, 8♥, 9♦, 10♦, J♣. Sadly, a near straight flush such as 7♠, 8♥, 9♥, 10♥, J♥ is just a straight, as is a regal-looking 10♠, J♠, Q♠, K♠, A♥.

---

[2]Perhaps the time has come to recognize such interesting, naturally occurring phenomena as gay flushes. Mathematically, they are just as valid as the straight ones.

Table 3.1 shows which types of hands are relatively common and which are quite rare, as well as those that are very rare indeed. There, "flush" means nonstraight flush.

A straight flush beats a full house, and three of a kind beats two pairs. Note that a full house is about 100 times more likely than a straight flush, which is, in turn, about 10 times more likely than a royal flush.

In summary,

royal flush > straight flush > four of a kind > full house > flush > straight > three of a kind > two pairs > one pair.

It may be advisable to have a copy of Table 3.1 handy, even as a prop in performance, for those who need reminding about the relative values.

Jonah Cards

Now we have some fun with a terrific idea of unknown origin[3] that is often referred to as the "Ten Card Poker Deal". It seems to have first appeared in print (simply as "Poker") in Arthur Buckley's *Card Control* [Buckley 46, page 103].[4] Over the decades, a whole sub-industry of improvements and extensions has developed, which we recommend exploring; we only consider the basic plot here.

Consider ten cards consisting of three sets of three of a kind, such as three 3s, three 8s, and three Kings, together with one nonmatching card, perhaps a Jack, which is known as the Jonah card. The relative values play no role, as they are never compared. The suits are arbitrary.

These cards are mixed up, face down, and a spectator picks any five of them for a poker hand, while you retain the other five as your hand. Amazingly, the outcome is somewhat predictable.

Jonah Card Principle
If ten cards consisting of three sets of three of a kind and one nonmatching "Jonah" card are divided into two poker hands, then whoever has the Jonah card loses, without fail.

To see why this is true, let's suppose that the cards in question are 3♣, 3♦, 3♠, 8♣, 8♥, 8♦, K♦, K♥, K♠, and J♠. Focus on the hand with the Jonah card; it's not difficult to see that one of the following three cases must hold.

Case I: This hand also contains three of a kind, and one other card. Then the other hand wins, being a full house containing three of a kind and a pair, such as the hands in Figure 3.1.

---

[3]Max Maven opines, "It is presumed to have come from the world of gambling cheats, but no citations are known."

[4]Some words of caution: Buckley got it backwards, and it's not in the Dover reissue.

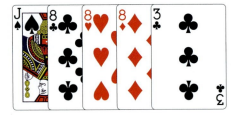

Figure 3.1. Typical losing and winning hands in Case 1.

Case 2: The hand with the Jonah card also contains two pairs. Then the other hand still wins, containing three of a kind and two unmatched cards, such as the hands in Figure 3.2.

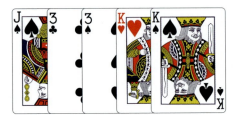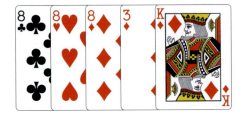

Figure 3.2. Typical losing and winning hands in Case 2.

Case 3: The hand with the Jonah card also contains one pair and two nonmatching cards. Then the other hand wins again because it contains two pairs, such as the hands in Figure 3.3.

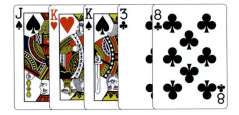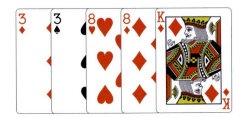

Figure 3.3. Typical losing and winning hands in Case 3.

The trick, so to speak, is to control who gets the Jonah card. There are not-so-honest ways to do this (e.g., card marking), but we consider two ingenious and fair-seeming mathematical methods that have numerous other applications. (A third method is discussed, along with these, in the June 2012 *Card Colm* "Something's Old, Something's True, Something's Borrowed, Something's New" [Mulcahy 12_06].)

Position Parity

In 2006, Martin Gardner shared the following delightful effect, adapted from a coin puzzle. It suggests a position parity method of controlling how to split up ten cards into two poker hands.

> Here's a curious card effect I based on the first problem in Peter Winkler's book *Mathematical Puzzles* [Winkler 04]. Eight cards are dealt face up from a shuffled deck to form a row. Players take turns removing a card from either end of the row.
>
> The person whose cards have the higher sum is the winner. To speed up the game, face cards count as 10. You always go first. It seems like a fair game, but you have total control, not only over who wins, but who wins by a precise number of points! It's what carnies used to call a "two-way" game, because you can always allow your opponent to win. Moreover, you can always predict the exact number of points by which a game is won! For example, you can say, "I'll win this game by n points." Or "This time you will win by n points." Or "This game is sure to be a draw."
>
> The technique is simple. As you deal the 8 cards, sum the cards in odd positions. Then sum the values in even positions. Assume the evens are higher. You first take the card at the even end. Both ends are now odd. This forces your opponent to take an odd card. You can then take an even. If the odd cards have the higher sum, you start by taking an odd card, forcing her to take an even. Then you take another odd, and so on ... With only 8 cards, the games go fast until the deck is exhausted.

The gist of this is that, provided you get to go first, you can control which cards you end up with: either the ones in the odd positions or the ones in the even positions. Specifically,

Position Parity Principle
Take turns selecting cards, always choosing from one of the two exposed ends, then if you go first and start by taking the first card in the row, you'll end up with all of the cards in the odd positions. If you start by taking the last card in the row, you'll get the cards in the even positions.

For instance, with the row of six cards depicted in Figure 3.4, you'll definitely end up with the Red cards in positions 1, 3, and 5, provided

Figure 3.4. The Red cards are yours for the taking.

that you go first and start by taking the first card. The spectator is doomed to get the Black cards, despite the illusion of three (okay, two) free choices.

Note that other than your first selection, each time you must take the exposed card next to the one that the spectator takes. This observation saves you trying to keep track of odd/even positions as the row shrinks. Don't make your strategy look too obvious, however; feel free to take your time and appear to give the cards at each end equal consideration before deciding on which one to remove.

By combining the Jonah Card and Position Parity Principles, you can control whether you or a spectator has the winning hand when splitting up ten cards for two poker hands, while giving the illusion that the spectator makes all of the decisions, as we now demonstrate.

3♣ Ditch the Dud

How it looks: *Ask for a spectator who likes poker, as you shuffle the deck. Have ten cards dealt out into a face-down pile, and have that pile further mixed. Pick up the cards and glance at their faces briefly, remarking on how random they are, and yet how they may result in two interesting poker hands. Announce which of you will win. Deal the cards into a face-down row, and alternate with the poker fan in taking cards from one end of the row or the other, until you both have five cards. Compare and see who has the winning poker hand. Your earlier prediction turns out to be correct.*

How it works: A ten-card stock (see page 10) is planted and left at the top throughout some shuffles. Use any three sets of three of a kind, and one nonmatching card as a Jonah card. Values and suits play no role here.

After the ten key cards have been thoroughly mixed by the poker fan, scan the faces, remarking how random they are, and yet how they may result in two interesting poker hands. Announce which of you will win.

Of course, you have simply noted whether the Jonah card is in an odd- or even-numbered position, counting the top card as position 1.

Let's assume that you then deal the cards into a face-down row so that the top card is on the left.

If the Jonah card is in an even position, you can avoid it by first selecting the card in the first position, then alternating with the spectator in taking cards from one end or the other of the row, as explained above, until you both have five cards. If the Jonah card is in an odd position, you can avoid it by first selecting the card in the last position, then proceeding as before.

Finally, compare and see who has the winning poker hand. Needless to say, your prediction turns out to be correct.

Remember always to select the exposed card next to the one the spectator has chosen, but only after pausing and appearing to consider each end of the shrinking row equally. Don't have any cards turned over until the end.

Source: Original, from roughly 2006, but hardly new. Published online at MAA.org in the June 2012 *Card Colm* "Something's Old, Something's True, Something's Borrowed, Something's New" [Mulcahy 12_06].

Presentational options: You can casually jumble the cards as you scan the faces, stopping any time you wish, as long as you note the parity of the final position of the Jonah card.

You can repeat this with the same cards, or perhaps with a similar selection also planted in the top half of the deck.

You can also control from which end you start selecting, if you think this adds to the effect upon repetition. If you want to start with the last card, for instance, then deal the packet from left to right, as usual, if the Jonah card is in an odd position in the spread. Otherwise, simply close up the face-up spread, turn it over, and respread it face-down on the table in such a way that the former top card is now at the end of the row.

Rather than pretend that those ten cards came off the top of a shuffled deck, you may opt to be open about their identities from the very start. It can increase audience interest in card mixing!

Once you've mastered Simon's 64 Separation Principle, explained next, you'll have a new way to control the Jonah card to yield impressive results. This is discussed online—along with a third method—in the June 2012 *Card Colm* [Mulcahy 12_06]. Then, we present an application in a different vein.

Enter Bill Simon

Here's a diabolically fair-seeming distribution procedure due to magician Bill Simon, who was the best man at Martin Gardner's wedding. It's really another controlling method, and is borrowed from Bill's *Mathematical Magic* [Simon 64], which was arguably the second book for the general public on the topic of the title, *Mathematics, Magic and Mystery* [Gardner 56] having been the first.

Simon's 64 Separation Principle
It is possible to control how to split up a packet of eight cards into two piles of four, while giving a spectator the illusion of multiple free choices.

You retain control of the division in one key sense: the top four cards end up in the first pile (the spectator's), and the bottom four in the second pile (yours). It's really a separation principle. Hence, if you start with four Red cards on top of four Black ones, the piles maintain that color separation, with the Reds in the first pile. In fact, the best way to see how Simon's separation works in practice is to follow along with a face-up packet of four Reds on top of four Blacks (in practice, the eight cards would be face down). The application of most interest to us is separating the cards according to their potential contribution to losing and winning poker hands, rather than based on their color.

Before explaining how it's done, we note that while it is possible to have a spectator correctly do the distribution, so that you never touch the cards from a certain point on, it first requires very careful demonstration by you, as people who are new to it are apt to get confused and make mistakes. We move forward under the assumption that you handle the cards throughout.

For the purpose of explaining what is going on, let's consider a face-up packet with four Red cards on top of four Black ones. Figure 3.5 shows the kind of thing that happens if we start with the face-down packet 4♣, 9♠, K♠, 2♣, 5♦, A♥, J♦, 8♥, flip it face up, and apply Simon's ideas.

Hold the packet in one hand, and take off the top two cards with the other. Ask the spectator which card he wants to start pile A on the table; the other one gets tucked underneath the six cards remaining in the packet. The upshot is that one of the first two Red cards starts pile A, and the other goes to the bottom of the packet. Give the spectator the exact same free choice for the next pair of cards. Pile A now contains two Red cards, and the retained packet has four Blacks on top of two Reds.

Next, ask the spectator to make similar choices to determine one of two Black cards to start a new pile B (to the right of the pile of Red

Figure 3.5. Simon's 64 Separation Principle in action.

cards in Figure 3.5). The Black card not selected is tucked underneath the rest of the cards in hand, and the next two Black cards are offered in the same way. As a result, two Blacks start that pile and the retained packet consists of two Reds on top of two Blacks. At this stage we have a scaled-down version of the original packet.

Now, the spectator picks one card for pile A in the usual way, and finally one for pile B. These choices also maintain the color separation.

Again, in practice, this is done with all cards face down, so that the spectator has no idea his hand is being forced. In performance, this is a good time to recap what has happened, brazenly claiming, "Six times, I gave you independent free choices. That's two to the power of six, or sixty-four, different things that could have happened."

The last two cards in our example are a Red on top of a Black; you must add the first to pile A, and the second to pile B. This is best done casually, perhaps while making the above comments, or you can use some magician's force (Bill Simon suggested one in his book).

When all is said and done, pile A contains the original four top cards, in some order, and the four bottom cards are in pile B, also mixed up.

This procedure can be extended to packets of size sixteen (or any power of 2), if suitable modifications are made. For instance, starting with eight Reds on top of eight Blacks, the first step is to offer the spectator one card from four pairs in turn, thus determining four of those eight Reds for pile A. Similarly, the spectator thinks he freely decides on four cards for pile B, but, of course, they are four of the eight Blacks. Now there are eight cards left—four Reds on top of four Blacks—and we are back to familiar territory.

As the following effects make clear, there's an easy way to extend the eight-card version to packets of ten cards, which is just what is needed to determine two poker hands. After learning one of these effects, it will be clear how to adapt the method to work in a "Ditch the Dud" (3♣) type scenario (details can be found online in the June 2012 *Card Colm* [Mulcahy 12_06]).

Next is a shocking case of a mathemagician deliberately leading an innocent spectator astray. While it's well known that probability can be counterintuitive, this one muddies the waters in a new way.

3♥   Worst-Case Scenario

How it looks: *Shuffle the deck, and deal ten cards face down. Ask for a spectator who likes poker. "Everybody knows that poker hands are ranked based on how likely they are to show up," you say brightly. "The more rare a hand is, the more highly it's valued. That's why a flush beats a straight: there are about 5,000 ways to get a flush, compared to about 10,000 straights, out of over two and a half million possible poker hands."*

Pause to let that sink in before continuing. "I'll give you a chance to get an even rarer kind of poker hand. There are only about 4,000 of these, which makes it almost as rare as a full house. Are you willing to play?" Hopefully, by now you have an eager spectator chomping at the bit. You may wish to have the table of poker hands at the ready or on display.

"I'll make you a deal. I promise I'll do no better than three of a kind, which can happen in about 50,000 ways. I'm not being greedy, as you'll get a much more unlikely hand. But what am I thinking? You get to make all of the decisions, of course! What's the worst that can happen?" If you wish, you can jot down a precise prediction at this point.

Give the spectator numerous choices as to how to distribute the packet of ten cards among the two of you. It transpires that the spectator does indeed get a very rare, but unfortunately useless hand, whereas you get three 2s, in each case matching your prediction.

In fact, the losing hand is about as bad as it could be: the worst-case scenario has indeed happened, despite the spectator making all the choices.

How it works: Needless to say, the deck is set up in advance. Start with these cards on top: 2♦, 2♥, 3♠, 4♠ (in any order), then 3♣, 4♣, 5♣, 7♠ (in any order), and finally 2♣, 2♠ (in any order). The initial shuffling should not disturb them. Note that the order of these is reversed when they are dealt out, yielding something like the configuration shown in Figure 3.6.

 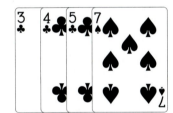 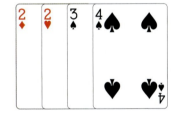

Figure 3.6. The cards on offer.

Ultimately, you'll end up with one of the two Black 2s at the top and all of the second set of four cards shown in Figure 3.6. The spectator gets the first set of four cards shown and the other Black 2.

As a result, you end up with three 2s, a 3, and a 4, and the spectator gets a 2, 3, 4, 5, and 7, in mixed suits, which does the trick nicely. The use of a written prediction is optional. In any case, say, "Congratulations! In spite of the fact that we both ended up with very low valued cards, it all came out just as I predicted. You got a very rare kind of hand—I believe there are exactly 4,080 of them, hence there's only a 1 in 637 chance of getting this lucky—whereas I got the much more common three of a kind. There are 54,912 hands like that, which means it happens about 1 time in 47, if my arithmetic is correct."

Ignore the look of disgust on the spectator's face, and conclude, "You don't believe me? Just enter the phrase 'worst poker hand' in Wolfram Alpha[5] and you'll see what I mean. It confirms the 4,080 count I just

---

[5]The computational knowledge engine www.wolframalpha.com.

mentioned for hands with 'no two cards of the same value, high card 7, neither a straight nor a flush,' which is exactly the kind of select scarce hand you picked for yourself. I guess it's the worst-case scenario; or the best, from my perspective."

If you've opted to provide the table of poker hands, you'll need to apologize for the absence of the spectator's hand in the listing. You could add insult to injury by saying, "You got a hand so rare they didn't even list it here."

Most of the choices offered to the spectator use Simon's 64 Separation Principle, thus ensuring that the desired sets of four cards end up in yours and the spectator's hands, respectively. Again, we suggest that you handle the cards throughout, to ensure that no mistakes are made, while emphasizing the apparent freedom of choice given to the spectator.

After the top ten cards are dealt out, pick them up, and offer the spectator one of the first two cards, pushed off the top, saying, "Pick one of these as your hole card, I get the other one." Of course, each of you gets a Black 2. Put the chosen 2 face down on the table in front of the spectator, to start his hand, and put the other 2 face down in front of you to start yours.

Eight cards remain in your hand, in a face-down packet. Push off the top two, saying, "One of these is for you, the other one goes underneath the rest. Which one would you like?" The selected card is put on top of the spectator's hole card, and the other one goes underneath the packet in your hand. Repeat with the next two cards, so that the spectator now has three cards to your one. Next, the spectator is asked to make similar choices to determine two more cards for you, working with the top two cards in the packet in your hand each time, so that both of you end up with three cards on the table in front of you. Now, the spectator is asked to choose a fourth card, deciding between the top two of the four cards in your hand, the rejected one being tucked underneath. You have three cards left. Offer the top two to the spectator, saying, "Please select one more for me," being careful then to tuck the other one underneath. Finally, give the top card to the spectator and the last one to yourself, without comment.

The worst-case scenario has come to pass for the spectator, whose use of free will led to something even rarer than three 2s, but nevertheless represents a spectacularly poor poker hand.

Source: Original, June 2012. Inspired in part by Dave Solomon's "Power of Poker," as gleaned from John Bannon's *Dear Mr. Fantasy* [Bannon 05].

For a statistical application of Simon's 64 Separation Principle, using mean, standard deviation, and skewness—with hints of kurtosis—see the online December 2007 *Card Colm* "Plurality Events, Standard Deviations and Skewed Perspectives" [Mulcahy 07_12].

Unquestionably, the most breathtaking application of Simon's 64 Separation Principle is the aforementioned "Power of Poker" from *Dear Mr. Fantasy* [Bannon 05], which pits a full house against a straight flush to great effect.

The next concoction runs with that idea, reinforcing the superiority of straight flushes, having first distracted the spectator with thoughts that some full houses are better than others. Along the way, everyone has cause to remember that four of a kind isn't too shabby a hand either. We suggest a "double whammy" below, but you may wish to settle for just one of the two parts until you feel comfortable with the way it works.

3♠ Full House Blues

How it looks: *Shuffle the deck, and deal ten cards face down. Ask for a spectator who is very familiar with the values of various poker hands. You could provide Table 3.1 from earlier as a prop. Pause to say that you are writing down what you yourself would like to receive as a poker hand. Briefly discuss the relative values of various full houses, raising the possibility that you might both get one. Have the spectator wish aloud for a pretty desirable full house, and write that down too.*

Now give the spectator numerous choices as to how to distribute the packet of cards among the two of you. The spectator always ends up with a full house of Sevens over Aces. Flip over the spectator's prediction, then have those cards shown to all, and say, "Congratulations, you got a full house, perhaps not exactly the hand you wished for, but impressive nonetheless. For your sake, let's hope I didn't get a better full house." Glance at your cards, as you confirm, "No, no full house at all for me."

Turn over your cards to reveal that you actually got four Twos, which not only beats the spectator's full house, it matches what you wrote earlier. Set all ten cards aside, and say, "Let's try again."

Shuffle the remaining part of the deck, and deal out ten more cards face down. Once more, pause to write down what you would like to receive as a poker hand. Ask the spectator to announce an even better full house than before, if not the best one possible. Once more give that person numerous choices as to how to distribute the new packet of ten cards among the two of you. This time, it turns out that the spectator ends up with a full house, specifically two Aces and three Kings. Say, "Congratulations, you got an astounding full house! That's as good as it gets, since the other two Aces turned up earlier and were set aside. Can that be beaten? Let's see how I did." Turn over your cards to reveal a straight flush in Spades, handily beating the spectator's full house. Now turn over your prediction. It says, *"I'll get a straight flush in Spades. You'll get a full house of Kings over Aces."* Verbally add, *"I took the liberty of predicting your cards also, I hope you don't mind."*

How it works: Needless to say, you set up a significant chunk of the deck. The early shuffling should not disturb the top twenty cards, and the second shuffling should not disturb the ten remaining top cards.

At the outset, these ten cards are stacked at the top, in three groups: 2♣, 2♥, 2♠, 2♦ (in any order), then A♦, A♠, 7♦, 7♠ (in any order), then 7♥, 7♣ (in any order). Once these are dealt out face down, the card order gets reversed. If these were fanned face up, they would look something like Figure 3.7, from left to right.

 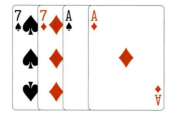 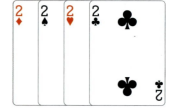

Figure 3.7. The first packet of ten cards.

However, the order of the cards within each group displayed is not important. For instance, any sandwiching of the two specified Aces between the four Twos and four Sevens works.

The next ten are also in three groups: first 9♠, 10♠, J♠, Q♠ (in any order), then K♠, K♦, K♥, A♥ (in any order), then A♠, 8♠ (in any order). Hence, after the first part of the effect has been concluded, and this second packet of ten cards is dealt out face down, if it were fanned face up, from left to right, it would look something like Figure 3.8, with the same flexibility as before within each group of cards displayed.

  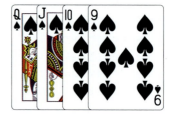

Figure 3.8. The second packet of ten cards.

Most of the choices offered to the spectator use Simon's 64 Separation Principle, in each case ensuring that the first group of four cards as displayed above will end up in the spectator's hands, and the second group of four will end up in yours. Also, each of you ends up with one of the original top two cards.

For phase one, once each desired poker hand has been committed to paper, pick up the first packet of ten dealt out, and offer the spectator either of the first two cards, saying, "You pick one of these as your hole card, and I get the other one as mine." Of course each of you ends up with a Seven.

Eight cards remain in your hand, in a face-down packet. Pick up the top two, saying, "One of these is for you, the other one goes underneath. Which one would you like?" Whichever one is selected is put on top of the spectator's hole card, and the other one goes underneath the packet in your hand. Repeat with the next two cards, so that the spectator now has three cards.

Next, the spectator is asked to make similar choices to determine two cards for you, working with the top two cards in the remaining packet in your hand each time, so that you too end up with three cards in front of you on the table.

Now, the spectator is asked to select on a fourth card, deciding between the top two of the four cards in your hand, the rejected one being tucked underneath as usual. You have three cards left, offer the top two to the spectator, saying, "Please select one more for me," tucking the other one underneath.

Pause to recap what has happened, "I believe I've been scrupulously fair, giving you so many totally free choices for your cards, and for mine too. Now, each of us has four cards. Please look at yours." The spectator's eyebrows should noticeably rise upon seeing what a good hand has been accumulated so far. Next, you move in for the kill. "I want you to get the best possible poker hand you can, so I'm going to let you see the last two cards here before you pick the one you want."

Show the spectator the faces of the two remaining cards in your hand, as you turn your head away. Whichever card is not selected, plunge it into the middle of your four cards on the table. Pick up your five cards, and say, "I'm going to shuffle these one more time, for good luck."

Ask to see the spectator's hand. It will be a full house, consisting of two Aces and three Sevens. Look crestfallen, and say, "Very impressive, that's going to be hard to beat." Then, turn over your cards to reveal that you got four Twos and a Seven.

To add insult to injury, the wishes written down earlier can now be read to reveal that the outcome was completely predetermined in your case.

In phase two, should you opt to do that too, start as above but also show the spectator the faces of the top two cards of the second packet of ten cards as you ask for one to be chosen as a hole card. You are counting on the spectator being sensible and selecting the Ace rather than the Eight.

The rest works as above: the spectator ends up with the other two Aces and the K♠, and hence a most (but not *the* most) impressive full house, whereas you end up with an even better hand, namely this straight flush: 8♠, 9♠, 10♠, J♠, and Q♠. Again, the note you made earlier can now be read to reveal that you predicted the outcomes of both hands.

You can add more drama to the proceedings as follows. Once the spectator's cards have been turned over, look at yours and say, "I see you have the King of Spades, if only I'd got that, I'd have had a King-high straight flush in Spades." Pause, and add, "I guess I should be happy with what I got," as you allow all to see your winning hand.

Source: Original. May 2012. Inspired by Dave Solomon's "Power of Poker," as learned from John Bannon's *Dear Mr. Fantasy* [Bannon 05].

Presentational options: Now that you are conditioned to work with a partially rigged deck, why not go all the way, especially if you've also mastered some complete-deck false shuffles? If this sounds like you, consider having five carefully planned ten-card stacks as below, with the two remaining cards (the 3♠ and 4♥) at the bottom.

Find five poker fans, and work through them (and the stacks we now list) one by one, for different dramatic effect.

Stack 1: 2♦, 2♥ on top, then 2♠, 2♣, A♦, A♠ for the spectator, and 5♦, 5♥, 5♠, 5♣ for you.

   Punchline: a full house for the spectator, and a winning four of a kind for you.

Stack 2: 8♠, K♠ on top, then K♥, K♦, A♣, A♥ for the spectator, and 9♠, 10♠, J♠, Q♠ for you.

   Punchline: two pairs (Kings and Aces) for the spectator, but a superior straight flush in Spades for you.

Stack 3: 6♥, 10♦ on top, then 9♦, 8♦, 7♦, 7♠ for the spectator, and 7♥, 8♥, 9♥, 10♥ for you.

   Have the spectator pick the first card based on seeing both available.

   Punchline: a straight for the spectator, but a superior straight flush in Hearts for you.

Stack 4: 6♣, J♣ on top, then J♦, J♥, 6♦, 6♠ for the spectator, and 7♣, 8♣, 9♣, 10♣ for you.

Punchline: a full house for the spectator, and a better straight flush in Clubs for you.

Stack 5: K♣, 3♣ on top, then 3♥, 3♦, 4♣, 4♠ for the spectator, and 4♦, Q♣, Q♦, Q♥ for you.

Start by letting the spectator see the faces of the top two cards, and having one of those selected.

Punchline: two pairs for the spectator, and even better, three of a kind for you.

Birthday Card Matches

The above effects are heavily dependent on partially rigged decks. Actually, something quite worthy in the poker realm can be pulled off with a genuinely shuffled deck, courtesy of a probabilistic observation.

Table 3.1 shows that half of all possible poker hands contain a pair or better. Hence, readers may be able to guess the answer to this question:

> How many cards, picked randomly from a standard deck of fifty-two, do we need so that there is a greater than 50% chance of getting at least one pair (i.e., two cards with the same value)?

This problem can be tackled by using the usual approach to the classic birthday problem, which asks how many people are required to ensure a greater than 50% chance of having at least two people born on the same day of the year, that is, at least one birthday match. The counterintuitively small answer to that question is twenty-three people if we assume that all birthdays are equally likely.

The key to estimating such probabilities is to turn things around, and focus on the chances of there being no such match, noting that

$$P(\text{some match}) = 1 - P(\text{no match}).$$

If k cards are randomly picked from a full deck, where $3 \leq k \leq 13$, then since there are four cards of each value, it can be shown, using an argument like the one on page 22, that

$$P(\text{some match}) = 1 - \left(\frac{52}{52} \times \frac{48}{51} \times \frac{44}{50} \times \cdots \times \frac{52 - 4k + 4}{52 - k + 1} \right).$$

Plugging in various values for k, we find that for five cards we can be about 49.29% sure of a "birthday card match"[6] and for six we can be about 65.48% sure; so we need to pick at least six cards to be at least 50% sure of such a match. For seven, eight, or nine cards, the probability of a match goes up to about 79%, 89%, or 95%, respectively. Throwing in one more card allows our confidence to soar.

Birthday Card Match Principle
In any ten randomly selected cards, there is a 98% chance of at least one value match.

We often get a lot more than a matching pair, as experimenting with ten randomly chosen cards confirms. Pairs of pairs, or even triple pairs, or instances of three of a kind, are far from unusual.

♣   ♥   ♠   ♦

We can use the Position Parity Principle together with this card match principle for an impromptu poker effect. Have the deck shuffled and take off the top ten cards. Glance at the card faces, then assuming that there is indeed at least one matching pair, drop clumps of cards so that there are two interwoven poker hands, one occupying the odd positions, and the other the even positions. If you are very unlucky and get no matches at all, simply show all of the cards and say, "Not the best cards for a game of poker, please shuffle and deal again," and all should be well.

Let's assume that as a result of the clumps you drop, the even positions contain the matching pair, or whatever good cards are among the ten. Lay out all of the cards face-down, being careful to note silently which end of the row includes a card from the winning hand.

As long as you go first, and start by selecting that card, and from there on always pick the card next to the one the spectator chooses, you are certain to end up with the victorious hand. It's important that no cards are turned over until the end.

♣   ♥   ♠   ♦

There aren't many poker effects that can be done with a genuinely shuffled deck. Combining the Birthday Card Match and Simon's 64 Separation principles—neither of them easy for onlookers to see through at first—yields an even better effect than the one just suggested. In our experience, it catches audiences totally offguard.

---

[6]As already seen in "Convention Center" on page 23.

3♦   Poker with Any Ten Cards

How it looks: *A spectator shuffles the deck, and deals ten cards face down. You glance at the card faces and pause to write down a prediction. Then the spectator is given numerous choices for distributing the cards between the two of you. Yet, without fail, not only do you end up with the winning poker hand, but your specific prediction, made in writing in advance—which may say something like, "You will have a pair of Jacks, but I'll have three Fives"—also comes true.*

How it works: Glance at the faces of the cards you are given. As explained earlier, there is a high probability that among these ten cards, being from the top of a shuffled deck, there will be at least one pair. We proceed under the assumption that this is the case; if it isn't, request a reshuffle, having first shown the card faces around and made remarks about how nobody can get a decent poker hand from those.

(The "any" of the title refers to cards from a shuffled deck. It is, of course, possible to get ten useless cards, it's just not that likely.)

The idea is to ensure that by innocently dropping clumps of cards face down on the table, the "winning cards" (i.e., a matching pair or better) are among the bottom four (i.e., in positions 7–10, counting from the top), and also the "losing cards" are in positions 3–6. It's important that the top two cards are such that it does not matter which of you gets which one as your hole card—they must not affect the balance of power. For instance, if you get two pairs, arrange it so that the better pair is among the last four cards, and the weaker pair is among the four cards above those.

To understand the seemingly fair separation procedure the spectator goes through, first forget about the (low-impact) top two cards. Each of you gets one of them, but it won't change "the balance of power." We focus on the remaining eight cards. Once again use Simon's 64 Separation Principle, so that you retain control of the division in this key sense: the top four cards of those eight end up in the spectator's pile and the bottom four in yours.

Suppose the cards you are handed are K♣, 8♠, J♥, 10♣, 5♣, 7♠, 5♥, 9♠, 4♦ and A♠, as shown in Figure 3.9.

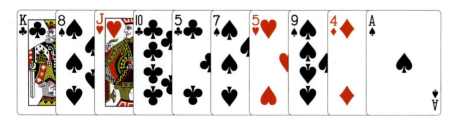

Figure 3.9. One possibility for the cards you are given.

There is a pair of 5s, separated only by one card, and cutting the 9♠, 4♦, and A♠ from one end of the packet to the other puts the Fives where you want them: among the bottom four cards when the fan is closed up and the packet is face down once more. There are no other pairs present, so that no matter which of you gets the K♣ or 8♠ on top, you're sure to get the winning hand.

In contrast, suppose the cards you are handed are 2♠, 9♣, 6♥, 2♥, K♠, J♦, 2♦, 4♠, A♦, and J♣, as shown in Figure 3.10.

Figure 3.10. Another possibility for the cards you are given.

There are three Twos and a pair of Jacks, so some fast decisions must be made. Options include winning with the three Twos, while letting the spectator have the Jacks, or winning with the Jacks, giving the spectator only two of the Twos, or even winning with a pair of Twos, making sure that neither of you gets both Jacks. Or you could be very greedy, and award yourself two Twos and both Jacks, with a 50% chance of clinching the third Two as well. Whichever winning strategy you decide on requires some care in what happens next. It's up to you how specific you wish to make your prediction.

The easiest way forward here is to proceed so that you get both Jacks, and the spectator gets at most two Twos. One way to achieve this is to pull out either the 4♠ or A♦ from the fan, making sure nobody sees its face, while making a comment such as, "Very interesting, I have a Two here, and there are two more where that came from. If you play your cards right, those could all be yours, which would guarantee a win for you." As you say this, close up the fan face down, and plunge the card in your hand face down somewhere in the top half of the packet. The upshot is that the bottom four cards contain both Jacks and 2♦. Were the packet fanned face up once more, it would look like Figure 3.11. Now you can't lose. You can even tease, by saying, "Try to get as many Twos as you can. There are three up for grabs here."

You'll end up with one of the first two cards and all of the last four shown. The spectator gets A♦, 6♥, 2♥, K♠, and, if he's lucky, the 2♠ as well. If he gets a pair of Twos, say, "Excellent! But it's a pity you didn't get all three of them," throwing the 2♦ on the table. Then, as an

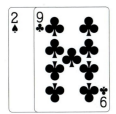 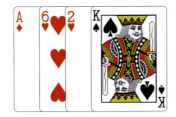 

Figure 3.11. The cards in Figure 3.10 as you think of (and rearrange) them.

apparent afterthought, quietly add your pair of Jacks. If you get the 2♠ as well, you can really gloat. "Too bad you only got one Two, because that means I got the other two," as you put all of your cards on the table.

Source: Original, inspired by Bill Simon's separation technique, learned from his 1964 book *Mathematical Magic* [Simon 64], where it was part of an effect called "The Four Queens." Published online in the June 2006 *Card Colm* "Better Poker Hands Guaranteed" [Mulcahy 06_06].

Presentational options: There is another way to split up the ten cards here to give the desired result—by using the Monge shuffle. (See *Magical Mathematics* [Diaconis and Graham 11] and the online October 2008 *Card Colm* "Monge Shuffle Cliques" [Mulcahy 08_10].) This is mentioned in the June 2012 *Card Colm* "Something's Old, Something's True, Something's Borrowed, Something's New" [Mulcahy 12_06].

Above we've focused on pairs, threes of a kind, and so on. By chance—or by design, if a bloody-minded spectator tries to trip you up upon a repeat performance and handpicks the cards to be used—you may find yourself presented with ten cards of different values. Regardless of the suits present, you now have ten of thirteen possible values, and it turns out that the odds of having six values in a row, such as 2, 3, 4, 5, 6, and 7, are even. In such cases, if cards are arranged so that the 2 and 7 are the first two, and the 3, 4, 5 and 6 are the last four, you will certainly get a straight. Of course the spectator may end up with a better straight, if the other four cards are an 8, 9, 10 and Jack, but at least it's still interesting.

Parting Thoughts

- Is it true to say that even if you are dealt 100,000 poker hands over the course of a gambling-intensive lifetime, you're still likely to have held less than 5% of all possible hands?

- Come up with better stacks, using most of the deck, for multiple demonstrations of the futility of free choice, as seen in "Full House Blues" (3♠).

- We claimed in "Poker with Any Ten Cards" (3♦) that if you are handed ten random cards, it's not unusual to find more than one pair, or even three of a kind. Can you quantify those claims, with probability estimates? For instance, what is the probability of getting at least one three of a kind? Similar questions are asked in the "Parting Thoughts" at the end of Chapter 10, and hints are given there as to how to do the analysis.

- We also claimed in "Poker with Any Ten Cards" (3♦) that if you have ten cards of different values, the odds of having six values in a row are even; in other words, the probability that this happens is about 50%. It's a more challenging exercise to show why this is true.

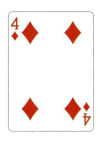

More Additional Certainties

Many, if not most, of the mathematical card effects popularized over the decades depend on nothing fancier than addition and subtraction, cleverly dressed up to hide that fact. As such, their inner workings are not hard to figure out for those with a little background in mathematics. Here, we present a selection of more recent addition- and subtraction-based effects that may surprise and amuse.

Chapter 2 considered two families of additional certainties: effects for which something definite could be said about the sum of the values of neighboring beads on bracelets of cards. In the effects that follow, we consider some more sum-rich experiences, but without the adjacency assumption. In essence, we'll learn how to *unadd* numbers. Given the sum of two (or maybe more) numbers, under appropriate circumstances, we'll soon be able to deduce what the individual numbers must be.

Our chapter opener here is hard to top. It's a perfect follow-up to one of the effects from Chapter 2, assuming you have a second deck of cards set up in advance, as required. You can even introduce it by saying, "You're probably thinking there's something special about cards in a circle. Not really. This time, you can break the circle and mix them up all you want."

4♣ Little Fibs

How it looks: *Give the deck several shuffles, then deal six cards face down to the table, setting the rest aside. Turn away, requesting that those six cards be thoroughly mixed up. Have any two cards selected by two spectators, who then compute and report the total of the two card values. From that information alone, you promptly name each card.*

How it works: It really is true that when several numbers are added up, each may be deduced with certainty from the sum, provided that the available numbers are somewhat restricted!

While totally free choices of two cards are indeed offered, they are from a carefully controlled small subset of the deck. The possibilities are narrowed down by having half a dozen key cards at the top of the deck at the start, in any order, and keeping that top stock there throughout some fair-looking shuffles. That's the first secret: you have memorized the suits and values of the half dozen key cards that start at the top. The second secret is the following little-known fact.

Zeckendorf Representation

Every positive whole number is a sum of numbers in the list 1, 2, 3, 5, 8, 13, 21, 34, ... *Moreover, there is only one possible sum if we avoid using any number twice, and we don't use any numbers that are adjacent in this list.*

The numbers listed are *Fibonacci numbers*, generated by starting with 1, 1, then adding those to get 2, adding the 2 to the 1 before it to get 3, adding the 3 to the 2 before it to get 5, adding the 5 to the 3 to get 8, and so on. Each term is the sum of the two before it. The list continues indefinitely: 1, 1, 2, 3, 5, 8, 13, 21, 34, 55, 89, 144, ... The numbers get large quite fast. For our effect, we restrict ourselves to values within the realm of a deck of cards, and use only one 1.

Hence, the Fibonacci numbers are essentially the building blocks of all positive whole numbers by using addition in an interesting way: we never have to use two consecutive Fibonacci numbers or the same one more than once. This decomposition is named after Edouard Zeckendorf, an amateur mathematician who was a doctor (and later a dentist) in the Belgian army, and who, in 1939, seems to have been first to observe this still-not-well-known property of Fibonacci numbers.

For instance, $6 = 5 + 1$ only (we don't allow $3 + 2 + 1$ or $3 + 3$), and $20 = 13 + 5 + 2$. Given any n, it's not difficult to find the appropriate breakdown, by using a greedy algorithm: first peel off the largest possible Fibonacci number as one summand, and repeat for what remains, until you are left with the smallest Fibonacci component. For instance, $50 = 34 + 16 = 34 + 13 + 3$.

Let's fix our attention on any Ace, 2, 3, 5, 8, and King, for instance, those displayed in Figure 4.1 with suits in standard CHaSeD order. These are the cards secretly at the top of the deck, before and after some seemingly convincing shuffling. The Zeckendorf payoff here is as follows.

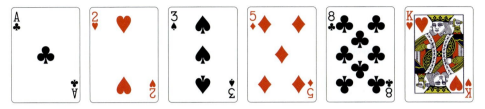

Figure 4.1. Little Fibs.

Little Fibs Principle

If two cards are selected from the A♣, 2♥, 3♠, 5♦, 8♣, and K♥, their individual values can be deduced from the sum of their values. In other words, any possible total can arise in only one way.

Even better, it's easy to decompose a given sum into its pieces. For instance, if a total of 11 is reported, subtract the biggest Fib you can, namely 8, to get 3. As long as you have the suits memorized, you can then reveal that the chosen cards are the 3♠ and 8♣.

What we have here is another sum-rich situation, earlier incarnations of which were encountered in Chapter 2. This time, however, it's for sets rather than bracelets or rows of numbers. The *additional certainty* concept carries over with more freedom: the cards really can be mixed up thoroughly because the order in which they are summed is irrelevant.

Source: Original. Inspiration arrived one day in January 2008 while lecturing at the board in an honors introductory class at Spelman College, using the generally excellent *The Heart of Mathematics* [Burger and Starbird 04]. Published online at MAA.org as the February 2008 *Card Colm* "Additional Certainties" [Mulcahy 08_02]. Fibonacci numbers have many applications to mystery and magic, from geometrical vanishes [Gardner 56, page 134] and instant addition stunts [Gardner 56, page 158], to Stewart James modular predictions [Diaconis and Graham 11, pages 187–189]. See the June 2007 *Card Colm* "Gibonacci Bracelets" [Mulcahy 07_06] for other recent modular effects.

Presentational options: If you are doing this as a follow-up to one of the first two effects in Chapter 2, you may wish to start by having the planted cards at the top dealt in a circle, as was done there, asking for two adjacent cards to be selected, then appear to change your mind and indicate that it can be *any* two cards. Say, "Let's break the circle," vigorously mixing up the six cards on the table. "Take any two, and I'll try to guess them from their value total."

Even mathematically savvy audiences tend to miss the fact that small Fibonacci numbers are at the heart of this effect. Feel free to have a little fun, saying, "Magicians often lie, to gain your confidence and distract you from what they are really up to. Mathemagicians would never dream of doing that, of course. At worst, we might sneak in a few little Fibs."

You can request that four spectators pick cards and reveal the total. Since you know that all six card values sum to 32, subtraction allows you to deduce which two were *not* chosen, and hence which four were.

It's even fairly safe to ask three spectators to participate. The unique summand property holds true in all but two cases: if a total of 16 is reported, the cards could be either the Ace, 2, and King, or the $3, 5$, and 8. Then you'll need to fish a little to determine which has occurred, perhaps by commenting, "I'll do the Hearts first, please raise your hand if you have a Heart." Assuming everyone is truthful, you can now proceed successfully to the finish line. (If no hand is raised, you can also make a crack about the difficulty of working with a heartless audience.)

A simpler way to ensure that you can work with three spectators and not risk the ambiguity just mentioned is to deal only five of the six key cards to the table at the outset. It doesn't matter which five are used.

Other sets of numbers work too, and can be used for a repeat performance. For instance, the cards in positions 7–11 could also be set up in advance, perhaps using an Ace, 3, 4, 7, and Jack. These are derived from the Lucas sequence, $2, 1, 3, 4, 7, 11, 18, \ldots$, omitting the initial 2 to avoid the undesirable $2 + 3 = 1 + 4$ possibility. Any list of increasing generalized Fibonacci numbers—formed by starting with any two numbers, and continuing by having each new term be the sum of the two before it—also has the desired property. (You'll also need to memorize whatever suits you use to go with these values.) When all is said and done, we don't even need segments of generalized Fibonacci sequences to pull off the above kind of effect. The set $1, 2, 4, 6, 10$ works. So does $1, 2, 5, 7, 13$. We'll have more to say about this observation in the final effect of this chapter.

Note that the just-approved sets suggest equally usable dual sets, which also work well for decks of cards, taking as values the "complements in 14"; for instance, {King, Queen, Jack, 9, 6, Ace} = 14−{Ace, 2, 3, 5, 8, King}. The use of the greedy algorithm needs to be modified accordingly, however. Any of those may be used to throw off an audience familiar with the Fibonacci version.

It's not difficult to find two nonoverlapping sets of five cards of the desired type with a regular deck of cards, which allows for two stacks to be planted at the top at the outset, permitting a second performance with different cards. For instance, one set could be any Ace, 2, 4, 6, and 10, the other being its complement in 13, namely, a Queen, Jack, 9, 7, and 3. There's just room to squeeze in two nonoverlapping sets of size six: any Ace, 2, 3, 5, 8, and Queen, along with its complement in 12, namely, a Jack, 10, 9, 7, 4 and Joker (which you'll need to explain has value 0).

Balanced Ternary Representations

Long before the unique summand of Fibonacci numbers property just exploited was noticed, it was known that every positive whole number n is a sum of powers of 2. This leads to the binary (base 2) representation of n. For example, $13 = 1 \times 10^1 + 3 \times 10^0$ can also be written $1 \times 2^3 + 1 \times 2^2 + 0 \times 2^1 + 1 \times 2^0$, so that 13 is 1101 in base 2. In the next two effects, we exploit a well-known ternary take on this.

> ### Balanced Ternary Principle
> *Every whole number can be written as a sum of signed distinct powers of 3, and this representation is unique apart from cancelations. Also, the whole numbers between −13 and 13 inclusive are precisely the ones that can be written by using* ±1, ±3, *and* ±9.

For example, $10 = 9 + 1$, $7 = 9 - 3 + 1$, $-5 = -9 + 3 + 1$, and $13 = 9 + 3 + 1$. It's also true that $10 = 9 + 1 + 3 - 3$, but in what follows we suggest a way to deal with this in the card context.

4♥ Consolidating Your Cards

How it looks overall: *Deal five or six cards face down from a shuffled deck. Turn away, and have the cards mixed up. A spectator selects and looks at any two or three of them. Under the guise of "consolidating credit and debit card rates," the total of the selected card values is reported, with the convention that Red card values are negative. From that information alone, you promptly name each value and suit.*

How it looks in detail: Shuffle a deck several times and then deal five or six cards to the table, face down. Address one spectator, saying, "Let's perform a little experiment here. No doubt you've heard about the credit card reforms that kicked in recently. A little-known aspect of those is a novel way to compute your credit rating, that is based on simply consolidating all of your cards. Of course, debit cards really have negative interest rates; and they too must be counted, in their own way."

Have the cards on the table mixed up. Ask the spectator to select two or three of the cards, without seeing their faces, and without telling you how many are chosen, saying, "I'm going to avert my eyes and say a little prayer to Plutus, the Greek god of credit, that you choose wisely." Then turn away while the selections are made. Request that the selected cards be held so that you can't see them, and have the unused cards stuffed into the rest of the deck, which is then set aside. Now turn back.

"You have selected several cards at random. I have no idea which ones, or even how many. Actually the cards you selected are probably a mixture of debit and credit cards. Each one has an interest rate. Don't look just yet, but you can tell whether it's debit or credit by checking its color. Of course, it's very important to know the difference! For instance, a Red Queen is a debit card with twelve percent interest, whereas a Black five is a credit card with five percent interest.

"Now, when I turn away again in a moment, I'd like you to look at your cards and add up their interest rates, and just tell me the total. Basically, you are going to consolidate all of your cards into just one card with an overall rate computed according to the new rules. Just tell me the overall rate. For instance, if you have a Red Ace and a Red four, which are debit cards at one and four percent, respectively, and a Black Jack, which is a credit card at eleven percent, that's minus one minus four plus eleven, hence overall a six percent interest credit card, if you were to consolidate all of your cards. Are you ready? I'm going to offer up another little prayer that this all works out." Turn away once more.

Suppose that an overall rate of five percent is announced once the spectator has looked at, and again hidden, his cards. Turn back one last time, and continue. "They say you can tell a lot about a man from simply knowing his credit rating. I can tell a lot about you. You've revealed that you came up with five percent when you consolidated your cards. In that case, having just attempted to communicate with the credit god, I'd say you have two debit cards and one credit card. Am I right?" That should get some reaction. But, wait, there's more.

"Actually, if I were a betting man, I'd say your debit cards have rates of one and three percent, respectively, and your credit card has a rate of nine percent. How did I do?"

At this point the spectator will likely move to reveal the cards he is holding, and you may wish to add quickly, "That's because the cards you selected at random were, in fact, the Ace of Diamonds, the Three of Hearts and the Nine of Clubs. Please show everyone your cards." You are right on all counts.

How it works: For this effect, six key cards must start at the top of the deck, and stay there through some casual shuffles of yours. The cards are shown in Figure 4.2, the suits being in classic CHaSeD order. You must have these committed to memory.

Following the convention that Red cards represent negative values (see page 14), the Balanced Ternary Principle is exactly what is needed to make this performable. If a five percent credit rate is reported, you

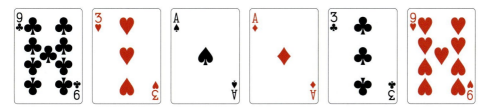

Figure 4.2. Debit and credit cards.

think of the decomposition $5 = 9 - 3 - 1$, realize that the resulting Ace and 3 are Red, while the 9 is Black; assuming you remember the suits of the six cards in question, all is well. If a seven percent debit rate is reported, you instead think of the decomposition $-7 = -9 + 3 - 1$, realize that the resulting Ace and 9 are Red, while the 3 is Black, and in due course reveal that the cards in question are the 3♣, A♦, and 9♥.

Two of the cards may cancel each other out, for instance if three cards are picked and a consolidated credit or debit rate of three percent is reported. In such cases, there is no way to tell which other identical powers of 3—one Red, one Black—are involved. When the consolidated rate is (plus or minus) a power of 3 like this, you can side-step this by commenting, "You have three cards, but I believe that two of them cancel each other out, so let's focus on the one that remains." However, if two or four cards are picked, and a consolidated rate of zero is reported, due to one or two canceling pairs, then you have less to work with; for this reason, you may wish to insist that three cards be taken in the first place.

Source: Original. Published online in the February 2010 *Card Colm* "Tighter Ascertainments: Matching Interest Rates" [Mulcahy 10_02].

Presentational options: There are other cards that may be used if you worry that the powers of 3 used above are "too obvious." For instance, all numbers between -13 and 13 can also be represented using just $-9, -3, -1, 2, 3, 8$. The cards in Figure 4.3 could be the ones stacked at the top of the deck at the outset.

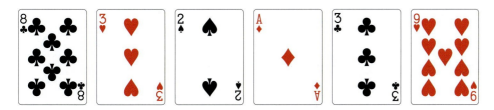

Figure 4.3. Other debit and credit cards.

However, with these values, uniqueness may fail, since, e.g., $5 = 2 + 3 = -3 + 8$ and so $6 = 1 + 2 + 3 = 1 - 3 + 8$. Hence, if a consolidated rate of five (or six) percent credit is reported, you'll need to fish a little to figure out which of two possible card combinations gave rise to it.

Three of the six cards in the previous Fibonacci-based effect are Red, and it's a pleasant surprise to note that if they are interpreted as debit cards, then the consolidating effect still works for two freely selected cards from those six. (Can you think of a reason why this must be so without checking the details?) In fact, if any two numbers from 1, -2, 3, -5, 8, and -13 are chosen and added, we obtain one of the distinct sums -18, -15, -12, -10, -7, -5, -4, -2, -1, 1, 3, 4, 6, 9, or 11. Note that in most cases it suffices to know the absolute value of the sum to deduce the summands; in other words, the overall rate is all that is needed, so that you could take a gamble on asking, "Debit or credit? No, don't tell me, I'll try to guess." If we switch red and black—changing suits to match, natch—we are still in luck. The same holds for the "alternating Lucas" numbers 1, -3, 4, -7, 11 and the alternating versions of some of the other lists mentioned at the end of "Little Fibs," such as 1, 2, 4, 6, 10. However, we cannot modify 1, 2, 5, 7, 13 in this way, since $1 + 5 = -7 + 13$.

The binary principle that inspired the last effect also underlies an old chestnut from the 1920s, which appears as "Findley's Four Card Trick" in Martin Gardner's *Mathematics, Magic and Mystery* [Gardner 56, pages 6–7].[1] What happens there, basically, is that an Ace, Two, Four, and Eight of different suits are hidden in a known order in a pocket, and a spectator calls out the name of any card in the deck. If it is one of the four in the pocket, whip it out and declare victory. Otherwise, take out those cards whose values sum to the desired value, if possible starting with (and commenting on) a card that also matches suit. For instance, if the cards are A♣, 2♥, 4♠, and 8♦, and the 10♠ is named, produce the 2♥ and 8♦.

A balanced ternary twist on this follows. It can be performed as an alternative to the last effect, but don't try both for the same audience.

4♠   Matching Interest Rates

How it looks overall: *Have a spectator freely remove a card from a shuffled deck. You rapidly pull out a few cards to match it in suit and value, using the usual conventions for debit (Red) and credit (Black) cards.*

---

[1]Max Maven reports that Arthur Findley wasn't the only one to invent it, Charles Jordan came up with it independently.

How it looks in detail: Take out a deck and shuffle it a few times, while mentioning the new financial regulations governing credit cards. Say something like, "Have you heard? They've really tightened up the rules on credit cards." Fan the deck so that the audience can see the card faces, but you can't. Say, "Pick a card, any card!" and have somebody remove a card. Close up the fan, and place the deck out of sight, in your pocket perhaps, or in a bag on the table. Have the chosen card placed face down on the table.

"From a shuffled deck, you've selected a card," you state. "I didn't influence your choice in any way, did I?" Once it has been acknowledged that a completely free selection has been made, add, "I couldn't possibly know what it is, but I will in just a minute."

After a little pause for effect, continue, "You have in fact chosen a new credit or debit card, whose interest rate is its numerical value. Of course, as you're a fiscally responsible person, you have chosen wisely. Now, about these new rules. On balance, they're a bit complicated, but I can help you since I understand the mathematics involved. The best news is that from now on, there aren't supposed to be any surprise rate hikes. Actually, I'm going to match your rate!"

While that information is sinking in, announce, "I need to swipe your card," as you quickly lean forward and brazenly take the card off the table. Innocently ask, "Debit or credit?" Regardless of the answer (if any) pretend to swipe it, as if it were an actual debit or credit card. Then casually glance at its face; suppose it's the 10♠. Modestly declare, "I think your card is the Ten of Spades, am I right? Give me a little credit, please."

Press on without delay: "My pledge to you is that I will match that interest rate, in my own way. Ten percent? It could be worse. Have you ever considered the advantages of several credit cards? Each has its own perks, some of which you might want to avail yourself of."

Reach into your pocket (or bag, as the case may be), continuing: "The match of the Ten of Spades would be the Ten of Clubs, correct?" Pull out one card and place it on the table, face up. "The Nine of Clubs, that was close! Only off by one. Aces represent one percent interest, so if I can find an Ace credit card to go with that, I'll declare victory." Sure enough, the next card you pull out is the Ace of Spades.

If the chosen card is the Seven of Clubs, you first pull out the Nine of Clubs. Look surprised and exclaim, "Silly me, how can I get the Seven of Clubs when you already have it!" Reach for the deck again, saying, "Let's see," before pulling out the Ace of Spades. "Oh dear, Aces have value one, so now we're up to ten percent interest. This isn't adding up as I'd hoped it would, and that's a much higher rate than the one you selected. But mathematics can get us out of this mess. If we subtract three percent, all will be well, so all we need is a card that represents three percent debit. We want to be three percent in the red, so to speak, so obviously a Red three is called for, that would be a debit card to go with the two credit cards here." Quickly pull out the Three of Hearts.

"There, all is well! It really doesn't matter how many credit or debit cards one has, as long as the combined interest rates add up correctly. I've heard that under the new rules, nobody will ever need more than six cards, three debit and three credit, but I'm not sure I believe that."

If the selected card is Red, you'll probably want to explain first that this has negative value, and hence is really a debit card, before proceeding accordingly. (But read on for an alternative presentational possibility that works about half of the time in this case.)

How it works: It should come as no surprise that once again the deck is not as shuffled as is claimed: the top three and bottom three cards are set up in advance and are kept there through several riffle shuffles that genuinely mix (most of) the rest of the deck. As in the last effect, the cards in Figure 4.2 are used, their suits, values and exact placements having being memorized in advance.

Figure 4.4—spread over two pages so that more cards can be seen (ten cards not shown)—gives a sense of what the deck would look like fanned if these cards were concealed at the bottom and top: we recommend that the Black 9, 3, and Ace be stashed at the bottom top of the deck, in that order, and likewise the Red Ace, 3, and 9 be stashed at the bottom, so that the 9s are the top and bottom cards overall. As the cards near the face of the deck might make a visual impression on your audience when fanned, we suggest adding a decoy in the form of a Red royal card at the very (visible) bottom.

Once the cards are out of sight, this decoy can be removed and plunged into the middle of the deck; that way it won't confuse you when you have to "make withdrawals" later. Make sure that you practice reaching in and quickly extricating several of the cards displayed above.

Following the convention that Red cards represent negative values, the second part of the Balanced Ternary Principle says that the value of

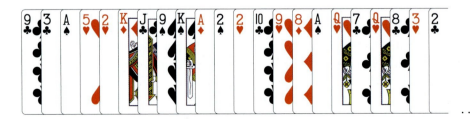

Figure 4.4. Hidden debit and credit cards.

any card in the deck can be represented with these cards. The negative associations for Red cards can be ignored if the selected card is a Red Ace, $3, 4, 9, 10$, Queen, or King. In such cases, you may wish to simply match it by producing one or more of the three Red cards in use.

One's patter must be adapted depending on the suit of the selected card; sometimes it's best not to dwell too much on the suit.

Source: Original. This too was published online in the February 2010 *Card Colm* "Tighter Ascertainments: Matching Interest Rates" [Mulcahy 10_02].

Presentational options: As noted at the end of the previous effect, other cards may be used if you worry that utilizing only powers of 3 is too obvious. The numbers between -13 and 13 can also be represented by using $-9, -3, -1, 2, 3, 8$. Hence, 8♣, 3♥, 2♠, A♦, 3♣, and 9♥ could be planted at the outset, three at each end of the deck.

However, with these values, sometimes four cards are needed to match certain interest rates, such as $6 = 8 + 2 - 1 - 3$. (The possible failure of uniqueness here is not a problem.) One could opt to place all cards with such troublesome values close to the ends of the deck at the start, thereby reducing the chances that any of them will be chosen.

The kind of inversion of the unique (balanced ternary) summand principle just explored can also be applied to the cards with Fibonacci (or related) values considered at the start of this chapter.

Two-Summers

Can an effect like "Little Fibs" (4♣) be pulled off with a genuinely shuffled deck? Yes, it turns out, with high probability. However, it's not for the faint of heart. It requires some quick thinking and fast memory work, but it can be repeated if you are feeling brave and like a challenge.

We need to take a closer look at some mathematics. Let's begin by studying sets of positive whole numbers $\{a_1, a_2, ..., a_k\}$ with the property that all possible $\frac{k(k-1)}{2}$ unordered pairs have unique sums.

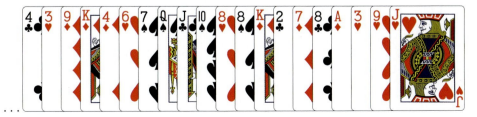

Figure 4.4. (continued)

We refer to such two-sum rich sets as *two-summers*, since our concern is all possible sums of two numbers in these sets. In the literature, these are called Sidon sets. For example, 1, 2, 4, 8, 13, 21, 31, 45, ... are the first terms of one such set, constructed by using a greedy algorithm [Mian and Chowla 44].

Of course, the a_i must all be different, and there is no harm is assuming that they are also listed in ascending order. Examples of two-summer sets include: $\{1,2,3,5\}$, $\{1,2,3,7,100,2013\}$, $\{1,2,4,6,10\}$, $\{1,2,5,7,12\}$, $\{1,3,4,5,9\}$, and $\{4,5,7,100\}$.

Since $\{1,2,4,6,10\}$ is a two-summer, it's not hard to see that its complement in 14 (see page 92), namely, $14-\{1,2,4,6,10\} = \{13,12,10,8,4\}$, is also two-summer. Moreover, there's nothing special about 14 here, it can be replaced by any other number. Indeed, the complement of $\{1,2,4,6,10\}$ in 13, namely, $13 - \{1,2,4,6,10\} = \{12,11,9,7,3\}$, is not only a two-summer, it has no numbers in common with the set from which is was derived. We'll put that example to good use in "Unadditional Love" (6♠) in Chapter 6.

Clearly, any $\{a_1, a_2, a_3\}$ for which $a_1 < a_2 < a_3$ yields a two-summer of size three. Given any collection of four or more numbers, it's easy to tell whether it forms a two-summer, based on this elementary but key observation:

Same Difference Principle

$$a + b = c + d \quad \text{if and only if} \quad a - c = d - b.$$

Hence, coming up with two-summers is equivalent to coming up with sets that avoid equal differences for distinct pairs.

Hence, we can be sure that $\{1,2,5,7\}$ is a two-summer—without computing a single sum—since it is obvious that no two differences are the same. Similarly, $\{1,2,4,7\}$ is two-summer, despite the fact that $7 - 4 = 4 - 1$, as is $\{1,2,4,7,10,14\}$. In contrast, $\{1,2,3,6,11,13\}$ is not a two-summer, since $13 - 11 = 3 - 1$ (corresponding to the undesirable $13 + 1 = 3 + 11$).

Suppose $\{a_1, a_2, ..., a_k\}$ is two-summer of size k. How can we extend it, by appending one more number a_{k+1}, to yield a two-summer of size $k+1$? What is needed here is that all new two-sums $a_{k+1} + a_i$ (for $i < k + 1$) be distinct from any two-sums that arose earlier. One way to guarantee this is to make sure that the smallest possible new two-sum exceeds the largest earlier two-sum. Hence,

Starting with a two-summer $\{a_1, a_2, ..., a_k\}$, we can extend it to a longer two-summer by appending one term a_{k+1} if $a_{k+1} + a_1 > a_{k-1} + a_k$, or equivalently, $a_{k+1} > a_{k-1} + a_k - a_1$.

Hence, if $a_1 = 1$, we can take $a_{k+1} = a_{k-1} + a_k$.

In this way, $\{1,2,3\}$ can be extended to $\{1,2,3,5\}$, and $\{1,2,4\}$ can be extended to $\{1,2,4,6\}$.

The condition above is sufficient, but not necessary; for example, $\{1,2,5,7\}$ can be extended to $\{1,2,5,7,12\}$, by the above method, but it can also be extended to the "tighter" $\{1,2,5,7,9\}$.

With these preliminaries, we are ready for our grand finale.

Disclaimer: What we describe next can be done with a genuinely shuffled deck only by those who can casually yet confidently glance through the card faces seeking five that enjoy a special relationship to one another, and can memorize them instantly, while chatting nonchalantly to the audience about how randomly shuffled the deck is.

4♦   Any Two Cards (No Fibbing)

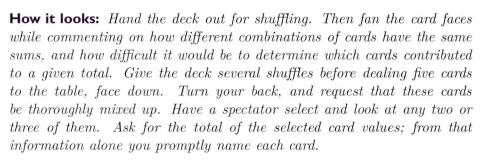

How it looks: *Hand the deck out for shuffling. Then fan the card faces while commenting on how different combinations of cards have the same sums, and how difficult it would be to determine which cards contributed to a given total. Give the deck several shuffles before dealing five cards to the table, face down. Turn your back, and request that these cards be thoroughly mixed up. Have a spectator select and look at any two or three of them. Ask for the total of the selected card values; from that information alone you promptly name each card.*

How it works: What you are actually doing while blathering on like this is calmly scanning the faces seeking five adjacent two-summer cards. It's okay to let the audience see some card faces, as long as they are not the ones you're about to zero in on!

We claim that the following holds, even though we offer no proof.

> ### Probabilistic Two-Summer Principle
> *If the card faces in a shuffled deck are scanned, then then with probability over 99%, there is a run of five adjacent cards whose values are a two-summer.*

For instance, Figure 4.5 shows such a run (put in ascending order to make it easier to verify its suitability).

A good way to locate such a run is to search for Aces and see what values are nearby. In the rare cases where this does not yield the desired result, repeat your scan, searching for Twos. The idea is to "accept" the

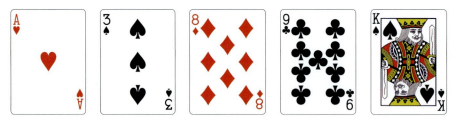

Figure 4.5. A lucky find: a two-summer.

next two lowest values within some five-card run containing an Ace or a Two, and then just worry about whether the addition (so to speak) of the two other cards in the run leads to a two-summer set. If you don't have the kind of luck we claim is likely, settle for a four-card two-summer run instead. Another possibility, keeping in mind the complement in 14 principle (see page 92), is to take a "top down" approach, and consider what values are near Kings or Queens.

Let's assume that the five cards shown in Figure 4.6 have been located together. Surreptitiously cut these to the top and do some riffle shuffles that do not disturb them, all the while chatting casually. The key now is to sear into your brain the card total (26 here) as well as a clear image in ascending order as follows. You may find it easier to remember the values {1, 3, 5, 8, 9} and suit order separately.

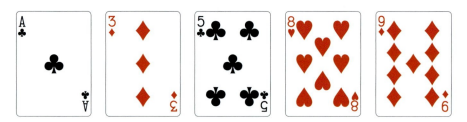

Figure 4.6. An unforgettable two-summer.

In general, you've memorized the top (four or) five card values and suits, and the resulting total t. Once those cards are dealt face-down to the table, and further mixed, a spectator selects two of them and tells you the sum of their values.

For instance, in the case of the five cards above, with total value 26, if the spectator reports a sum of 17, you can quickly deduce that the selected cards are an 8 and a 9. Assuming that you haven't forgotten the suits, you can now conclude the effect as desired.

Source: Published online as the June 2009 *Card Colm* "Two Summer Difference Certainties" [Mulcahy 09_06].

Parting Thoughts

- What property of small sets of whole numbers makes them suitable for effects like the ternary ones presented above? Can such an analysis suggest other sets of cards to use that are less obviously related to powers of three?

- Readers with some experience with probability may enjoy estimating the probability claimed in the Probabilistic Two-Summer Principle on page 101.

- Which n-element sets S of numbers have the property that all of their $2^n - 1$ nonempty subsets have distinct sums? Such sets are said to be *dissociated*, and these generalize the kind of sets used in this chapter. Examples include $\{2, 8, 16, 64, 256\}$ (or any collection of distinct powers of 2), $\{1, 27, 81\}$ (distinct powers of 3), $\{s^1, s^4, s^{11}, s^{101}\}$ (for any whole number $s > 3$), as well as the more interesting $\{1, 3, 6, 13\}$. For such a set, given the sum of any of its subsets, we can tell which elements were used, without advance knowledge of how many were involved.

  The last example above, in which each value is one more than twice the previous one, lends itself to cards but is too small to be impressive. If we allow the use of negatives, as seen in the second and third effects in this chapter, we know that we can find two-summers of size six. Can we find sets of size six (or larger) such that given the sum of any of its subsets, we know which subset it is?

- Here's a deceptively innocent-sounding question shared by mathematician Neil Calkin whose solution has proved elusive.

  > For a collection S of k numbers from $\{1, 2, \ldots, n\}$, there are $\frac{k(k-1)}{2}$ ways to subtract one of them from a larger one. Hence if all of the differences are distinct, n must be at least as big as $\frac{k(k-1)}{2}$. Solving for k gives $k < \sqrt{2n}+$ constant. Chowla, Erdős, and Turan improved that crude inequality by showing that $k < \sqrt{n} + n^{1/4} + 1$, and so the constant in front of $\sqrt{n}$ has to be 1: the question then is to determine the size of the remainder term.
  >
  > Erdős conjectured that there exists a constant C such that $k < n^{1/2} + C$ and offered \$500 for a proof. It was one of his favorite problems. Isn't it amazing: we're only talking about sums, and sums of two numbers at that, but already we've reached a problem that even the best mathematicians don't know how to answer!

Off-Centered COATs

"Never Forget a Face" (5♥) is a good example of mathematics in action behind the scenes. It makes use of an extension of the main idea of Chapter A.

A deck is handed out for shuffling, and a spectator is invited to count out any nine cards in a pile, face down, the rest of the deck being set aside. Have a number between one and nine called out. The card at that position is shown around by the spectator, while you avert your eyes, then it is replaced where it came from.

Take the pile of cards behind your back, or hold it under the table, claiming that you're feeling the face of the selected card with your extra-sensitive fingers, so that you'll be able to recognize it later.

Now put the pile down, and explain a spelling and transferring routine, based on the name of the selected card. Demonstrate with cards from the rest of the deck and the name of a random card, say, the Four of Clubs.

Turn away again, and have the spectator do the spelling and transferring to the pile of nine cards, twice, using the name of the selected card.

Take the pile back and place it out of view again, as you comment, "Your actions have mixed these cards well, and there's no way that I could know what card you selected." Bring one card forward.

*"But my fingers never forget a face. What was your selected
card?" As soon as it is named, turn over the card in your
hand to reveal that you have indeed located it, sight unseen.*

Double-Dealing with a Difference

We start with a nonsymmetric version of the Low-Down Triple and Quadruple Dealing (or COATing) of Chapter A. Recall that COAT stands for Count Out (hence reversing card order) And Transfer (behind what remains). The 2008 print incarnation [Mulcahy 08] of the inaugural *Card Colm* [Mulcahy 04_10] ends by pointing out that the Low-Down Triple and Quadruple Dealing principles hold with minor modifications under more general conditions. Martii Sirén, long-term editor of the Finnish magic magazine *Jokeri*, had corresponded the year before to note that instead of COATing k cards three (or four) times, where $k + k \geq n$, we can COAT s cards, then t more, then s cards again (and if desired, another t), provided that $s + t \geq n$ [Sirén 08]. Austrian magician Werner Miller also made use of such an observation in "Hold the Line!" from his manuscript *Enigmaths* [Miller 10].

For example, starting with a packet of ten cards, if we COAT four cards, then seven, then four again, we end up with the original seven bottom cards on top, in reversed order, followed by the original three top cards in their original order. One more COAT of seven cards restores the whole packet to its initial order. Try it out with any Ace to 10 in order.

If we COAT seven cards first, then four, then seven again, we end up with the original four bottom cards on the top, in reverse order, followed by the original six top cards in order. One more COAT of four cards restores the whole packet to its initial order. The results for that second example are in some sense the dual of those for the first example.

Likewise, starting with thirteen cards, if we COAT six, then nine, then six again, we end up with the original nine bottom cards on the top, reversed, followed by the original four top cards, and one more COAT of nine cards restores the packet to its starting order. Hence, in order to get the bottom four cards of a thirteen-card packet to the top, in reverse order, there's an alternative to blatantly counting all of the cards into a pile. Select s and t with $s \geq 4$ and $s + t \geq 13$, then COAT three times: first s cards, then t, finally s again. Done in-hand while chatting, this gives the illusion of mixing the cards. This all generalizes as follows.

Off-Centered Bottom to Top Principle

*Starting with a packet of n cards, and s, t with $s + t \geq n$, i.e.,
s and t are on average at least $\frac{n}{2}$, if we COAT s cards, then
t more, and then s cards again, then the original bottom card
ends up on top.*

Off-Centered Low-Down Triple Dealing Principle ⋈

In fact the original t bottom cards end up on the top, in reversed order, followed by the original $n - t$ top cards in their initial order.

Four OverCOATs On Average Principle ⋈

Furthermore, one more COAT of t cards restores the entire packet to its initial order.

Since s can be less than $\frac{n}{2}$ here, provided t compensates for that by being at least $\frac{n}{2} - s$, the Save at Least 50% Principle of Chapter A doesn't generalize *per se*. However, we do get a new false shuffle (or two).

Off-Centered Quad False Shuffle Principle ⋈

If $s + t \geq n$, then two applications of the following restores a packet of size n to its initial order: first run off s cards from the top of the packet into a waiting hand, thus reversing their order, transfer them behind the remainder—or just drop the rest on top—then do the same with t cards from the top.

Done casually to a small packet, say of size ten, e.g., twice running off five then seven cards, this gives the illusion of mixing the cards (see page 10). As noted above, skipping the last running off of seven cards just moves the original bottom stock to the top, reversed.

The Low-Down Deal Packet Separation Principle of Chapter A doesn't hold for underCOATs either, but something is salvageable.

Two UnderCOATs around One OverCOAT Separation Principle ⋈

Starting with a packet of n cards, and $s < \frac{n}{2} < t$ such that $s + t \geq n$, if we COAT s cards, then t more, and finally s cards again, then the top and bottom halves switch places, subject to some internal re-ordering.

This also suggests a useful in-hand shuffle with a notable property.

The Top, Midriff, Bottom Decomposition

To see why the above principles hold, it once again pays to do some bookkeeping. Recall that after COATing k cards, a packet arranged $\{1, 2, \ldots, k - 1, k, k + 1, k + 2, \ldots, n - 1, n\}$ assumes the order $\{k + 1, k + 2, \ldots, n - 1, n, k, k - 1, \ldots, 2, 1\}$. Just as in Chapter A, there are three portions of the packet to keep track of. They move around intact, subject at most to some internal reversals, but the one between the top and bottom may not be in the middle, so we rename it accordingly, as the midriff.

We alternately COAT s and t cards, where $s + t \geq n$. So $n - t \geq 0$, $s + t - n \geq 0$, and $n - s \geq 0$, and those three numbers sum to n.

> **Adhuc Gallia est omnis divisa etiam in partes tres**[1]
> *Writing $n = s + (n - s) = (n - t) + (s - (n - t)) + (n - s) = (n - t) + (s + t - n) + (n - s)$, we see that the packet of n cards breaks up naturally into subpackets T, M, B (top, midriff, and bottom) of sizes $n - t$, $s + t - n$, and $n - s$, respectively.*
>
> *For the packet $\{1, 2, \ldots, n\}$, we have*
>
> $$T = \{1, 2, \ldots, n - t\},$$
> $$M = \{n - t + 1, n - t + 2, \ldots, s + t - n\},$$
> $$B = \{s + t - n + 1, s + t - n + 2, \ldots, n\}.$$

Some of the three pieces may be nonexistent for certain values of s and t (relative to n), but that's not a problem. In Table 5.1 on page 111, these are denoted by the empty set $\{\}$. Unless $s = t$, this decomposition is not symmetric about the middle, as it was in Chapter A.

If $n = 13$, $s = 6$, $t = 9$, then $\{1, 2, \ldots, 13\}$ breaks into $T = \{1, 2, 3, 4\}$, $M = \{5, 6\}$ and $B = \{7, 8, \ldots, 13\}$. Figure 5.1, like the others to follow, shows what such a *face-down* packet of Diamonds looks like if fanned face up: it should be read from left to right. The three key parts have been spaced out to ease with tracking the COATs to follow. The midriff is not in the middle this time.

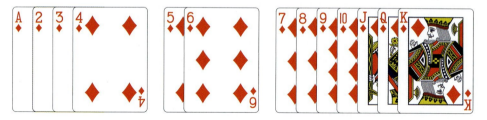

Figure 5.1. *T, M, B before COATing six, then nine, cards.*

The next image, Figure 5.2, shows what this packet—first flipped over face down—looks like (face up) after COATing six cards: counting out the top six cards (i.e., T and M together) and transferring them under the rest yields $\{7, 8, 9, 10, 11, 12, 13, 6, 5, 4, 3, 2, 1\}$, which is B followed by M reversed, followed by T reversed.

---

[1]Another Roman Empire reference; courtesy of Klaus Peters.

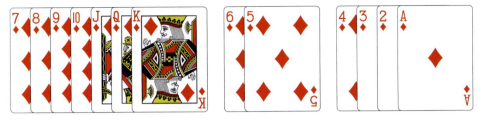

Figure 5.2. After COATing six cards.

COATing nine additional cards (namely B and the reversed M) yields $\{4, 3, 2, 1, 5, 6, 13, 12, 11, 10, 9, 8, 7\}$, which is T reversed followed by M, followed by B reversed, as shown in Figure 5.3. Just as in the symmetric case explored in Chapter A, the midriff is back where it started, although here it may be off-centered rather than in the middle.

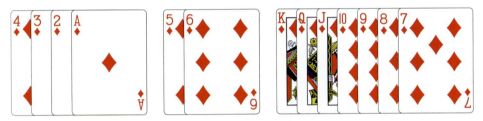

Figure 5.3. After COATing nine more cards from A–K♦.

Next, COATing six cards again (namely T reversed and M), yields $\{13, 12, 11, 10, 9, 8, 7, 6, 5, 1, 2, 3, 4\}$, which is B reversed followed by M reversed, followed by T, as shown in Figure 5.4. As claimed, the original bottom card ends up as the new top card. Finally, one last COAT of nine cards (namely B reversed and M reversed) clearly returns us to the starting configuration.

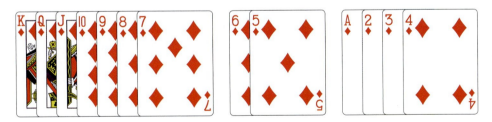

Figure 5.4. After another COAT of six cards.

The above works in general. COATing s cards (T and M together), leads to B followed by M reversed, followed by T reversed. In fact, just as in Chapter A, each COAT can be represented in the same fashion, as $X, Y, Z \to Z, \overline{Y}, \overline{X}$, where the bar over a letter indicates a complete subpacket reversal.

Figure 5.5 shows another visualization of two rounds of double COAT-ing in the case of $n = 13$, $s = 6$, and $t = 9$, for which $T = \{1, 2, 3, 4\}$, $M = \{5, 6\}$, and $B = \{7, 8, 9, 10, 11, 12, 13\}$, analogous to the "proof without words" for the symmetric case from page 39. As before, we represent a pile of thirteen cards by a vertical strip of gray-scale panels in decreasing order of brightness, from white for the top card to black for the bottom card, as depicted in the leftmost strip in Figure 5.5.

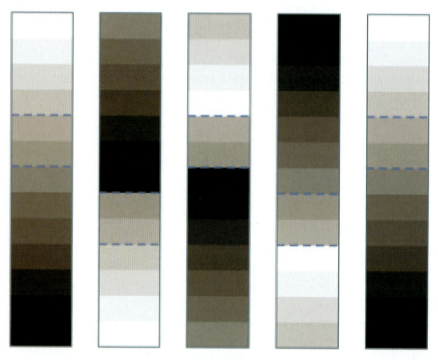

Figure 5.5. Off-centered proof without words.

The results of the double COAT of six then nine cards is given by the successive vertical strips. The last strip shows a fully restored packet, as expected, since the double COAT in question has period two. Notice how the midriff moves around from COAT to COAT.

We can also see why the original bottom card (represented as a black panel) has risen to the top after three of the four COATs involved, in preparation for its final journey back to the bottom under one more COAT. Moreover, the nine bottom cards become the nine top cards, suitably reversed, after three COATs. The three key portions of the packet—of sizes four, two and seven here—move around intact, subject at worst to some predictable internal reversals.

The only relationship between 13, 6, and 9 required to make this sequence of images generalizable is $6 + 9 \geq 13$, so in theory a similar succession of strips can be constructed to illustrate any case of interest to us. However, in practice, one of the key portions of the packet may be vanishingly small, which makes this approach tricky, as the midriff part behaves differently from the others.

Examine Table 5.1 to get a sense of how $\{1, 2, \ldots, n\}$ breaks up into T, M, and B, for $n = 11$ and various values of s and t with $s + t \geq n$.

n	s	t	T	M	B
11	11	2	$\{1,2,3,4,5,6,7,8,9\}$	$\{10,11\}$	$\{\}$
11	11	3	$\{1,2,3,4,5,6,7,8\}$	$\{9,10,11\}$	$\{\}$
11	11	4	$\{1,2,3,4,5,6,7\}$	$\{8,9,10,11\}$	$\{\}$
11	10	2	$\{1,2,3,4,5,6,7,8,9\}$	$\{10\}$	$\{11\}$
11	10	3	$\{1,2,3,4,5,6,7,8\}$	$\{9,10\}$	$\{11\}$
11	10	4	$\{1,2,3,4,5,6,7\}$	$\{8,9,10\}$	$\{11\}$
11	9	2	$\{1,2,3,4,5,6,7,8,9\}$	$\{\}$	$\{10,11\}$
11	9	3	$\{1,2,3,4,5,6,7,8\}$	$\{9\}$	$\{10,11\}$
11	9	4	$\{1,2,3,4,5,6,7\}$	$\{8,9\}$	$\{10,11\}$
11	9	5	$\{1,2,3,4,5,6\}$	$\{7,8,9\}$	$\{10,11\}$
11	8	3	$\{1,2,3,4,5,6,7,8\}$	$\{\}$	$\{9,10,11\}$
11	8	4	$\{1,2,3,4,5,6,7\}$	$\{8\}$	$\{9,10,11\}$
11	8	5	$\{1,2,3,4,5,6\}$	$\{7,8\}$	$\{9,10,11\}$
11	8	6	$\{1,2,3,4,5\}$	$\{6,7,8\}$	$\{9,10,11\}$

Table 5.1. The breakup of 1–11 into top (T), midriff (M) and bottom (B) parts.

B is small if s is close to n, and M is small if $s + t$ is close to n. An extended table would show evidence that T is small if t is close to n. Of course, T and B would have the same size only when s and t are equal.

We have now shown how knowledge of n, s, and t determines T, M, and B. It works the other way around also; in fact the Midriff alone, which is back where it started after any even number of COATs, determines everything.

Actually, any desired Midriff can be achieved. If we want $\{1, \ldots, 13\}$ to break up into $T = \{1, 2, 3, 4, 5\}$, $M = \{6, 7\}$, $B = \{8, 9, 10, 11, 12, 13\}$, then we have $n = 13$, and so $s = 7$, $n - t = 5$ and $t = n - (n - t) = 13 - 5 = 8$. Similarly, to break $\{1, \ldots, 12\}$ up into $T = \{1, 2, 3\}$, $M = \{4, 5, 6, 7, 8\}$, $B = \{9, 10, 11, 12\}$, we have $n = 12$, and so $s = 8$, $n - t = 3$, so $t = n - (n - t) = 12 - 3 = 9$. Using these COATs, perhaps in conjunction with the spelling of two appropriate words over and over, a desired royal flush in order can be maintained in positions 4–8 in a packet of twelve cards.

Finally, let's look at a more extreme example: to break up $\{1, \ldots, 10\}$ into $T = \{\}$, $M = \{1, 2, 3, 4, 5, 6\}$, and $B = \{7, 8, 9, 10\}$, we have $n = 10$, and so $s = 6$, $n - t = 0$, and $t = n - (n - t) = 10 - 0 = 10$.

The above considerations open up a new world of possibilities. For instance, if an audience member declares a liking for *mint chocolate chip* ice cream, then since $4 + 9 \geq 11$, given any eleven cards, the bottom card will rise to the top after a single such triple spelling of words of different lengths: first COAT four cards as you spell *mint*, then another nine for *chocolate*, and finally four more for *chip*.

Here's an effect where the spectator peeks at the top card of a packet before using one secretly chosen name to mix up the cards, yet you have no trouble locating her card.

5♣ Celebrity Selection

How it looks: *A spectator is invited to shuffle a deck of cards and deal any ten to the table. Have her look at and remember the top one, before placing them all on top of the remainder of the deck. You then present her with a list of well-known people from the worlds of literature, music, film, sports and politics, who are known by three names, such as Francis Ford Coppola, and have her secretly select one of these people.*

Deal out a pile of cards to the table and hand it to her, saying, "Here are your ten cards, please hold them in your hand. While I look away, I want you to use the three names of your selected celebrity, one by one, to spell out cards to the table, each time dropping the rest on top. Let me show what I mean with some other cards from the deck." Once you are sure she understands the directions, turn away.

When she is finished, turn back, take her cards from her and combine them with the remainder of the deck, then shuffle freely. Scan the card faces, fan the deck to the audience, and say, "I thought so," as you set the deck down. Have her name her card, and turn over the top card of the deck. They match.

How it works: The list provided can include names such as: Hillary Rodham Clinton, Andrew Lloyd Webber, Francis Ford Coppola, Helena Bonham Carter, Kareem Abdul-Jabbar, Helen Gurley Brown, Keenan Ivory Wayans, George Washington Carver.

After the spectator has peeked at the top card of her packet of ten, place those cards on top of the remainder of the deck and shuffle, being careful not to disturb the top card.

Then deal out eleven cards this time, while claiming it's only ten. This puts the selected card at the bottom of the resulting pile. No matter which person's name the spectator has selected to spell out, her card will rise to the top after the three names have been used. Reassemble the deck by putting her cards on top.

One more round of convincing looking shuffling will leave the selected card in place for the final revelation.

Why it works: In each case the first and last names have the same number of letters, and the first and middle names combined have at least eleven letters. Hence, by the Off-Centered Bottom to Top Principle, the bottom card comes to the top as desired.

Source: Original. December 2012. Thanks to magician Max Maven for suggesting some of the suitable names.

Presentation options: The sneaky addition of one extra card to the spectator's ten is to make it harder for the audience to reconstruct why it works, should they later try.

Another option, pointed out by Max Maven, is to only deal nine cards the second time, in which case the list of celebrities can be lengthened with any of Joyce Carol Oates, Louisa May Alcott, David Allen Grier, Lesley Ann Warren, and Evan Rachel Wood. Of course, Stevie Ray Vaughn, Carol Bayer Sager, June Carter Cash, and James Earl Jones are also contenders here. The number of cards used can also be adjusted downward if you wish to take advantage of shorter names such as Tommy Lee Jones or Jerry Lee Lewis.

People may notice that your list shows only people whose first and last names have the same length. For a repeat performance, you could offer a second list of names such as Arthur Conan Doyle and Kristin Scott Thomas, where the last name is one letter shorter than the first name. Working with nine or ten cards, the selected card will end up one from the top at the conclusion of the three COATs, and minor adjustments allow you to produce it later on.

Double-Dealing and Fixed Points

For select values of n, s, and t, Table 5.2 documents what happens to a packet of size n if we COAT s cards, then COAT t more, and stop.

n	s	t	double COAT	two double COATs	period
6	2	2	{5,6,2,1,4,3}	{4,3,6,5,1,2}	3
6	2	3	{6,2,1,5,4,3}	{3,**2**,6,4,**5**,1}	6
6	2	4	{2,1,6,5,4,3}	{**1,2,3,4,5,6**}	2
6	3	2	{6,3,2,1,5,4}	{4,**2,3**,6,**5**,1}	3
6	3	3	{2,3,1,6,5,4}	{**1,2,3,4,5,6**}	2
6	3	4	{2,1,**3**,6,**5**,4}	{**1,2,3,4,5,6**}	2
6	4	2	{4,3,2,1,6,5}	{**1,2,3,4,5,6**}	2
6	4	3	{3,**2**,1,4,6,5}	{**1,2,3,4,5,6**}	2
6	4	4	{2,1,**3,4**,6,5}	{**1,2,3,4,5,6**}	2
7	2	2	{5,6,7,2,1,4,3}	{**1**,4,3,6,**5**,2,**7**}	6
7	2	3	{6,7,2,1,**5**,4,3}	{4,3,7,6,**5**,1,2}	3
7	2	4	{7,2,1,6,**5**,4,3}	{3,**2**,7,4,**5,6**,1}	6
7	2	5	{2,1,7,6,**5**,4,3}	{**1,2,3,4,5,6,7**}	2
7	3	2	{6,7,**3**,2,1,5,4}	{5,4,**3**,7,6,1,2}	3
7	3	3	{7,3,2,1,6,5,4}	{4,**2,3**,7,**5,6**,1}	6
7	3	4	{3,**2**,1,7,6,5,4}	{**1,2,3,4,5,6,7**}	2
7	3	5	{2,1,**3**,7,6,5,4}	{**1,2,3,4,5,6,7**}	2
7	4	2	{7,4,**3**,2,1,**6**,5}	{5,**2**,3,4,7,**6**,1}	6
7	4	3	{4,3,2,1,7,**6**,5}	{**1,2,3,4,5,6,7**}	2
7	4	4	{3,**2**,1,4,7,**6**,5}	{**1,2,3,4,5,6,7**}	2
7	4	5	{2,1,**3,4**,7,**6**,5}	{**1,2,3,4,5,6,7**}	2
7	5	2	{5,4,**3**,2,1,7,6}	{**1,2,3,4,5,6,7**}	2
7	5	3	{4,3,2,1,**5**,7,6}	{**1,2,3,4,5,6,7**}	2
7	5	4	{3,**2**,1,4,**5**,7,6}	{**1,2,3,4,5,6,7**}	2
7	5	5	{2,1,**3,4,5**,7,6}	{**1,2,3,4,5,6,7**}	2

Table 5.2. The effect of double COATing s then t from n cards.

It lists the order of $\{1, 2, 3, \ldots, n\}$ after one and two double COATs of s and t cards, respectively, as well as the period of the double COAT itself. Fixed points are indicated in bold.

As usual, we consider the results only when $2 \le s \le n-2$ and $2 \le t \le n - 2$. For completeness' sake, we include the cases where $s = t$, noting that the overCOATing versions of those results follow from Chapter A.

In the overCOATing case where $s + t \ge n$, we obtain a result that, if duplicated, restores the packet to its original order. In other words, double overCOATing has period two.

Hence, using the language introduced in the "Convention Center" (page 19), the cycle decomposition of a double overCOAT is in terms of 2-cycles (transpositions, or card pair switches), and maybe some fixed points too (namely, cards that effectively don't move at all).

For example, if $n = 9$, $s = 4$, $t = 7$, then double overCOATing converts $\{1, 2, \ldots, 9\}$ to $\{2, 1, 3, 4, 9, 8, 7, 6, 5\}$, which has cycle decomposition

(1 2)(5 9)(6 8). The pairs indicated in parentheses denote switches, and the unmentioned 3, 4, and 7 effectively do not move.

As Table 5.2 shows, in the double underCOATing cases, where $s + t < n$, there are no such guarantees. However, several patterns are suggestive. For the values listed, at least one of the last two cards is almost always fixed. Also, for a particular value of n, switching the values of s and t leads to predictable results: for instance, the $s = 3$ and $t = 5$ row of the table allows one to predict the row for $s = 5$ and $t = 3$ (this is the duality hinted at on page 106).

Let's consider values of s and t that permit another family of spelling effects of interest, namely, the nine cases where, $(s, t) = (3, 5)$, $(3, 6)$, $(3, 8)$, $(4, 5)$, $(4, 6)$, $(4, 8)$, $(5, 5)$, $(5, 6)$, or $(5, 8)$. These are of interest because every card in the deck has a value and suit where the former is a word of three, four, or five letters (Ace, Two, Six, Ten; Four, Five, Nine, Jack, King; or Three, Seven, Eight, Queen), and the latter is a word of five, six, or eight letters (Clubs, Hearts and Spades, or Diamonds, respectively).

What happens if we double COAT $\{1, 2, \ldots, 9\}$ twice, using those s and t values? Since $s + t \geq n$ in eight of the nine cases, we generally get back the same order as we started with. In the remaining case, when $(s, t) = (3, 5)$, we're COATing three, then five, cards from nine, twice. Starting with a face-down A–9♣, the first double COAT yields a packet that, when fanned face up, is illustrated in Figure 5.6.

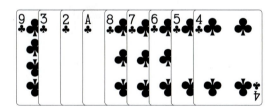

Figure 5.6. The effect of COATing three, then five, cards from A–9♣.

The second double COAT yields a packet that, when fanned face up, is illustrated in Figure 5.7.

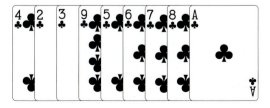

Figure 5.7. The effect of again COATing three, then five, cards from A–9♣.

It's a pleasant surprise to note how many cards are fixed. As a result, we have the following principle.

> **Nine-Card Double COATing Spelling Principle** ▷◁
> *When double COATing a packet of nine cards twice, using the nine possibilities for card value and suit name to determine s and t, if $(s, t) = (3, 5)$, the cards in positions 2–3 and 5–8 end up back where they started; in the other eight cases, all of the cards return to their original positions.*

It's time to put these observations to work, for our chapter highlight.

5♥ Never Forget a Face

How it looks overall: *A deck is handed out for shuffling, and a spectator is invited to count out any nine cards in a pile, face down, the rest of the deck being set aside. Have a number between one and nine called out. The card at that position is looked at and remembered by the spectator, and then shown around while you avert your eyes. The spectator then replaces it where it came from.*

Take the pile of cards behind your back, or hold it under the table, claiming that you're feeling the face of the selected card with your extra-sensitive fingers, so that you'll be able to recognize it later. Add, "My fingers never forget a face."

Now put the pile down, and explain a double spelling and transferring routine, based on the name of the selected card. Demonstrate with cards from the rest of the deck and the name of a random card, say, the Ten of Clubs.

Turn away again, and have the spectator do the spelling and transferring to the pile of nine cards, twice, using the name of the selected card.

Take the pile back and place it out of view again, as you comment, "Your actions have mixed these cards well, and there's no way that I could know what card you selected." Bring one card forward.

"What'd I say? My fingers never forget a face. What was your selected card?" As soon as it is named, turn over the card in your hand to reveal that you have indeed located it, sight unseen.

How it looks in detail: We recommend that in your demonstration you use the shortest possible card name, such as the suggested Ten of Clubs. The spelling and transferring routine is a double COAT that is done twice, using card names to determine the numbers of cards COATed each time. For instance, when the spectator does it, perhaps using the Nine of Spades, then four cards are first transferred from one hand to the other, thus reversing their order, while silently spelling out *N-i-n-e*, before tucking them behind the rest of the pile in the first hand. The spectator does likewise for *Spades*, spelling out and transferring six cards this time. The spectator does all of this twice.

How it works: Since you asked for a number *between* one and nine, the worst that can can happen is that four is called out. If so, simply move the fourth card to any other position other than the top or bottom, remember which position that is, and there you will find it later. In all other cases, the card the spectator notes ends up in the same position it started in. Either way, you know where it is at the conclusion of the double spelling, and so it's easy to extract it from there at the end.

The upshot is that Steinmeyer's effect can be duplicated in the following sense: the third card of nine will always be in a predictable position— the third—after two rounds of suit then value spelling and dropping. Alternatively, the fifth card of nine will always be back in the fifth position after two rounds of suit then value spelling and dropping.

Source: Original, from December 2012. Published online as the April 2013 *Card Colm* "Never Forget a Face (Double-Dealing with a Difference)" [Mulcahy 13_04]. Inspired by Jim Steinmeyer's landmark Nine Card Problem [Steinmeyer 93, Steinmeyer 02] (also known today as Nine Card Speller), which also works with any nine cards. There, the original third card ends up in the middle position after a single spelling of any card name, where the spelling proceeds in three stages (e.g., *Queen, of,* and *Spades*), so that the cards are effectively COATed each time—that is, the card order is reversed with the spelling of each word, before the rest are dropped on top.

Presentational options: As in the case of popular presentations of Steinmeyer's effect, this can be repeated while encouraging the spectator to try to throw you off by lying and spelling the name of another card.

Of course, if the cards available to the spectator somehow exclude those for which the value is three letters long and the suit is five letters long, i.e., A♣, 2♣, 6♣, and 10♣, then the selected card will automatically end up where it started in the packet.

At the end, with the cards out of view again, instead of extracting the selection as we've suggested, you could announce that your extra-sensitive fingers were able to determine the card's identity at the outset, and you've figured out that it's now in position 7 (or any other position you wish to name). Simply move it there from its known position, then bring the cards forward and ask the spectator to name his card. You will be proved correct.

It's also possible to have the suit spelled out before the value. In that case, we need to invoke a natural dual of the Nine-Card Double COATing Spelling Principle: all cards will return to their original positions after two such double spellings, except possibly when the suit is Clubs and the value is one of Ace, Two, Six, or Ten, in which case only the cards in positions 2–5 and 7–8 are back where they started. Hence, you only need to make an adjustment here if six is called out. Likewise, if four is called out you can switch to this version if you like. If neither is the number called out, you can say, "Spell either the suit or the value first, I don't care," stressing that whichever method is opted for, it is the one used throughout.

Low-Down Single Dealing

We can say something interesting about a single application of Low-Down Dealing or COATing any k cards from a packet of size n, no matter how small k is, when the two numbers have opposite parity (see page 16). Note that this is equivalent to the oddness of $n - k$, the number of cards dropped on top of the rest (or the number behind which the counted out cards are transferred) after counting out and reversing; hence the name we give this principle.

> ### First Odd Drop Principle ▶◀
> *If k is even and n is odd, or* vice versa, *then under any number of applications of dealing out k cards from n, and dropping the rest on top, an alternating packet remains alternating.*

In other words, COATing preserves alternation in an alternating packet provided $n - k$ is odd.

Specifically, if k is even and n is odd, the cards in odd-numbered positions remain in odd-numbered positions, and the cards in even-numbered positions remain in even-numbered positions. On the other hand, if k is odd and n is even, then the cards in odd-numbered positions and the cards in even-numbered positions exchange places.

The easiest way to see why this holds is to first arrange six Red and five Black cards in a face-up alternating packet, say, Red, Black, Red, Black, etc., and take $k = 4$—in effect, moving that many cards from the face to the back, and *watch* what happens; then repeat with six Red and six Black cards, taking $k = 3$ or 5.

Figure 5.8 shows an alternating packet of size eleven before and after COATing four (so that seven are dropped on top of the four reversed).

Alternating packets can thus be subjected to repeated deals of various numbers of cards, dropping the rest on top, provided the parity of the number of cards dealt is different from that of the packet size each time.

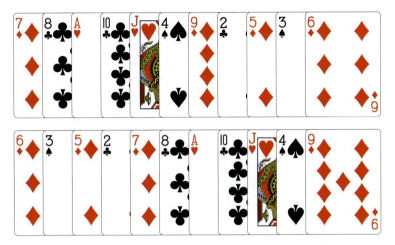

Figure 5.8. A face-up alternating packet before and after COATing four cards.

For instance, if deals of two, four, or six cards are done from an odd-sized alternating packet, in any order or combination, the rest being dropped on top each time, then the packet will still be alternating.

Similarly, if deals of three, five, or seven cards are done from an even-sized alternating packet, in any order or combination, the rest being dropped on top each time, then the packet will still be alternating. The nature of its alternation switches with each such deal.

Alternating the cards of a first-rate poker hand with five or six boring cards gives rise to effects in which a winning hand can be teased out without fail at the conclusion of numerous mixing or spelling steps, perhaps all done in a spectator's hands.

The smallest number of cards that we consider COATing is $k = 2$; we refer to this as *minimal underCOATing*. What happens to a packet in this situation depends on the parity of its size n, as Table 5.2 earlier suggests.

Let's focus on the case where n is odd. As just seen above, we might as well COAT any even number of cards k, not just 2.

Minimal UnderCOAT Parity Principle ▶◀

If n is odd and k is even, then when k cards are COATed, the cards in odd positions remain in odd positions, and the cards in even positions remain in even positions.

The next effect is one simple application of this principle.

5♠ Oddly Enough

How it looks: *Have a pile of cards dealt to the table from the deck until a spectator calls out, "Stop." The remainder of the deck is set aside. Have the resulting card looked at and remembered, and then put on the bottom of the pile. While you turn away, have the spectator repeatedly count out two or four cards from one hand to the other, reversing their order, then tucking those behind the rest. This is done as often as is desired.*

Turn back, and have the cards dealt into two new piles on the table, alternating left and right. Each of you takes one pile. Request that the spectator seek their chosen card in their pile. When they fail, remark, "Oddly enough, that's because I have it right here," as you throw it face up on the table.

How it works: The top half of the deck alternates prime and composite values, starting with a prime: the cards in odd positions have only values 2, 3, 5, 7, Jack, or King, in any order, and likewise the cards in even positions have only values 4, 6, 8, 9, 10, or Queen.

When dealing, it's vital to end up with an odd-sized pile. If an odd number of cards have been dealt when the spectator says to stop, then the last card on the pile is the one you have somebody memorize, and then put on the bottom of the pile. If an even number of cards have been dealt, say, "Very good, I'd like you to look at the next card, and remember it, then put it underneath the pile dealt to the table." Either way, there is now an odd-sized pile on the table, with one more prime than composite, that alternates except for the final card. It starts with a composite, and ends with two primes (the last of which is the chosen card). The odd positions are all occupied by primes, and the even positions are occupied by composites, except for one "imposter" card.

If any number of COATs is performed, moving an even number of cards each time, the upshot is that the even positions will be occupied by primes, and the odd positions by composites and the one out-of-step prime card.

When the packet is subsequently dealt into two piles, take the larger one for yourself; that's the one on which the first and last cards were dealt. The chosen card will be easy to spot: it will be the lone prime value in a run of composites.

The use of alternative primes and composites is recommended over odd and even values, or Black and Red cards, to make the inner workings less obvious to onlookers who may later inspect the two piles.

Source: Original. June 2003.

Presentational options: Another way to reveal the selected card, without having cards dealt out at the end, is to have the pile turned face up and the card names read out in order. The chosen card will be the middle one in the only run of three primes in a row, in an otherwise alternating sequence of primes and composites. (Cycle from the end back to the start, if necessary.)

The Second Odd Drop Principle

As seen in "Low-Down Single Dealing," when the number of cards dropped is odd, a single COAT yields a position parity principle. Something else of interest is true in this case: there is a pair of symmetrically positioned cards that always switch places. Just like the First Odd Drop Principle, this works for underCOATs as well as overCOATs.

Figure 5.9 shows what happens to a particular face-up packet of size eleven if we COAT six cards: the cards in positions 3 and 9 (namely, the 4♣ and K♣) switch places.

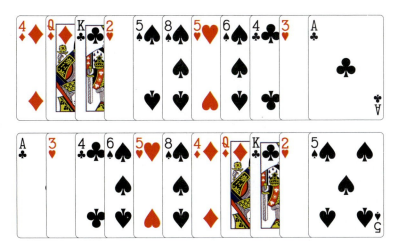

Figure 5.9. A face-up packet of size eleven before and after COATing six cards.

Similarly, if we COAT three cards from a packet of eight, then the cards in positions 3 and 6 switch places, and if we COAT four cards from nine, it's the cards in positions 3 and 7 that switch places. Likewise, if we COAT five cards from ten, the cards in positions 3 and 8 switch places.

Note that it's always the cards three in from each end of the packet that switch positions. It also works for overCOATing: COATing eight cards from thirteen leads to the cards in positions 3 and 11 being switched (the same is true for the cards in positions 6 and 8, as seen in Chapter A).

Also, if we COAT five cards from a packet of size twelve, the cards in positions 4 and 9 switch places, and if we COAT five cards from fourteen, it's the cards in positions 5 and 10 that switch.

Table 5.3 shows more values of n and k and the corresponding special transpositions. When overCOATing, where $k \geq \frac{n}{2}$, there are often other interesting transpositions, such as pairs of cards near and symmetric about the center, but these are ignored here.

n	k	transposition
5	2	(2 4)
8	3	(3 6)
9	4	(3 7)
10	5	(3 8)
11	6	(3 9)
12	7	(3 10)
13	8	(3 11)
11	4	(4 8)
12	5	(4 9)
13	6	(4 10)
14	7	(4 11)
15	8	(4 12)
14	5	(5 10)
15	6	(5 11)
16	7	(5 12)
18	8	(5 13)

Table 5.3. COATing k cards from a packet of size $n = k + (2v + 1)$.

These observations generalize as follows.

Second Odd Drop Principle

If we COAT k cards from a packet of size $n = k + (2v + 1)$, then the card $v + 1$ from the top switches places with the card $v + 1$ from the bottom, provided that $2k \geq \frac{n}{2}$.

The reason we need $2k \geq \frac{n}{2}$, or $k \geq \frac{n}{4}$, is to have a fighting chance. Otherwise two COATs of k cards leaves more than half of the packet "intact" (though shifted).

We're now ready for our final fling.

5♦ My Lucky Number Is Seven

How it looks: *"What's your lucky number?" you ask a spectator brightly, handing out a deck of cards for shuffling. "Mine's seven! But four is lucky for me, too."*

Have seven random cards handed to you. Glance at their faces briefly, adding, "Amazing, I don't know why it always works," then place the cards in a neat face-down pile on the table.

"You said eight, right? Take any eight cards for yourself, and while I turn my head I want you to note the fourth one from the bottom. You can even change your mind about your lucky number. Would you rather use five cards, not eight? It's up to you."

When the spectator has settled on a lucky number, selected that many cards, and looked at the fourth one from the bottom, have those placed in a second pile on the table. Turn back, "What number did you settle on? Six? Excellent. Please place my seven cards on top of your six. That will bury yours under mine. Take that combined packet in your hands, and watch what I now do."

Now demonstrate COATing six cards from another packet, hastily selected from the rest of the deck. When you are convinced that the spectator understands, set those aside, and turn away, saying, "I want you to do that six-card count-out-and-transfer move over and over with the packet of cards in your hands. Do it as often as you wish, quietly so I won't know how many times, and let me know when you're finished."

At the conclusion, have the packet of cards put in your pocket. Comment, "You don't know where your card is, and neither do I. Not only that, I have no idea what it is! How could I?" Pause for effect. Say, "Earlier I mentioned that seven is my lucky number," before producing seven cards from your pocket.

The last card you produce is the spectator's card.

How it works: You need to know the middle card of your seven, that is, the fourth one from either end. Casually glancing at the faces of your cards works, but we suggest an alternative below if you want to be less blatant.

No matter what number k is decided upon as the spectator's lucky number, the result is a packet of size $7 + k$ whose fourth card you know, and the spectator's card is fourth from the bottom. Those two cards are switched repeatedly under all of the COATs of k cards to follow.

Upon having the packet put in your pocket, remind the audience that the spectator did all of the shuffling and you couldn't possibly know where any card is. Start by producing the three top and bottom cards from the packet, slowly and steadily, as if you are concentrating on making some difficult choices. The spectator's card is either the next one on the top or bottom of the remaining packet in your pocket. Bring one of these forward as your seventh card, face up. If you don't recognize it as your card, then it must be the spectator's card, which you announce triumphantly. If it is your card, keep cool and bring the other one forward as well, saying,

"Seven is my lucky number because after seven failures I always succeed. *This* is your card!"

Source: Original. Conceived at the Joint Mathematics Meeting in San Diego, in January 2013.

Presentational options: A less blatant way to gain knowledge of the middle card in your packet is to take the seven cards you are given, face down, and peek at the bottom one while squaring them up on the table, before running off four cards from one end to the other.

It's possible to do this effect without needing to peek at, or even handle, any of the cards the spectator has. First, demonstrate how to COAT by using cards from the remainder of the deck. Later, when the packet is in your pocket, isolate the cards four in from either end, relocating one to the top and the other to the bottom. Bring the packet forward sneaking a peek at the bottom card. Ask the spectator what the selected card was. If it's the bottom card, have the packet turned over. Otherwise, have the top card turned over.

If you'd rather use five as your lucky number, that can be arranged.

Top Visitors and Orbits

We close with some observations in search of an entertaining application.

In Chapter A, when studying quadruple overCOATing, we found four key card positions that were intimately linked. They marked both the top visitors and the orbits of any of those four cards. With double overCOATs, those two concepts diverge notably. The top visitors are well-behaved, but the orbits of those four cards hold some surprises.

If $n = 10$, $s = 6$, and $t = 5$, we find that by double COATing $\{1, 2, \ldots, 10\} = \{1, 2, 3, 4, 5\}, \{6\}, \{7, 8, 9, 10\}$ twice, the top visitors, in sequence, are the cards originally in positions $1, 7, 5$, and 10, before the original top card returns to the top. These four cards are the boundaries of T and B, just as in the $s = t$ case. Likewise, if $n = 13$, $s = 9$, and $t = 6$, by double COATing $\{1, 2, \ldots, 13\} = \{1, 2, 3, 4, 5, 6, 7\}, \{8, 9\}, \{10, 11, 12, 13\}$ twice, the top visitors, in sequence, are the cards originally in positions 1, 10, 7, and 13, before the original top card returns to the top. Again, these four cards are the boundaries of T and B. This generalizes as follows.

> ### Off-Centered Top Visitors Principle ▶◀
> *Under a double overCOAT of s then t cards, the top visitors of an n-card packet are the cards originally in positions 1, $s + 1$ (the first card in B), $n - t$ (the last card in T), and n (the last card in B), before the original top card returns there.*

The journey made by the original top card of a packet is less obvious.

Let's return to the first example above, where $n = 10$, $s = 6$, and $t = 5$, with $\{1, 2, \ldots, 10\} = \{1, 2, 3, 4, 5\}, \{6\}, \{7, 8, 9, 10\}$ and top visitors 1, 7, 5, and 10. It can be checked that the top card in turn orbits to positions 10, 5, and 6, before returning to the top, so that the orbit of 1 isn't just the top visitors rearranged.

The card in position 5 orbits to positions 6, 1, and 10, in turn, before returning to position 5. However, the card in position 6, being in the midriff subpacket, can't get near positions 1 or 10: it just oscillates between positions 5 and 6. The card in position 10 orbits to positions 4, 7, and 1, in turn, before returning to position 10.

There are more surprises in the second example above, where $n = 13$, $s = 9$, and $t = 6$, in which case $\{1, 2, \ldots, 13\} = \{1, 2, 3, 4, 5, 6, 7\}, \{8, 9\},$ $\{10, 11, 12, 13\}$. The top visitors are 1, 10, 7, and 13. It can be checked that the initial top card (1 here) orbits to positions $13, 7$, and again to 7, before returning to the top. Note that 7 didn't move in the third step because that position is the middle card of 13 and the third COAT (like the first) is an overCOAT. Meanwhile, 7 orbits to 7, 1, 13, and back to 7; and 13 orbits to 4, 10, 1, and back to 13. Now that position 4 has just come to our attention, we note that it's an oscillator, as is 10; in fact, each of these is in the two-element orbit of the other.

Is there any new magic to be derived from all of this?

Parting Thoughts

- The classic Nine Card Problem (a.k.a. Nine Card Speller) of Jim Steinmeyer (see page 117) is based on the fact that given any nine-card packet, the card that starts in position 3 always ends up in position 5 after a single spelling of *any* card name, where the spelling proceeds in three stages (e.g., "Three," "of," and "Spades"). In our language, this corresponds to three COATs, using different numbers of cards each time.

  Prove that the Steinmeyer spelling effect is unique in the sense that if n, s, t are three numbers such that, given any packet of n cards, the one which starts in position s always ends up in position t when the spelling principle in question is applied (using the possible names of cards in a deck, in English), then we must have $n = 9$, $s = 3$, and $t = 5$. Not only is this a killer effect, it can't be topped.

- Are there other Steinmeyer-like possibilities in other languages?

- Minimal underCOATing from an even-sized packet: show that if $k = 2$ and n is even, then $2 \times \frac{n}{2} = n$ COATs are required to restore a packet of size n to its original order. Hint: first consider the case

$n = 12$, and the packet A♣, A♥, 2♣, 2♥, 3♣, 3♥, 4♣, 4♥, 5♣, 5♥, 6♣, 6♥, from the top down. See what happens after six COATs.

- The above minimal underCOATing result generalizes as follows.

  ### Divisor UnderCOAT Principle ▶◀
  If k divides n, say, $n = dk$, then $2 \times \frac{n}{k} = 2d$ COATs of k cards are required to restore the packet to its initial order. Also, the original top card will not return to that position under fewer COATs.

For instance, $2 \times \frac{12}{3} = 8$ COATs of 3 cards from 12 are required to restore the packet to its initial order, but only $2 \times \frac{12}{4} = 6$ COATs of 4 cards from 12 are required to have the same effect.

The result follows from the result of mathematician Jay Cummings given at the end of Chapter A, but a direct proof may be given as follows: Consider $\{1, 2, \ldots, n-1, n\}$ decomposed into the subpackets $P_1 = \{1, 2, \ldots, k-1, k\}$, $P_2 = \{k+1, k+2, \ldots, 2k-1, 2k\}, \ldots, P_d = \{(d-1)k+1, (d-1)k+2, \ldots, dk-1, dk\}$. For instance, if $n = 12$ and $k = 3$, we have the decomposition of $\{1, 2, \ldots, 11, 12\}$ into the four subpackets $P_1 = \{1, 2, 3\}$, $P_2 = \{4, 5, 6\}$, $P_3 = \{7, 8, 9\}$ and $P_4 = \{10, 11, 12\}$. After a single COAT, $\{1, 2, \ldots, n-1, n\}$ becomes $P_2, P_3, \ldots, P_d, \overline{P_1}$, where the bar denotes a complete subpacket reversal. After a second COAT, we have $P_3, \ldots, P_d, \overline{P_1}, \overline{P_2}$. After d COATs, we have $\overline{P_1}, \overline{P_2}, \ldots, \overline{P_d}$, and a further d COATs restores the packet to its initial order. Then, and only then, will the original top card be back on top.

- Can the period results of Jay Cummings (see page 45) be extended to give the period of any COAT of s cards from n followed by a COAT of t more?

- At the end of Chapter A, in "Parting Thoughts," we introduced the TACO(LM) move as a natural dual of the basic COAT move. Adapt the above results in the off-centered context—COATing any s cards from n followed by COATing t more—to work for TACO(LM)s. You should end up with numerous Off-Centered TACO(LM) Principles.

Gilbreath Variations

On the cover of this book, there is a cascade of cards, one of whose faces is not visible. You'll be able to figure out what card it is once you've read "Lucky Number between One and Thirteen" (6♥), which revisits well-known turf. Among the newer effects, "Unadditional Love" (6♠) seems like a miracle to those with no inkling of the underlying mathematical principles.

Take out a deck, and say, "Have you heard of the power of unadditional love? Here's how it works: you get to shuffle, and I get to demonstrate my unadditional love."

Have cards from the deck dealt to a pile on the table, asking for somebody to say when to stop. Invite another spectator to twirl the resulting two piles into rosettes, and mush them together, before squaring up the deck once more.

"Excellent: the cards are totally mixed up—and I have not touched them, agreed? Now I want you to take off as many cards you wish—I suggest any number between two and ten—and add up their values. I'll then unadd them, sight unseen, and tell you what you have just from knowing the total. I just love unaddition."

Suppose the spectator takes off some cards and tells you that their values sum to sixty-eight. You take the remainder of the deck, count it out to the table, and say, "Forty-four cards. So you must have eight. That's called subtraction. Now comes the tricky part—the unaddition." You think hard, soon cor-

rectly revealing that the selected cards are two Jacks, a Nine, a Seven, two Queens, and two Threes.

"Let's do it again," you continue cheerily, "Take some more cards; it doesn't have to be the same number." The spectator picks more cards and tells you that the values add up to nineteen. You again count out the remaining cards, and conclude, "Forty—so you have four." In due course you correctly announce that the spectator has an Ace, a Six, a Two, and a Ten.

Gilbreath Shuffling

Recall that riffle shuffling a packet of cards refers to dividing it into two not necessarily equal piles, and then dovetailing those together—perhaps using the thumbs to release the cards—with no particular regularity (see page 3). This type of shuffling is popularly believed to randomize the cards, but the truth isn't so simple.

Here we are not concerned with the probability or statistics of these shuffles, nor with the fascinating but difficult question of how often such a shuffle needs to be done to a full fifty-two-card deck to really randomize it. All that matters is that the cards within each original "half" maintain the same order, relative to each other, after the shuffle. Think of shuffling a Red Ace–King pile into a Black Ace–King pile. If the Red cards were then extracted, they would still be in order, as would the Black cards left behind. So it's just a matter of knowing which of the thirteen positions are occupied by the Red cards, and once that is nailed down, everything else is determined. Hence, there are $\frac{26!}{13!13!}$, or about 10 million, possible results for such a shuffled half-deck of cards, starting with two quarter-decks (see page 16). From that perspective, as seen on page 4, a riffle shuffle is equivalent to the rosette shuffle of Swedish magic maestro Lennart Green.

Rosette Shuffle
Take a packet of cards and cut it into two piles, side by side. Use your fingers to "twirl" each pile into a rosette. Now, push the rosettes together, and square up the resulting packet.

Figure 6.1, reproduced here for convenience from the "Convention Center," shows the basic steps of a rosette shuffle. Effectively, the packet has been riffle shuffled. This works particularly well with small packets. While rosette shuffling is optional now, it will be a required skill by the end of Chapter 8. We note the following internal coherence result.

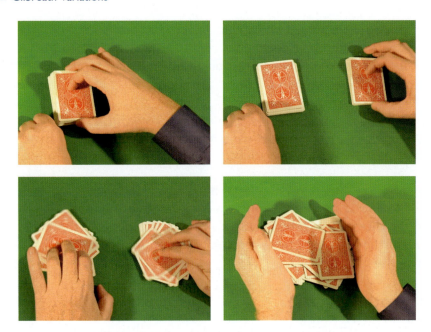

Figure 6.1. Lennart Green's rosette shuffle.

Solitaire Principle

Suppose two face-up packets of cards, one the A–K♥, the other the A–K♣, are riffle or rosette shuffled together. If the cards are now dealt one at a time into two face-up piles separated by color, then the original two packets will be reformed, in reverse order.

In other words, we'll never be faced with a situation such as dealing the 4♣ onto a Black card other than the 3♣; by the time we get to the Four, we can be sure that the Ace, Two, and Three of Clubs will already have been dealt, in that order.

This holds for any two packets of arbitrary size. Figure 6.2 shows one possibility when the face-up packet A–10♥ is shuffled into the face-up packet A–10♣, and the cards are then fanned. In practice, such packets would likely be face down throughout.

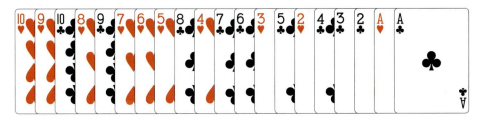

Figure 6.2. Hearts and Clubs.

Our goal is to explore the now well-known fact that a single riffle shuffle applied to a packet with cyclic prearrangement leads to quite predictable results when one of the shuffled "halves" has first been dealt into a pile.

Gilbreath Shuffling
This refers to the process of dealing some of the cards in a packet to the table, hence reversing their order, and then riffle shuffling those into the rest of the packet.

For our purposes, the two piles that are riffled together here—one reversed compared to how it started—may just as well be rosette shuffled. There is no need for these piles to be equal in size, which would be the case if exactly half of the cards had been dealt, although in practice they are often of similar sizes.

The Basic Gilbreath Principle

In 1958, mathematician Norman Gilbreath published the following result in *The Linking Ring*, under the title "Magnetic Colors" [Gilbreath 58]:

Magnetic Colors Principle
If an alternating color packet of even size is split into two parts, with cards of different colors at the bottom, then after a single riffle shuffle, the packet will consist of pairs of Black and Red cards, or Red and Black cards.

(We'll see shortly how this relates to Gilbreath shuffling, in which one of the packs arose from dealing from a larger alternating packet.)

Prolific magic author and Gilbreath shuffle expert Karl Fulves, in his presentation of "Magnetic Colors" [Fulves 84], remarks, "This routine was independently devised by Gene Finnell, Norman Gilbreath and others."

For instance, if we riffle or rosette shuffle the face-down packet A♥, A♣, 2♥, 2♣, 3♥, 3♣, 4♥, 4♣, 5♥, 5♣, into the face-down packet 5♠, 5♦, 4♠, 4♦, 3♠, 3♦, 2♠, 2♦, A♠, A♦, and fan the results face up, we obtain something like what is shown in Figure 6.3.

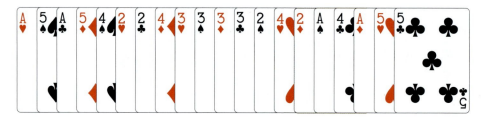

Figure 6.3. Basic Gilbreath in action.

What's being asserted here is that if we pull pairs off the top, i.e., those cards in positions 1–2, 3–4, 5–6, etc., then we will never get two cards of the same color. The same is true if we pull pairs off the bottom (face) instead.

The packet may now start Red, Black; Black, Red; Black, Red; Black, Red, Red, Black; ..., as the one in Figure 6.3 does (reading from left to right), in which adjacent cards *do* sometimes have the same color—but that doesn't violate the claim, as those emerge in different pairs, such as positions 2–3 and 8–9 (counting from the right in the image). Note the use of semicolons above to delineate pairs as peeled off the shuffled packet; in the image, we've inserted larger gaps for the same purpose.

Of course the result is just as valid for other types of card alternation, such as even or odd values, and primes or composites.

The following observation is critical, and will allow for numerous extensions in due course.

> When splitting an even-sized alternating packet into two parts to riffle shuffle together, the proviso that the two parts end with cards of different types is equivalent to having formed one of the piles by simply dealing out cards from the packet into a pile, thus reversing the alternating pattern for those cards.

That being said, we can now recast Gilbreath's original result more formally as follows.

Basic Gilbreath Principle
If an alternating packet of $2m$ cards is Gilbreath shuffled, the result is a packet of m pairs of alternating cards.

The packet can be cut any number of times before the shuffle—the result does not change.

So if we start with twenty-two cards alternating Black and Red and deal off about nine or ten of them into a pile, riffle shuffling those into the remaining thirteen or twelve cards, we will end up with eleven pairs of cards of different colors. That is to say, the first two cards will definitely be one of each color—and the same holds for the next two, and the two after those, down to the last pair. Each card in an odd position has the opposite color of the card right after it.

As already noted, we're not saying that the colors alternate perfectly again—that would hold only if each card in an even position still had the opposite color of the card after it. It's possible, but doesn't happen often.

Let's formally record this, along with a second observation of note that foreshadows material to come in Chapter 8.

If we do encounter two Red (or Black) cards in a row after such a shuffle, we can be sure of the following:

1. They occur as we switch from one natural pair to the next (separated by semicolons in some of the examples above), so that the first one is in an even position and the second one is in a odd position,

2. The cards will then alternate colors perfectly again until we run into two in a row of the opposite color, and so on.

The moral is that only certain patterns of deviation from an alternating pattern can occur in such a riffle-shuffled packet.

Note that it also is permissible to cut the packet anywhere, as often as desired, before it is Gilbreath shuffled, and the given outcome is still guaranteed.

A lingering and important question remains: what if we start with an alternating packet and split it into two parts that have the same type of card (say, both Red) on the bottom, and riffle those? We could well end up with a packet that starts Black, Black; Red, Black; Red, Red; ....

The good news is that a shuffled packet in that state is easily converted into one in a post-Gilbreath shuffle state. Even better, the method of handling this situation also works if we split and shuffle without first checking whether the bottom cards match.

Fan Restoration Principle

Assume that an alternating deck of even size has been split into two parts and those are riffle shuffled together. Fan the card faces under the guise of showing the audience how well mixed they are. If the top and bottom cards are of the same type (e.g., color), then all is well; the bottom cards of the two shuffled packs must have been of opposite types (colors) before the riffle. If the top and bottom cards are of different types (colors), speedily scan to find two adjacent cards of the same type and simply cut the fan between those two; then switch the fan parts as you reassemble the packet the other way around.

Few audiences will question this move if you do it steadily and without fuss. You don't know which state the packet was in before your split and cut; it may already have been exactly as you wanted it! Either way, the packet is in the desired state after this move.

If you cannot find such an adjacent pair of the same type, then the shuffled packet is still alternating, and you have two possible courses of action. You might opt to acknowledge this miracle, and break out the champagne, congratulating the shuffler on an amazing achievement. (Just don't mention the fact that the packet started out alternating.) Alternatively, you could say nothing and have the cards shuffled a second time, "to really mix them up," and proceed as you had originally planned.

Here's another way to express the impact of Gilbreath shuffling an alternating packet:

Gilbreath Duality Principle
If an alternating packet of 2m cards is Gilbreath shuffled and then dealt out alternately into two piles of m cards, then these two piles are dual in the sense that the card in any particular position in the second pile is the "opposite" of the corresponding card in the first pile.

For instance, if a twelve-card packet that alternates Red/Black is Gilbreath shuffled and dealt out as indicated, then, if the first pile runs Red, Black, Black, Black, Red, Black from the top down, we can be sure that the second runs Black, Red, Red, Red, Black, Red.

J. W. Sarles milked that fact to good effect in his "Posi-Negative Cards" [Fulves 68]: Have a spectator do the post-shuffle dealing, one pile being dealt face up. You take the face-down pile, and ask the spectator to separate the face-up pile into Reds and Blacks. All you do is mimic the spectator's every move, perhaps with your pile held out of view, to achieve a similar separation, even though you have not seen the faces of any of your cards.

The Gilbreath Principle is a dramatic confirmation of what card experts have known for a long time: contrary to most people's intuition, riffle shuffling isn't all it's cracked up to be when it comes to mixing cards. Fulves has written that the shortcomings of a single irregular riffle shuffle as a deck randomizer were explored as far back as the early twentieth century by O.C. Williams, and then by Charles Jordan in the 1920s and 1930s [Fulves 92, pages 114-129].

Martin Gardner included "Magnetic Colors" in his June 1960 *Scientific American* column, available in *New Mathematical Diversions from Scientific American* [Gardner 66]. An elegant explanation of why the principle works can be found there in the addendum to Chapter 9.

What follows is based on de Bruijn's exposition in his paper "A riffle-shuffle card trick and its relation to quasicrystal theory" [de Bruijn 87].

There are two cases to consider, depending on whether the bottom cards of the piles shuffled have opposite colors. If both of those cards have the same color, just ignore one of them entirely, thinking of the shuffled packet as "out of sync by one." Moving the bottom card of this packet to the top switches its state to that of one derived from piles that started with cards of opposite color on the bottom. Cutting the packet between two like-colored cards has an equivalent effect.

There are three things to keep track of: the initial two piles, held in the left and right hands, which we assume consist of alternating cards with bottom cards of different colors; and the pile of shuffled cards, which starts off empty. Consider the situation after the first two cards have fallen. If they both came from the same hand, then the pile of shuffled cards consists of a pair of oppositely colored cards (in what order we cannot say), and the remaining left and right packets still alternate in color with bottom cards of different colors. The very same observation is true if the two fallen cards came from different hands. This argument continues to hold for each successive pair of fallen cards; hence the shuffled cards consist of unmatched pairs as claimed.

Time for a Little Tête-à-Tête

We now introduce a more visual approach, which will be used throughout the rest of this chapter as well as in Chapter 8. It's a kind of head-to-head argument, which we refer to as a *tête-à-tête*.

Let 0 represent Red and 1 represent Black (or whatever two types of card are alternating), and for the sake of concreteness, suppose we have ten cards, so that the packet runs 0 1; 0 1; 0 1; 0 1; 0 1 from the top.

Gilbreath shuffling such a packet starts with dealing a certain number of cards to a table, which amounts to deciding where to insert dividing bars ‖ to mark the break between the dealt and undealt cards. There are two possibilities, depending on whether an odd or even number of cards is dealt—in other words whether an original pair is split or not, yielding something like

(top) 0 1; 0 1; 0 ‖ 1; 0 1; 0 1 (bottom)

or

(top) 0 1; 0 1; 0 1; 0 1; ‖ 0 1; 0 1 (bottom).

The cards to the left of the dividing bars are then dealt (and hence reversed) to form what we call pile D, and the cards to the right form the remainder of the packet, which we denote by pile C. The pile D reversal means that the top of the original packet is now the bottom of D, and the top of D is the card to the left of the divider bars.

So, in essence, when we consider the pending riffle shuffle, which mixes cards from the tops of each pile, we are working "from the inside out" within the next displays, starting with cards beside the dividing bars.

We have either

(bottom) 0 1; 0 1; 0 (top of D) ‖ (top of C) 1; 0 1; 0 1 (bottom)

or

(bottom) 0 1; 0 1; 0 1; (top of D) ‖ (top of C) 0 1; 0 1 (bottom).

Now it's time for a little tête-à-tête. Upon riffling, the top two cards of the shuffled packet must be either the top two from one of the piles or the top card from each one. Either way, we are removing an adjacent 1 and 0 (in some order) from a region of one of the above displays that includes, or is beside, the divider bars. We are left with an eight-card version of one of the displays, and hence the same result is true for the next pair in the shuffled packet, and so on down to the final two cards.

The style of analysis just given will prove to be very useful later when we examine various extensions to other shuffles.

Now let's look at a playful effect that applies the Basic Gilibreath Principle to show how a shuffled deck of cards can "reveal" some surprising mathematics (or maybe somebody's phone number). As presented, several decimals of π emerge dramatically from a shuffled deck that is out of view. Any decimal (or binary) string can be produced just as easily.

6♣ Easy as Pi

How it looks: *"We're going to do an experiment here and see if this deck of cards knows any mathematics. Can somebody get a pen and paper, please?" Overhand shuffle and fan the cards, commenting that they are well mixed, some face up, others face down. Give them to a volunteer, who is invited to riffle shuffle. Take the cards back, fan them toward the audience again so that all can see how "random" they are, and then place them out of view.*

Bring forward three pairs, in quick succession, and throw them on the table, noting, "Look—three pairs all facing the same way! Amazing, isn't it?" Then drop some cards on the floor, clumsily. Reprimand yourself, saying, "Too late now—we'll never know which way those were facing. Let's try again."

The next pair proves to consist of two cards facing different directions, which you comment on. Then you produce four pairs facing the same way.

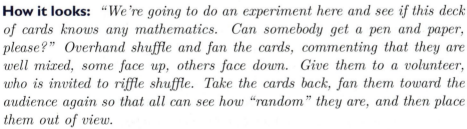

It's time to suggest that the second person keep track of these details on paper. Recap: "We started with three pairs facing the same way; then we got a pair facing opposite ways; then four more facing the same way. Write down 314, please." Continue, until three more numbers have been generated, namely one, five, and nine. Have 159 written down beside the 314. Then step back, and gasp, "I don't believe it! Remember that I dropped some cards after the first string of three pairs facing the same way? Let's put a marker, a period, after the 3 you wrote to denote that. Do you notice anything? Three point one four one five nine! It's as easy as pi!" Further confirming decimals can be produced if required.

How it works: Any numerical sequence, such as a telephone number, can be spelled out in this effect. At the outset, the deck is arranged so that it alternates face-up and face-down cards. Casually fanned to the audience, the cards will likely clump so that this arrangement will not be obvious. Proceed as above, this time cutting (if necessary) after the second fanning to ensure that the top and bottom cards are facing the same way.

With the cards hidden, take pairs off the top, one at a time. If brought forward as they are, all such pairs would consist of one face-up and one face-down card (in some order). By silently turning one of the cards over, you can convert any such pair to one whose cards face the same way. In this manner you can control the outcome for all pairs produced, to generate the digits of π or τ or any desired number. The decimal point gag depends on your "clumsily" dropping an even number of cards so as not to disturb the order that remains after the riffle shuffle.

For instance, in order to generate the digits 4, 2, 3 in a row, start by bringing forward four mixed pairs, one by one—namely, pairs with cards facing opposite directions. This is easy: they come off the packet that way. Next bring forward a pair of cards facing the same way, by secretly flipping over one card of the next pair. At this point pause and say, "Let's keep track here. We just got four mixed pairs facing opposite ways; then it switched to same way. Repeat, without much comment. Next, go back to producing pairs as they are, three times in total. After the first such pair, you can say, "Interesting, we got two matched pairs, cards facing the same way, before it switched back. Let's write that down." Have somebody take notes. Until you switch back to at least one more matched pair, you have no right to note that you got three mixed pairs in a row! Assuming an audience member is keeping track on paper, there should now be a written record something like OOOO, SS, OOO, S... (O for opposite direction and S for same direction). Now you have a good excuse to say, "Wait a minute—I'm seeing 423. Isn't that your office number?" or something appropriate to whatever gag you have planned.

Source: Published online in the August 2005 *Card Colm* "The First Norman Invasion" [Mulcahy 05_08]. This was inspired by "The Hustler" from Peter Duffie and Robin Robertson's *Card Conspiracy Vol. 1* [Duffie and Robertson 03a]. The idea of using face-up and face-down cards in place of Red and Black in the Gilbreath context goes back as least as far as magician Nick Trost in 1964.

Presentational options: The method outlined can be used to produce telephone numbers or numerical postal codes (such as ZIP codes), or short words using the usual numerical/letter equivalences (e.g., 414714 for "DADGAD" tuning, if you seek to impress a Celtic guitar player).[1]

The digits of ϕ, the Golden Ratio (which starts 1.6180339...) can also be summoned up, perhaps accompanied by a line about it being natural here since playing cards utilize "this most pleasing of ratios" (which isn't true, but then neither is much of the nonmathematical nonsense written about ϕ; see Keith Devlin's "The Myth That Will Not Go Away" (http://www.maa.org/devlin/devlin_05_07.html). You'll have to use your imagination there when dealing with the digit 0.

Since SOS came up earlier, this one also adapts naturally to Morse code.

Doubled-Up Alternation

A natural extension of alternating packets is doubled-up alternating packets; for instance, packets whose core consists of two Reds, then two Blacks, over and over.

One option is to start with a packet that runs Red, Red, Black, Black, repeated over and over from top to bottom, hence containing $4m$ cards for some m. If we deal off an even number of cards to the table, then the order reversal of those yields a pile that starts either with the same RRBB pattern or the similar BBRR pattern. In the first case, the remaining cards start BBRR; in the second, they start RRBB, just like the original packet. Now riffle or rosette shuffle those two piles together, and take cards from the top, four at a time. What can we say for sure about these quadlets?

It seems possible that all four cards could be the same color, perhaps not at first, but deeper into the packet. Actually, that can't happen in a quadlet: it is possible to get four Reds or four Blacks together, just not in any of the positions 1–4, 5–8, 9–12, and so on. (Like-colored runs of four cards must straddle two such quadlets.)

---

[1] A verbal pun springs to mind there, too, if you don't mind adopting a Southern U.S. drawl.

In the unlikely event that each pair of like colors remains together during the shuffle, then by our earlier analysis of alternating packets, each quadlet will consist of either two Reds followed by two Blacks, or the other way around. In particular, each quadlet will be color balanced.

That last property holds in general; in fact, each quadlet will either consist of two mixed pairs in a row (such as RBRB, RBBR, BRBR or BRRB) or two like pairs in a row (RRBB or BBRR). It's an easy exercise to check why this is so. It enables us to make predictions about the second pair in each quadlet, given information about the first pair, just as the first card in a pair predicts the second one in the basic Gilbreath situation. Karl Fulves has observed the same phenomenon in a related situation; see his "Backwards Bet" [Fulves 01].

For another option, we start with a packet running Red, Red, Black, Black, over and over as above, but deal an odd number of cards to the table before riffling or rosetting. This means we stop dealing between two cards of the same color. We can also ask all of those questions about a packet consisting of Red, Black, Black, Red, cycled repeatedly. It's the same pattern as before, with different boundary conditions.

It's interesting to ask what happens if we don't deal at all but simply cut the original packet in two, and riffle or rosette shuffle the resulting piles. If we don't get a result covered by our comments above, can we restore the shuffled packet to one that is in such a state, via an appropriate adjustment, such as moving a card or two from one end of the packet to the other?

For instance, consider two piles consisting of Red, Black, Black, Red, cycled repeatedly. Fulves has written about this situation at great length in his book *Riffle Shuffle Set-Ups*, pointing out that if such piles are riffled, and one card is moved from top to bottom, then ignoring pairs with one card of each color, pairs of two Reds must alternate with pairs of two Blacks [Fulves 68, page 4].

We invite readers to explore these options, some of which are discussed briefly in Chapter 8; a little tête-à-tête might be in order.

What we considered above is best analyzed by tracking with suits, not just colors. For instance, consider a packet that cycles Hearts, Diamonds, Clubs, Spades, over and over, or something similar. That conveniently opens the door to our next topic.

The General Gilbreath Principle

A far-reaching extension of the Basic Gilbreath Principle is variously known as the Second or General Gilbreath Principle. It applies to packets (or entire decks) prearranged in a cyclic pattern of a more general type than the binary alternating considered above. It first saw the light of day in another *Linking Ring* article, in June 1966 [Gilbreath 66]. A personal communication from Norman Gilbreath dating from 2006 suggests that the discovery predates the 1958 publication of the original Gilbreath principle.

> I am afraid I was the first to use the phrase Second Gilbreath Principle—this was to make it more understandable for magicians. At the time while I had realized the general principle I had not yet found an application that was interesting except for pairs and the whole deck. It is best in principle to just refer to the Principle. By the way, since it took over two years for the first application to appear in the *Linking Ring* I had already developed most of the routines that I later published before the first was in print. [Mulcahy 06_08]

The general principle deals with packets comprising repeated cycles of any length, whereas the basic principle assumes that the cycles have length two, namely, that we have an alternating packet.

To get a sense of what the general Gilbreath shuffle is all about, let's imagine a packet of cards consisting of repeated cycles of length nine from the top down, a pattern we can choose to represent as $G\,I\,L\,B\,R\,E\,A\,T\,H$.

$$(top)\ G\ I\ L\ \dots\ H;\ G\ I\ L\ B\ R\ E\ A\ T\ H;\ G\ \dots\ T\ H\ (bottom)$$

It's time for another tête-à-tête. Dealing out some cards to the table reverses their order, let's call that pile C. Denote what's left (the original bottom) by D. There is no loss of generality in assuming that the last card dealt to pile C is of type B, so that the dividing bars $\|$ below mark the break between the dealt and undealt cards.

$$G\ I\ L\ \dots\ T\ H;\ G\ I\ L\ \text{(top of C)}\ \|\ \text{(top of D)}\ R\ E\ A\ T\ H;\ \dots\ T\ H$$

Let's try another tête-à-tête argument: a subsequent riffle or rosette shuffle mixes cards from the tops of each pile. Where do the top nine cards of the shuffled packet come from? They come from selecting nine adjacent cards from the mid-region of the above display, starting with those on one or both sides of the dividing bars.

There are nine possibilities—depending on how many, say, k (a number between 0 and 9) come from pile C, to the left of the divider bars above; the other $9 - k$ cards must come from pile D, to the right of the bars. As a result, these $9 = k + (9 - k)$ cards were together in the original packet and hence must contain one card of each type. In other words, cards 1–9 in the shuffled packet consist of all nine types being considered (that is, the individual letters of GILBREATH in some order, with no repeats). The same holds for cards 10–18, 19–27, and so on.

The above tête-à-tête shows that determining the top nine cards of the Gilbreath shuffled packet can be viewed as freely deciding each time from which of two piles to pick a card, the piles in question connecting head-to-head to give an unbroken cycle of GILBREATHs. Phil Goldstein [Goldstein 02, page 22] remarks that such a perspective, for general stacked cycles, goes back at least as far as magician Nick Trost in 1969.

The most common application of the General Gilbreath Principle is to consider a deck arranged in cycling suits, ignoring values—for instance, having thirteen stacks of Clubs, Hearts, Spades and Diamonds in that order. If we deal about half of those to the table, and riffle shuffle them into the remainder of the deck, then the quadlets 1–4, 5–8, and so on are all guaranteed to consist of one card of each suit, in some order.

For instance, starting with a face-down deck A♣, 2♥, 3♠, 4♦, 5♣, 6♥, ..., 8♠, 9♦, 10♣, J♥, Q♠, K♦, dealing off twenty cards to a pile, riffling those into what remains, and fanning face up from left to right, the first sixteen cards from the top may look like those shown in Figure 6.4.

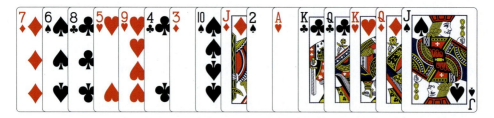

Figure 6.4. Generalized Gilbreath for suit cycling.

The spacing here highlights the quadlets (cards 1–4, 5–8, etc.); each does indeed consist of one card of each suit, although the same cannot be said for cards 2–5, 3–6, 4–7, 6–9, and so on.

Another popular option is to start with four stacked repeats of thirteen cards, say Ace to King in the usual order,[2] regardless of suits. If we deal some of those into a pile, and riffle shuffle them into the remainder of the deck, then cards 1–13, 14–26, 27–39, and 40–52, are all guaranteed to consist of cards of all thirteen values, in some order.

The Gilbreath shuffle just discussed is also an example of this, as it uses numerical order as well as CHaSeD suit order cycling, and the result can be reconsidered in that light. Figure 6.5 is just Figure 6.4 with the cards respaced to draw attention to the first thirteen of them. Cards 1–13 do indeed exhibit all thirteen values, but cards 2–14, 3–15 and 4–16 do not, having repeated Aces, 2s and 10s, respectively.

[2]Or perhaps Ace, 8, 5, 4, Jack, King, 9, Queen, 7, 6, 10, 3, 2, in alphabetical order.

Figure 6.5. Generalized Gilbreath reexamined.

There's no need to use an entire deck; indeed some smaller packet sizes have more interesting factorizations than does 52. We can, in general, consider a packet of ms cards consisting of m cycled repeats of a particular stack of s cards, each in the same order. Once again, Gilbreath shuffling such a packet refers to the process of dealing out some of these to a table, hence reversing their order, and riffle shuffling the two resulting piles together. The earlier Red/Black cycling and face-up/face-down options are, of course, special cases of this with $s = 2$, as are the examples mentioned above, for which $s = 4$ or 13, respectively.

What Norman Gilbreath noted about the resulting packet is this: from the top down—or the bottom up—counting s cards at a time, and disregarding order, we are sure to get sets that are identical in composition to the original stacks.

> **General Gilbreath Principle**
>
> *If we Gilbreath shuffle a packet of sm cards that consists of m repeated stacks of a particular set of s cards, each set being in the same fixed order with respect to some key characteristic, then we get a packet in which each stack of s cards pulled off the top (or bottom) contains exactly one card of each type, in some order.*

The result holds for the cards in positions 1 to s, $(s + 1)$ to $2s$, $(2s + 1)$ to $3s$, and so on, down to those in positions $(m − 1)s + 1$ to ms.

As was the case when $s = 2$, it's okay to cut the packet one or more times before it is Gilbreath shuffled. The cycles are just cycled around, and the stated outcome is unavoidable.

We've already noted that the case $s = 2$ permits extra flexibility: instead of dealing out cards to reverse their order, the packet can simply be cut anywhere to form two piles before shuffling, provided that the restoration move covered above is implemented after the shuffle. In Chapter 8, we consider a far-reaching extension of this property. In [Gardner 77, Chapter 7], another way to Gilbreath shuffle is discussed:

> Take cards singly from the top of the deck and push them into the pack, inserting the first card near the bottom, the

next anywhere above the previously inserted card (directly above it if you wish), the third above that, and so on until you have gone as high as you can.

Here's a quick justification of the General Gilbreath Principle. Assume that some cards are dealt out from the packet to form pile C, which is then riffle shuffled into the rest (pile D). Consider the top s cards of the shuffled packet: if k of them come from pile C, then they are intermingled with $s-k$ cards that can only have come from pile D. A moment's thought reveals that we then have $s = k+(s-k)$ cards overall, which were together in the original packet, and hence contain one card of each type. Similar reasoning applies to the next s cards, and so on, until we arrive at the bottom stack of s cards. We will present a more visual version of this argument shortly.

Note that suit cycling can be represented as a repeating 1 2, 3, 4 pattern (where 1 = Clubs, 2 = Hearts, etc.), and full-value cycling by 1, 2, 3, ..., 13 repeated. It is helpful (and consistent with the earlier treatment in the case of alternating colors), to consider repeating cycles of the pattern 0, 1, 2, ..., $s - 1$ in general, unnatural though that seems in the value case (unless we agree that 13 is the same as 0, and list it first).

A little tête-à-tête does wonders. No matter what kind of cycling we have, the initial packet can be represented by

(top) 0 1 2 ... $s - 1$; ... ...; 0 1 2 ... $s - 1$ (bottom).

Gilbreath shuffling such a packet starts with dealing a certain number of cards to a table, which amounts to deciding where to insert dividing bars ‖ to mark the break between the dealt and undealt cards. There are s possibilities, depending on where a cycle is broken when the dealing stops. Since everything repeats in a cyclic manner, there is really no harm in assuming that the break occurs between an $s - 1$ and a 0:

(top) ... ; 0 1 2 ... $s - 1$; ‖ 0 1 2 ... $s - 1$; ... (bottom).

The cards to the left of the dividing bars are then dealt (and hence reversed) to form what we call pile C, and the cards to the right form the remainder of the packet, which we'll denote by pile D. The pile C reversal means that the top of the original packet is now the bottom of C, and the top of C is the card to the left of the divider bars.

The riffle shuffle to follow, which mixes cards from the tops of each pile, can be thought of as selecting cards from the middle of this display, starting with cards on one or both sides of the dividing bars:

... ; 0 1 2 ... $s - 1$; (top of C) ‖ (top of D) 0 1 2 ... $s - 1$; ....

Where do the top s cards of the riffle-shuffled packet come from? There are $s + 1$ possibilities, depending on how many of them (say, k, a number between 0 and s) come from pile C, to the left of the divider bars. The other $s - k$ must come from pile D, to the right of the bars.

These $s = k + (s - k)$ cards were together in the original packet, namely, some cycled version of 0, 1, 2, ..., $s - 1$, and so contain one card of each type. The order in which they appear in the shuffled packet is variable, subject to the two internal order considerations; for instance, the first 0 to the right of the bars cannot show up under the 1 from the same pile. Once s consecutive cards are removed from the above display, the row looks identical, although shorter than it was, so a similar argument applies to the next s cards, and so on, until we arrive at the bottom stack of s cards.

Before continuing, we remark that COATing any number of cards from a packet, in the sense of Chapter A, may be thought of as a spectacularly bad example of a Gilbreath shuffle. A reversed packet is formed but is then merely placed underneath the remaining cards, as opposed to riffle shuffled into them. The resulting top cards all come from the bottom portion of the original packet, assuming that the cycle length does not exceed the number of cards COATed.

If you don't mind using cards from several decks, and you are careful to remove all unused cards from the scene of the crime before people start poking around, here's a simple miracle you can perform. Decide on two desirable poker hands, hand A (*admirable* in its own right) and hand B (even *better*). Fix a specific ordering of the cards in each hand. Prepare three card stacks: pile C has two cycles of hand B, as does pile D; pile E has four cycles of hand A. Now "build a DEC computer" by stacking D on top of E on top of C; that's forty cards in total.

If about half of these are dealt out and riffle or rosette shuffled into what remains, then the top five cards are guaranteed to form hand A, in some order. Give those to a spectator, who should be quite pleased. Take the bottom five cards for yourself; they will constitute hand B, in some order. Bottom-dealing pays off!

Here's a classic application of the General Gilbreath Principle, learned from mathematician John H. Conway circa 2000. In more recent times, he has embellished his own presentation as a crowd pleaser, called "This is the trick that I can't do," involving a dozen participants—but we cannot do justice to that here.

A deck is handed out, and a spectator invited to cut it as often as she wishes. She is then asked to deal between a third and a half of the cards to the table before riffle shuffling the resulting piles together. Take the deck back and remark that surely the cards are thoroughly mixed now, flashing some of the card faces to prove your point.

Take pairs of cards from the top of the deck, dropping them on the table face up, and remarking, "One Red, one Black." Once the audience is suitably impressed, drop clumps of four cards face up on the table, drawing attention to the fact that all four suits are represented. Finally, deal thirteen cards face up, saying, "Even more remarkably, there is one card of each possible value here." Turn over the remaining cards, saying, "The same holds for these cards."

The secret here is to have interwoven cycles of length 13 and 4: the values and suits simultaneously repeat in separate fixed orders. If we use CHaSeD order for suits and alphabetical for values, the deck initially runs: A♣, 8♥, 5♠, 4♦, J♣, ..., 10♥, 3♠, 2♦, from top to bottom.

After the Gilbreath shuffle, the deck has these properties, starting at the top (or bottom): each group of four cards has one of each suit, and each group of thirteen cards has one of each value. Of course, since the deck started with alternating colors, each pair of cards still consists of one Red and one Black (in some order).

The goal upon getting the deck back is, first, to turn over an even number of pairs, one at a time, verifying that there is always one Red and one Black card. Next, turn over quadlets, confirming that each suit is present, until twenty-four cards have been dealt. Turn over one more pair as an apparent afterthought, drawing attention to the opposite colors. The remaining twenty-six cards can now be dealt out face up, thirteen at a time, to reveal one of each possible value each time. Much of this is conveniently illustrated in the sixteen cards displayed earlier (post-shuffle) in Figures 6.4 and 6.5.

Once you understand that, you are all set to identify the mystery card on the front cover of this volume (also depicted in Figure 6.6).

6♥   Lucky Number between One and Thirteen

How it looks: *Hand the deck to a spectator, asking that it be cut several times. Have about half of the cards dealt to a pile on the table and then riffle shuffled into the rest. Take the deck back and ask the shuffler to call out a number between one and thirteen, inclusive, saying, "Some say thirteen is a lucky number; for me they're all lucky. I'm going to try to guess the card in the position you ask for."*

Suppose "Nine" is called out. Deal out about a third of the deck in an untidy overlapping row, all cards face up except for the ninth, which is face down. Stress the difficulty of determining the face-down card, since you can't see all of the fifty-one remaining cards. Nevertheless, in short order you correctly announce the identity of the ninth card.

How it works: Assume the deck starts in the order that intertwines both alphabetical-ordered values and CHaSeD cycling, namely,

A♣, 8♥, 5♠, 4♦, J♣, K♥, 9♠, Q♦, 7♣, 6♥, 10♠, 3♦, 2♣, A♥, 8♠, 5♦,
..., 6♣, 10♥, 3♠, 2♦.

After cutting several times, counting out twenty-one cards to a pile, riffle shuffling, and then dealing sixteen cards in a largely face-up row with the ninth card face down instead, something like Figure 6.6 will be seen.

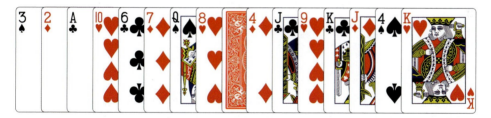

Figure 6.6. What's the ninth card?

To determine the suit of the face-down card, scan the cards in groups of four, starting with the 3♠, and see what suit is missing. In the case of the ninth card, you focus on the cards in positions 10–12. Here, a Spade is missing. To determine the value, scan the values of the first thirteen cards, and note which one is not visible. It's the Five: the ninth card must be the 5♠.[3] Note that while the colors of the part of the shuffled deck shown here alternate, there is no significance to that.

If "thirteen" is called out, and the corresponding card hidden, then you need to examine the cards in positions 14–16 to see which suit is missing. Except for this case, consideration of the first thirteen cards is always sufficient to determine the value and suit of the hidden card, but it's advisable to deal out sixteen each time so that some values are seen to repeat.

The above also works with the shuffled deck considered earlier, sixteen cards of which are displayed (post-shuffle) in Figures 6.4 and 6.5.

---

[3]This sequence of sixteen cards is also depicted on the cover of this book.

Source: Far from original. Published online in the August 2006 *Card Colm* "The Second Norman Invasion" [Mulcahy 06_08].

Presentational options: You could ask the spectator to deal sixteen or seventeen cards face down into a neat pile while you turn aside, and then turn one of those cards face up in place, before flipping the pile over and handing it to you face up. You quickly glance through the card faces and name the one face-down card.

You can even have a little fun, muttering something like, "I see two Kings with value thirteen ... raise that to the power of nine—you did say 'nine,' right? Now take the remainder upon division by fifty-one. Next use the Chinese Remainder Theorem ... let's see ... yes; your card is ...."

In truth, such an effect can be carried out without using Gilbreath ideas: if the deck is in any known order, structured or otherwise, and it is riffle shuffled (with or without first dealing out cards into a pile), then the identity of any card can be deduced from its neighbors by considering the two interwoven streams formed by the shuffle. Such approaches have been explored by many magicians on and off for almost a century.

Separated at Girth

Gilbreath shuffling a deck in which the colors alternate—such as in some of the effects presented above—facilitates a display of a "sixth sense" that can be used at the conclusion of one of those routines. Care must be taken in picking up the cards not to disturb the order they were in after the riffle shuffle.

The next presentation was suggested by Spelman College colleague Jeffrey Ehme. It's a twist on a standard flourish also learned from John H. Conway, made possible by the Duality Principle mentioned earlier.

Having already performed an effect such as "Lucky Number between One and Thirteen" (6♥), hand the reassembled deck face down to another spectator and say, "Let's do an experiment here. Please deal these cards alternately into two piles." Once this has been done, ask the spectator to point to either pile and pick it up, turning it over so that the card faces can be seen. You take the other pile for yourself and hold it behind your back.

Request that this spectator deal his or her cards into two face-up piles according to color, Reds to the left, Blacks to the right. Bring your own hands forward and say, "I asked you to pick a pile at random, and then to split those cards into Red and Black. I made it easy for you by letting you see the card faces. I tried to achieve the same result myself, but my cards were well out of sight. I used my sixth sense. A waist is a terrible thing to mind. Let's see how I did." Turn over your hands to reveal that the left one has all Red cards and the right one all Blacks.

Here's how it's done. With your hands behind your back, you mirror the spectator's actions as follows. Holding your pile face up in the left hand, collect cards in your right hand in two new piles, say a lower and an upper, using some of your right fingers as a natural separator, and anchoring everything with your right thumb and little finger.

Every time the spectator deals a Red card to the left on the table, you move a card from your left hand to the lower pile in your right hand, and every time the spectator deals a Black card to the right on the table, you move a card from your left hand to the upper pile in your right hand.

When it's all over, your lower pile is all Black and your upper pile is all Red. The only steps that remain are to transfer the upper pile back to the left hand and bring both hands forward.

The General Gilbreath Principle is very useful for forcing specific cards into the hands of a spectator who has just watched a deck being shuffled. For an application to ESP cards,[4] see "An ESPeriment with Cards" in the February 2007 issue of *Math Horizons* [Mulcahy 07_02a].

For our chapter highlight, we use that fact in conjunction with an unaddition result from Chapter 4.

6♠ Unadditional Love

How it looks: *Take out a deck, and say, "Have you heard of the power of unadditional love? Here's how it works: you get to shuffle, and I get to demonstrate my unadditional love."*

Have cards from the deck dealt to a pile on the table, asking for somebody to say when to stop. Invite another spectator to twirl the resulting two piles into rosettes, and mush them together, before squaring up the deck once more.

"Excellent: the cards are totally mixed up—and I have not touched them, agreed? Now I want you to take off as many cards you wish—I suggest any number between two and ten—and add up their values. I'll then unadd them, sight unseen, and tell you what you have just from knowing the total. I just love unaddition."

Suppose the spectator takes off some cards and tells you that their values sum to sixty-eight. You take the remainder of the deck, count it out to the table, and say, "Forty-four cards. So you must have eight.

---

[4]Those are also known as Zenner cards.

That's called subtraction. Now comes the tricky part—the unaddition."
You think hard, soon correctly revealing that the selected cards are two
Jacks, a Nine, a Seven, two Queens, and two Threes.

"Let's do it again," you continue cheerily, "Take some more cards;
it doesn't have to be the same number." The spectator picks more cards
and tells you that the values add up to nineteen. You again count out the
remaining cards, and conclude, "Forty—so you have four." In due course
you correctly announce that the spectator has an Ace, a Six, a Two, and
a Ten.

How it works: The deck is entirely rigged, of course. If you are willing to
settle for a single revelation, rather than the double whammy suggested
above, a much simplified version is possible.

Each time the spectator picks some cards, they must be from the
top of the packet available. The first tedious counting out on your part,
thinly disguised as a way to find out how many cards were selected (via
subtraction), actually serves a much more important purpose. It reverses
the card order, so that the second time the spectator picks cards, they
are really from the other end of the shuffled deck.

The second tedious counting out is for consistency, but can be circum-
vented to speed things up. You could clumsily drop the cards early in
the count, saying, "Oops, silly me. You know I can do subtraction; please
just tell me how many cards you took this time." This has the added
advantage of really mixing up the remaining cards so that all evidence of
the initial setup is destroyed, in case anyone gets curious later on.

The design of the deck is such that when about half of the cards
are counted out, we get two piles with the property that after a riffle or
rosette shuffle, the values—though not the order—of the top five (or ten)
and the bottom five (or ten) cards are totally predictable. The values of
any number of cards from either end can be deduced from their totals if
we use sum-rich numbers such as Fibonacci numbers or, more generally,
two-summers (see page 100).

We also saw on page 100 that $\{1, 2, 4, 6, 10\}$ is a two-summer, as is
its complement in 13, namely, $\{12, 11, 9, 7, 3\}$, and further that these sets
have no numbers in common. Thinking of these as card values, between
them they account for forty cards of the deck; only the Fives, Eights, and
Kings are left out. These two two-summers are shown in Figure 6.7 (with
arbitrary suits).

Note that the values in the first set sum to $1 + 2 + 4 + 6 + 10 = 23$,
and those in second set sum to $12 + 11 + 9 + 7 + 3 = 42$.

Stack the deck as follows, jumbling suits randomly:
- two stacks of Ace, Two, Four, Six, and Ten in that order;
- six of the Fives, Eights, and Kings in any order;

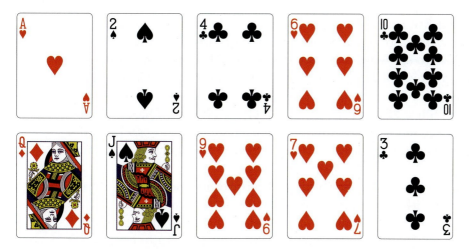

Figure 6.7. A two-summer and its complement in 13.

- four stacks of Queen, Jack, Nine, Seven, and Three in that order;

- the other six of the Fives, Eights, and Kings in any order;

- two stacks of Ace, Two, Four, Six, and Ten in that order.

- The two-summers can be in any fixed order.

It is important to have close to half of the deck dealt out before the shuffle, so don't ask for somebody to say "Stop" until twenty cards are dealt. After the riffle or rosette, the top ten (if not fifteen) cards consist of separated sets of Queen, Jack, 9, 7, and 3, each one in an unpredictable order. Similarly, for the bottom ten (and perhaps fifteen) cards, for complete sets of Ace, Two, Four, Six and Ten.

Let's focus on the top of the shuffled deck. The spectator takes off some cards, and tells you what the values add up to. You count out the remaining cards, so that the bottom and top of what remains are switched in preparation for the last part. Here's how the unaddition works.

If you find that five cards were taken, then you know that the selected cards are one of each type. The total is 42 here.

If fewer than five cards are involved, proceed as in "Any Two Cards (Little Fibs)" (4♣), using the above values, modifying the use of the greedy algorithm on page 90 accordingly. For instance, if it's three cards with a total of 22, then we note that $22 = 12 + 10 = 12 + 7 + 3$ and deduce that a Seven, Three, and Queen were selected. (We recommend naming the values in random rather than numerical order to deflect attention from the predictability of it all.) If it's four cards with total 30, then it's easiest to note that since all five card values here sum to 42, only the Queen is missing, so announce that a Nine, Three, Jack, and Seven were selected.

If more than five cards are involved, you can be sure that there's at least one of each type, so just subtract the total of those (namely, 42) and proceed as above to see which values are repeated. For instance, suppose the spectator gives you a total of 68 and your counting informs you that eight cards are involved. Noting that $68 - 42 = 26$, we seek three values with that sum. Since $26 = 12 + 14 = 12 + 11 + 3$, it's the Queen, Jack, and Three that are represented twice. So your announcement can be that the selected cards are two Jacks, a Nine, a Seven, two Queens, and two Threes.

Upon repetition, we're really working with the other end of the shuffled deck, and hence with the values Ace, Two, Four, Six, and Ten. If the spectator tells you that the values of the selected cards sum up to 19, and you know that four cards are involved, then you mentally compute, $19 = 10 + 9 = 10 + 6 + 3 = 10 + 6 + 2 + 1$, so that you can announce that the spectator has an Ace, a Six, a Two, and a Ten.

If the spectator tells you that the values of the selected cards sum to 33, and you know that six cards are involved, then it's particularly easy, since five of the cards sum to 23. That leaves an extra ten, so announce that the selected cards are a Six, an Ace, two Tens, a Four, and a Two.

Don't forget to shuffle all of the nonselected cards before anyone gets the bright idea to look over them.

Source: Original. Conceived in 2008.

Presentational options: A simpler version with only one revelation can be pulled off with Fibonacci numbers, or some of the variations suggested at the end of "Any Two Cards (Little Fibs)." For instance, a stack of twenty-four cards using four repeated sets of Ace, Two, Three, Five, Eight, and King, with random suits, can be used, planted at the top of the deck. Only those cards are used, either counted out, or innocently pulled off the rest of the deck via a pinky break (see page 11). This stack can be cut by the spectator at the start; then, after the subsequent Gilbreath shuffle and selection of some cards, be sure to shuffle the remaining cards into the others set aside earlier to hide all evidence of a setup.

As seen before, there is one possible problem when using six Fibonacci numbers: there are two sets of three cards with the same value sum, namely, the Ace, Two, and King and the Three, Five, and Eight. We already discussed a fishy work-around (see page 92). If nine cards are chosen, the same risk of ambiguity arises. The whole issue is easily avoided by starting out with a stack of twenty cards using any five of the six Fibonacci numbers above.

We close with a tapping-and-spelling effect stolen from Martin Gardner, made to appear more random thanks to Norman Gilbreath.

6♦   Tapped Out

How it looks: *The deck is handed out for riffle shuffling. Fan the card faces to demonstrate how mixed they are, and deal out the first five cards face-up. Have a spectator merely think of one, asking that a careful selection be made, one that you would not expect.*

Next, pick up the cards and arrange them face down in a circle, asking the spectator to turn away while this is taking place. Point out that the spectator couldn't know where the thought-of card is, you don't know which card that is, and the five cards being used were determined totally by chance. Explain that you will momentarily have the spectator "count out time" to their card. "We need a starting point. Imagine that these five cards form the face of a clock. Let's start counting at five o'clock." Indicate the card closest to the five o'clock mark on an ordinary clock. Now direct the spectator to tap the backs of the cards starting at that one, in a clockwise circle skipping over one card each time, *while silently spelling out the letters of the name of the thought-of card (such as the thirteen-letter "Ace of Diamonds"), tapping one card for each letter. Predict that when the spelling and tapping is complete, the thought-of card will have been arrived at. This turns out to be true. Congratulate the spectator for a job well done.*

How it works: This effect takes advantage of an old tapping principle, as implemented by Martin Gardner in 1939.

The Gilbreath principle is used to ensure that the lengths of the names of the five cards dealt out vary from 10 to 14, with exactly one of each length present. First arrange twenty cards such that their name lengths cycle 10, 11, 12, 13, 14 five times over, with the assistance of Table 6.1.

Place these twenty cards in the middle of the deck (i.e., with any other sixteen cards on top and the remaining sixteen on the bottom). To perform, hand the deck out—a few false cuts won't hurt—and ask that about half of the cards be dealt out into a pile. Once twenty cards have been dealt say, "Stop any time." Have the resulting packets riffle or rosette shuffled; the General Gilbreath Principle guarantees that the first five cards from the top consist of one each whose name has length 10, 11, 12, 13, and 14.

While the spectator is making a mental selection, scan the faces to see which card values have which lengths. You must arrange the cards face down in a circle with the length 10 ones at the top, and the others

10	Ace of Clubs, Two of Clubs, Six of Clubs, Ten of Clubs
11	Four of Clubs, Five of Clubs, Nine of Clubs, Jack of Clubs, King of Clubs, Ace of Hearts, Two of Hearts, Six of Hearts, Ten of Hearts, Ace of Spades, Two of Spades, Six of Spades, Ten of Spades,
12	Three of Clubs, Seven of Clubs, Eight of Clubs, Queen of Clubs, Four of Hearts, Five of Hearts, Nine of Hearts, Jack of Hearts, King of Hearts, Four of Spades, Five of Spades, Nine of Spades, Jack of Spades, King of Spades
13	Ace of Diamonds, Two of Diamonds, Six of Diamonds, Ten of Diamonds, Three of Hearts, Seven of Hearts, Eight of Hearts, Queen of Hearts, Three of Spades, Seven of Spades, Eight of Spades, Queen of Spades
14	Four of Diamonds, Five of Diamonds, Nine of Diamonds, Jack of Diamonds, King of Diamonds
15	Three of Diamonds, Seven of Diamonds, Eight of Diamonds, Queen of Diamonds

Table 6.1. Lengths of card names.

(running clockwise) in order 13, 11, 14, and 12. For instance, you might end up with what is shown in Figure 6.8, displayed face down. The numbers cycle 10, 11, 12, 13, and 14 clockwise, *skipping two at a time*. Starting at the card of length 11 in the five o'clock position results in the tenth card arrived at being the one of length 10, the eleventh being the one of length 11, and so on.

Figure 6.8. Circle of 5.

Source: Original. January 2007. The alternating tapping circle idea was gleaned from Martin Gardner's promotional items "Magic Tap-A-Drink Card" and "What Is Your Favorite Dessert?" as found in *Martin Gardner Presents* [Gardner 93, pages 338–339].

Presentational options: If you wish to do the tapping yourself, asking the spectator to stop you when the spelling is complete—then since you'll be tapping at least ten cards—the first nine are of no importance. Hence you can begin by tapping randomly, hamming it up and pretending to be "getting in tune" with brainwaves sent by the spectator. From the tenth on, you must tap the cards of length 10, 11, 12, 13, and 14 in that order. Having the circle arranged as above is not essential here; any order you can remember (avoiding anything too obvious) will suffice. Success is guaranteed.

Parting Thoughts

- Generalize the Gilbreath Duality Principle to the context of the General Gilbreath Principle. Hint: there may be poker-hand applications.

- There is now a vast literature on applications of the Gilbreath shuffle. Of particular note is the long-awaited *Beyond Imagination* [Gilbreath 13] by the namesake of the principle. It has a wealth of material spanning six decades.

- Diaconis and Graham's book *Magical Mathematics* [Diaconis and Graham 11] contains lots of analysis of Gilbreath shuffles. Their chapter "From the Gilbreath Principle to the Mandelbrot Set" should be enough to make any true-blooded mathematician drool. In it they present a result they modestly call the Ultimate Gilbreath Principle, before moving on to the headier matters suggested by the title.

- Can anything intelligent be said about "Gilbreath shuffling" three or more piles of cards? We return to this, with a little help from our friends, at the end of Chapter 8.

Word Row

"Twisting the Knight Away" (7♥) is a surprising revelation or location effect in which a low-intensity twist on reverse transferring is applied to a packet with just a hint of symmetry.

> *Give instructions to a spectator for randomizing a packet of mixed face-up and face-down cards, the goal being to lose the Jack. Upon receiving the deck back, you speedily locate it, face down.*

Like the other three effects presented here, it takes advantage of packets that display some kind of palindromic symmetry. A *palindrome* is a word or phrase that reads the same forward as backward, i.e., when the sequence of letters is reversed. Our chapter title, hijacked from Phil Goldstein's book *Redivider* [Goldstein 02], is an example, as are the single words *eye, sees, pip*, and the phrases *I prefer pi, never odd or even, air an aria, dog as a devil deified lived as a god*, and *was it a car or a cat I saw?*, not to mention *if I had a hi fi* and *aibohphobia* (the fear of palindromes).

If a palindrome has an even number of letters, then its second half can be viewed, to to speak, as a mirror image of the first half.

Of course, for us the focus is on arrangements of numbers or cards that display some kind of symmetry about their middle, including cases where items equidistant from the center merely share some property of note, being somehow complementary.

An obvious example is shown in Figure 7.1: A♥, 2♥, 3♥, 4♥, 5♥, 6♥, followed by the mirror images 6♠, 5♠, 4♠, 3♠, 2♠, A♠.

Actually, any packet containing $2m$ cards in which the last m cards in some sense match the first m cards, in reverse order, is fair game. Any

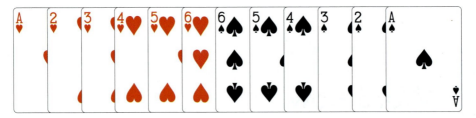

Figure 7.1. Palindromic display of Hearts and Clubs.

kind of buddying up, relative to the middle, is permissable.

> In a palindrome of length n, the first and last cards are buddies
> or matches, as are the second and second to last, and so on. In
> general, the cards in two given positions are buddies precisely
> when their position numbers sum to $n + 1$.

So if we were to follow the A–6♥ above with the repeated values A–
6♠, where the values of the corresponding pairs sum to 7, there is a sense
in which that too can be considered palindromic. Likewise if we were to
follow those Hearts with K♥, Q♥, J♥, 10♥, 9♥, 8♥ (can you see why?),
perhaps sticking 7♥ (or a Joker) in the middle for good measure, that
could also be considered palindromic.

As just seen, odd-length palindromes are obtained from even-length
ones by inserting a neutral card in the middle. Likewise, odd-length
examples can be turned into even-length ones by doubling up the middle
card. It's not uncommon upon breaking the seal on a new deck to find
that it's already palindromic—all cards of the same suit together, ordered
Ace–King, Ace–King, King–Ace, King–Ace.

Palindromic magic has a substantial literature, starting with the landmark work of Rusduck (J. Russell Duck) [Duck 57]. In the last decade
alone, offerings range from the aforementioned *Redivider* [Goldstein 02]
to British magician Stewart Murray's upcoming tome *Palindromic Magic
Revealed—Theory Techniques Tricks* [Murray to appear].

We only scratch the surface of this topic. A recurring theme is the
preservation of palindromic packets under moves that the audience assumes mix the cards. Such ultimately ineffective mixing can be thrown
in at will, in an effort to convince onlookers that there is no structure to
the packet, when in fact the exact opposite is true.

We mainly concentrate on two considerations: (1) how symmetric
packets fare under certain reversed transfers (the overCOATs of Chapter A); and (2) dealing into piles of equal size, something that has been
studied since the 1950s. First, we acknowledge some well-known basics.

Goldstein makes extensive use of the following, and variations thereof.

Cut Wrapped in Two Half COATs Principle

An even-sized palindromic packet is still palindromic after repeated applications of this sequence of moves:

1. *Run off half of the cards and drop the rest on top.*

2. *Cut anywhere.*

3. *Repeat 1.*

The first and last moves are what we call COATing exactly half of the packet. The first COAT converts the packet from palindromic to two-repeated-cycle form. The cut preserves that state and the last COAT restores it to palindromic form.

Here's another key property for packets of even size, as utilized in [Goldstein 02, pages 21–22].

Riffle Shuffle Palindrome Principle

When a palindromic packet is split exactly in the middle, and the two halves are riffle or rosette shuffled, then some order remains, in the sense that a new split into two halves preserves representation of the values involved.

For instance, if we start with Ace, 2, 3, 5, 8, King, King, 8, 5, 3, 2, Ace, ignoring suits, then after splitting in the middle and riffle shuffling, the top half of the packet (and likewise the bottom half) will contain an Ace, 2, 3, 5, 8 and King, in some order. This fact can, of course, be used to force card totals (32 if those cards are used), or even a modified "Little Fibs" (4♣) routine.

The Riffle Shuffle Palindrome Principle is a direct consequence of the General Gilbreath Principle of Chapter 6. There we saw what happens when cards are dealt to a pile from a packet consisting of repeated blocks, and that is riffle shuffled into the remainder of the packet. An even-sized palindromic stack, duly split in the middle in readiness for riffle shuffling, yields the same situation as fully dealing one duplicated block of two, thus reversing the card order, and so the desired conclusion must hold.

Palindromic packets also have interesting interactions with some other important types of shuffles, such as Faro shuffles (see [Duck 57], [Goldstein 02] and [Diaconis and Graham 11, pages 88–89, 94], the last source also discusses connections to the Monge shuffle).

Low-Down Double Dealing

In Chapters A and 5, we discussed Low-Down Triple (and Quadruple) Dealing. If you ever run into a low-down double-dealing type, then hand that person a palindromic packet! Here's a basic observation about packets with central symmetry, that we saved until now.

Low-Down Double Dealing Palindrome Principle   ><

A palindromic packet of cards remains palindromic if a fixed number of cards, representing at least half of the packet, is twice reverse transferred, that is to say, dealt off into a pile with the rest of the packet dropped on top.

In other words, in the language introduced in Chapter A, double over-COATing a palindrome produces another palindrome. This was first mentioned online in the February 2011 *Card Colm* "Low-Down Double Dealing with the Big Boys" [Mulcahy 12_02].

This holds because an overCOAT sends each card in the first half of the packet, say in position i ($\leq \frac{n}{2}$), to position $(n+1) - i$ in the second half (see page 43). Since those position numbers sum to $n + 1$, the cards match up as required.

See what happens when at least half of the cards in the twelve- or thirteen-card packets suggested earlier are low-down double dealt (or double overCOATed) as just explained, transferring the same number of cards each time. It's easier to track with the cards all face up.[1]

We already mentioned that one standard way to exploit palindromic packets is the apparent randomization of such cards, whether the initial setup is advertised or not. This permits the location of surprisingly matching or complementary pairs. A second idea is the hidden use of such structure to keep track of one or more cards of interest.

Our first effect utilizes the first approach just outlined. The second approach will be explored later.

7♣   The Biggest Names in Magic

How it looks overall: *A packet of cards is well mixed by a spectator using some of the biggest names in magic, freely selected from a list provided. Despite the fact that you have no knowledge of the names used or the final order of the cards, you are able to produce a suitable match for any card the spectator indicates.*

---

[1] Of course, as we saw in Chapter A, another two overCOATs restores the palindrome to its original order.

How it looks in detail: Ask audience members if they have had much luck with magic. Regardless of the responses received, say that with the help of some of the biggest names in magic, you're going to do an experiment.

Hand out a piece of paper on which is written the names Gardner, Goldstein, Diaconis, Benjamin, Boudreau, Gilbreath, Steinmeyer, and Copperfield, in any order. Say, "You've probably heard of Copperfield— he's certainly one of the biggest names in magic. Actually, all of these people are, and some of them have been very influential behind the scenes."

Shuffle a deck of cards and count off the top thirteen cards into a pile, saying, "Maybe these thirteen cards will bring us all some luck. In a moment, I'll turn away. Then I want you to mix up these cards using names on this list, as many as you like. You can use some of them more than once if you wish."

Grab about a third of the remaining cards, and demonstrate double overCOATing using the name Goldstein from the list. Now set those cards aside and turn away, making sure that the spectator works with the original pile dealt out. "Remember that for each magician's name that you use, you need to spell out that word, one card for each letter, and drop the rest on top, then do that again."

When the spectator has done as much of this mixing as desired, using different names from the list, turn back. Reclaim the packet of cards, saying, "These are totally randomized, based on magical actions. I honestly have no idea where any particular card is at this stage. How could I? Let's commence the experiment."

Fan the cards face down, and have the spectator take one, after which you place the others out of view. "Look at the card you just selected, and then put it face down on the table. What is its value? Don't tell me the suit." You claim that you are feeling the cards' faces, seeking some synergy, "Despite the unknown final state of the packet." Produce a card and put it face down on top of the spectator's first selection, commenting, "I think these match."

Repeat, having the spectator pick a second card from the eleven that are left, to generate a second face-down pair. Continue until all thirteen cards have been accounted for. At one stage, which could be at the start or finish, express difficulty and say, "I think that one is matchless," having it turned over on the spot, to reveal that it's a Joker.

At the end, six pairs of cards will be face down on the table. Have them turned over by the spectator, to reveal that all are perfect matches, such as the A♣ and A♠, the Q♦ and Q♥, and so on. Say, "I guess the experiment worked. It probably helped that we used thirteen cards and some of the biggest names in magic."

Needless to say, when working with that number of cards, smaller names in magic such as Vernon, Jay, Sands, or Hummer wouldn't have been at all suitable!

How it works: Don't draw attention to the fact that the names available have at least seven letters; it's crucial but should remain unnoticed.

Start by selecting six pairs such as A♣ and A♠, or Q♦ and Q♥. Arrange these in a palindromic packet of thirteen, with the Joker in the middle. Place those on top of the rest of the deck.

At showtime, maintain this stack in place throughout some casual shuffles. The pile you then deal out is still palindromic. By the Low-Down Double Dealing Principle for Palindromes, this property is retained no matter how many double overCOATs are done by the spectator, since the names provided on the list vary in length from seven to eleven letters.

When the spectator is finished mixing, and you turn back and reclaim the packet, fan it face down, and watch where the selected card is taken from. If it's from the middle, then it's the Joker; but don't reveal that immediately. With the cards out of view, feel around and then admit defeat, before proceeding as indicated above. If the card is selected from position 4 from one end of the fan, then you know that its match is in position 4 relative to the other end, and that's where you extract it. This logic applies in general: each time, just note how far the selected card is from the closest end of the fan. Then, out of view, pull out and bring forward the card in the same position relative to the other end.

Source: Original, December 2012. Published earlier online in a rather different two-stage version, with matchings of an algebraic nature, in the February 2011 *Card Colm*, "Low-Down Double Dealing with the Big Boys" [Mulcahy 11_02].

Presentational options: Many alternatives suggest themselves; for instance, you could be up-front about the final palindromic state of the packets, "milking" the top and bottom cards repeatedly (i.e., sliding them off in pairs with the thumb and matching fingers of one hand), to produce the desired pairs. This does, however, give away part of the secret.

The TOAFUH Alternative

If you're tired of dealing cards to the table—and that's understandable after we've inflicted so much of it on you—here's a low-intensity twist on the low-down dealing concept that works just as well in certain circumstances. It can be thought of as "low-down dealing without the dealing."

Before exploring the deal-free alternative, let's review the basics. Start with a packet of n cards, and a number k between $\frac{n}{2}$ and n. COATing k cards refers to dealing that many into a pile, thus reversing their order, and having the remaining $n - k$ cards dropped on top as a unit.

We now explain several ways to have a similar effect on a packet of cards without any dealing and dropping. The key is to note that overCOATing basically divides a packet into two subpackets, reverses the order of the cards in the larger one, and reassembles the packet by combining the resulting two subpackets in switched order. Among the other ways to achieve the same goal is reversing the order of the cards in the smaller subpacket.

COATing nine cards from any face-down packet, Ace, 2, ..., Queen, yields the face-down packet, 10, Jack, Queen, 9, 8, ..., Ace. Instead of doing that, simply thumb off the top nine cards as a unit—or isolate them by sliding away the bottom three—then flip the block of nine over, and tuck them underneath the remaining three cards, as shown in Figure 7.2. We also obtain a packet with the order, 10, Jack, Queen, 9, 8, ..., Ace, but the last nine cards are face up. Let's agree to denote that by 10, Jack, Queen, −9, −8, ..., −Ace.

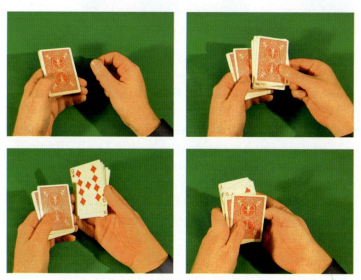

Figure 7.2. One alternative to COATing.

Here's another alternative with even less thumbing off. Again start with any face-down packet Ace–Queen. First turn the whole packet over, so it's now face up, and then thumb off three cards from the face as a unit, holding them to one side. Now take the nine cards remaining underneath, bring those forward and turn them over as a unit, dropping them on top of the three thumbed off cards. Finally, turn the whole packet over. Once more, we end up with 10, Jack, Queen, −9, −8, ..., −Ace. That sequence of moves is shown in Figure 7.3, the last image showing what the final state of the packet would look like fanned.

Even that maneuver can be streamlined: having thumbed off three cards as above, we can flip those over as a unit, and put them on top of the rest.

Figure 7.3. Another alternative to COATing.

Ignoring card orientations, these methods yield similar results. In what follows, we will apply such alternating moves to both ends of the packet, so we can dispense with any initial or final turning over of the whole packet.

These options are illustrated in three video clips linked from the April 2011 *Card Colm* "Twisting the Knight Away (No Big Deal)" [Mulcahy 11_04] at http://www.maa.org/columns/colm/cardcolm201104.html.

The key here is to **T**humb **O**ff (retaining card order) **A**nd **F**lip *Under*[2] *H*alf *(of the cards)*, which we abbreviate as TOAFUH.[3]

No big deal, you may say. But it does suggest the following easily verified and useful property.

Balanced TOAFUH Principle ▶◀

If a face-down palindromic packet of even size has one card of each matching pair turned face up, and is then subjected to an arbitrary number of double TOAFUHs, then it's still palindromic with one card in each matching pair facing the other way from its mate.

It's time to put all of this to work.

---

[2]In the sense of *fewer than*.

[3]Pronounced *tofu*—it's healthier sounding than the FAT (**F**lip **A**nd **T**ransfer) alternative.

7♥   Twisting the Knight Away

How it looks overall: *Give instructions to a spectator for randomizing a packet of mixed face-up and face-down cards, the goal being to lose the Jack. Upon receiving the deck back, you speedily locate it, face down.*

How it looks in detail: Hand out a deck of cards, requesting that a Jack and a dozen lower-valued cards be extracted. Have the selected cards well mixed, setting the rest of the deck aside. Take the chosen cards back and fan through them, face up, joking, "These could be mixed more, I can't help noticing that they're all facing the same way!"

Turn some cards over, and overhand shuffle the quarter-deck a little. Fan again, flipping the packet over if necessary, until you spot the Jack. Say, "The goal is to make this Knight vanish into the packet so that none of us knows where he is. There's a twist: you get to make him disappear, by thumbing off between one and six cards, and flipping them over like this. Do it as many times as you like."

Cut as needed until the Jack is face-up in the top half of the packet. Let's suppose it's in position 4. Thumb off four, five or six cards as a unit, and then flip them over, replacing them on top of the rest of the packet, now reversed. This time, the Jack will no longer be visible if the cards are fanned. Now turn the entire packet over and repeat, thumbing off and flipping over the same number of cards as before. Explain that you think of this thumb-and-flip routine as "twisting" and also that it must always be done to both ends of the packet—hence, the turning over of the packet and second thumbing and flipping, using the same number of cards as the first time.

Continue, "It doesn't matter if you know where the Knight is, as long as I don't have any idea. Please use this die here to randomize things. Roll it, and whatever number comes up, twist that many cards, as explained before. Then turn the whole packet over and twist the same number of cards again. Next, roll the die once more, and use that new number to determine how many cards you twist, twice, turning the whole packet over in between as I did just now. I'll watch the first time so that I know you've got the hang of it, then I'll turn away and you can do it as often as you wish. Stop when you're convinced that the Knight is well and truly lost. Finally, fan the cards to make sure he's not face-up—if so, simply turn everything over—then hand the packet to me."

Turn away as promised, and have the die rolled and the resulting double twisting done quite a few times. In due course, you turn back and are handed the packet. Fanning it quickly, you have no difficulty correctly identifying the face-down Knight.

How it works: The only preparation consists of silently noting where the Jack is when the cards are handed to the spectator—making sure it's not in the exact middle—and also paying attention to the identity of its match, namely, the card that starts out the same distance from the center. This match must be facing the opposite way of the Jack. For the sake of argument, let's suppose this card is the Six of Hearts.

No matter how much "double twisting" is done, as explained earlier, the Jack and Six of Hearts will remain equidistant from the center, and facing opposite ways. Upon reclaiming the cards, all you need to do is find the face-up Six of Hearts. The Jack (or Knight) will be in the matching position, counted out from the middle (or in from one of the ends).

Source: Original. Published online as April 2011's *Card Colm* "Twisting the Knight Away (No Big Deal)" [Mulcahy 11_04], with linked video clips.

Presentational options: The original packet could be constructed more purposefully. For instance, let's use a twelve-card packet, Ace–Queen in mixed and memorized suits, pairing the Ace with the Queen, 2 with 7, 3 with 9, 4 with 10, 5 with 8, and 6 with the Jack. The initial packet might run 3, Queen, 2, 4, 6, 5, 8, Jack, 10, 7, Ace, 9, from top to bottom. If one card in each pair, including the 6, is now turned to face the other way, we are all set for multiple revelations. At the conclusion of the spectator's die-directed mixing, each of the six face-down cards can be quickly identified if the above pairings are remembered. Their secret matches will be face-up by the Balanced TOAFUH Property, allowing you to identify them by position.

The logic behind the odd-seeming pairings is twofold: (1) to a casual observer there is no pattern suggesting an initial setup, but (2) they make sense mathematically. Apart from the Ace and Queen, the five products of the two numbers paired all have remainder 1 when divided by 13, that it to say, $2 \times 7 = 3 \times 9 = 4 \times 10 = 5 \times 8 = 6 \times 11 = 1 \pmod{13}$. The numbers in those pairs are each other's multiplicative inverses (mod 13).

Stay-Stack

The most famous palindromic property for cards is the Stay-Stack Principle. It has been studied since at least 1957, when Rusduck explored it in the context of Faro shuffles for two-pile deals [Duck 57].

The Faro shuffles can be replaced with the much easier dealing into two piles: for instance, if a palindromic packet of twelve cards is dealt out from left to right to form two piles of six cards, and the packet is reassembled by gathering the piles in either possible order, then it is still palindromic.

Actually, such a packet can be dealt out from left to right into three piles of four cards each, and the packet reassembled by gathering the piles in order (left to right or right to left, it doesn't matter which), and it's still palindromic. There's nothing special about the $2 \times 6 = 3 \times 4$ factorizations of 12 above.

Stay-Stack Principle

If r and c are two positive whole numbers, and a palindromic packet of rc cards is dealt out—from left to right in the usual way—into c piles of r cards each, then when the packet is reassembled by gathering the piles in order (left to right or right to left, it doesn't matter which), it is still palindromic.

From Rusduck in the 1950s, to Goldstein more recently, this is most often applied for two-pile deals, i.e., where $c = 2$.

To understand the effect of dealing a palindromic stack into piles, we first step back and consider dealing "ReCtangular" packets into piles in general. Imagine dealing out a packet numbered 1 to rc into c piles (or columns), from left to right, each one containing r cards (or rows). This is equivalent to dealing into a rectangular array, with r rows and c columns, row by row, building up each column one card at time. (Mathematicians use the word *matrix* for such an array.)

We can track what is going on by means of the rectangular array (or matrix) in Table 7.1, which indicates the positions of the cards in the original packet, reading left to right, row by row from the top down.

1	2	$\ldots$	j	$\ldots$	c
$c+1$	$c+2$	$\ldots$	$c+j$	$\ldots$	$c+c$
$2c+1$	$2c+2$	$\ldots$	$2c+j$	$\ldots$	$2c+c$
$\ldots$	$\ldots$	$\ldots$	$\ldots$	$\ldots$	$\ldots$
$(i-1)c+1$	$(i-1)c+2$	$\ldots$	$(i-1)c+j$	$\ldots$	$(i-1)c+c$
$\ldots$	$\ldots$	$\ldots$	$\ldots$	$\ldots$	$\ldots$
$(r-1)c+1$	$(r-1)c+2$	$\ldots$	$(r-1)c+j$	$\ldots$	$(r-1)c+c$

Table 7.1. The Matrix.

The general card numbered $(i-1)c+j$ is in both the ith row and jth column, For instance, if $r = 5$ and $c = 9$, then the card in the fourth row and third column is numbered $(4-1)9 + 3 = 30$. Note that this card is three rows from the top of the array, and two columns from the left. Now consider its "palindromic match," namely, the card that is three rows from the bottom of the array and two columns from the right. It's in row $5 - 3 = 2$ and column $9 - 2 = 7$, and is numbered $(2 - 1)9 + 7 = 16$. Observe how the sum of the numbers for the card and its match is $30 + 16 = 46 = 5 \times 9 + 1$, which is also the sum of the numbers for the very first card and its match, the last card. This is no accident; as remarked earlier, the sum of any card's number and that of its palindromic match is always $rc + 1$, and conversely, two cards are matches whenever their position numbers add up to $rc + 1$.

As already noted, the card in the ith row and jth column is numbered $(i - 1)c + j = ic - c + j$. Its match is in the $r - (i - 1)$th row and the $c - (j - 1)$th column and thus is numbered $([r - (i - 1)] - 1)c + [c - (j - 1)] = rc - ic + c - j + 1$. Summing the position number of that card and its match yields $ic - c + j + rc - ic + c - j + 1 = rc + 1$, as desired.

Note that if $r = 5$ and $c = 9$, then the cards with numbers 11 and 35 must be matches, since those sum to $46 = rc + 1$. It's now not too hard to verify that the Stay-Stack Principle is valid, using Table 7.1 and the fact that two card numbers sum to $rc + 1$ precisely when the cards are palindromic matches.

Let's have some circular fun. Tauists take note. This one is quite mathematical in flavor.

7♠   Easy as Tau (I Prefer Pi)

How it looks overall: *Twelve cards from a shuffled deck are repeatedly dealt into various face-down piles on the table, then reassembled, the number and sizes of the piles, as well as the frequency of the dealing, being determined at random by audience members. This packet is then split in two and riffle shuffled. You and a spectator each take six cards, and deal those to form two small face-down circles, seeking, but not finding, some numerical synergy.*

Finally the cards are dealt into a single large circle. Despite all of the randomization and shuffling, upon being turned over, the cards reveal two copies of a well-known approximation to π, *as befits a full circle.*

You comment, "In a sense it's twice as easy as pi, so I guess you could say it's as easy as tau. Tau is two pi. Have you heard of tauism? It's a movement to have all formulas using pi rewritten in terms of tau. I kid you not. While the idea has some merits, I must say: 'I prefer pi!' That's a palindrome. I wonder if there are any magic applications of palindromes?"

How it looks in detail: You take out a deck of cards and shuffle it, before dealing twelve cards into a pile on the table. Pick up this pile, and ask somebody to shout out any one of "two," "three," or "four." Deal the packet into the indicated number of piles, left to right, in the usual way. For instance, if "three" is called for, you will end up with three piles of four cards each. Now reassemble the packet by collecting the piles in order, and repeat, first requesting that a new number be shouted out to determine how many piles to deal into this time. Reassemble the packet as before.

Emphasize how mixed up "these randomly selected cards" must now be. "Just to make sure," you add, "Let's do this." Split the packet in two and riffle shuffle the halves together. "That should do it," you say triumphantly.

Split the packet in two again and invite a spectator to take either half. "Please look at the faces of the cards you have in your hand, and arrange them in numerical order. An Ace counts as one, a Jack as eleven, and so on. The lowest value card should be at the top of the face-down packet, and the highest one at the bottom."

Once that has been accomplished, announce, "I'm now going to deal my cards into a magical circle, and I'd like you do exactly as I do. Please stand beside me with your packet of cards ready, like mine, face-down." Slowly deal your cards into a face-down clockwise circle, starting at the twelve o'clock position, as indicated in Figure 7.4, allowing the spectator time to duplicate your actions with a second such circle beside yours.

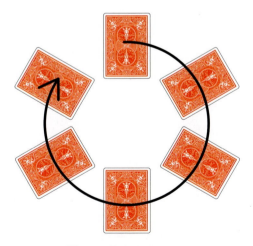

Figure 7.4. Half time.

Pause and comment, "It would be wonderful if we could generate some synergy between our respective circles of cards, despite the completely random methods used to determine them." Continue, "Something's not quite right. I think we need to combine our efforts. Let's try again. Once more, please copy my actions exactly."

Pick up the six cards in your circle, making four "NE to SW" diagonal sweeps, starting at the bottom right corner and ending at the top left corner, as indicated in Figure 7.5. Place the first card you pick up on top of the second, and those two on top of the third, and so on.

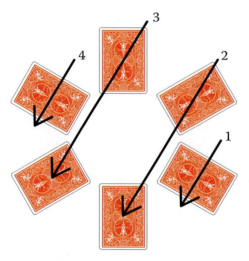

Figure 7.5. Pick up time

When the spectator has done the same, place those six cards on top of yours. Say, "I believe one larger circle would be better, I was confusing circumference and area before. Also, I should deal starting from the three o'clock position, and in the other direction, going through the quadrants in the usual order." Deal out this twelve-card packet into a single face-down larger circle, starting at the three o'clock position this time, but going counterclockwise.

"Maybe we've got a perfect clock here," you say, hopefully. Turn over the card in the three o'clock position, which is indeed a 3. Express satisfaction at this. Next, turn over the card in the two o'clock position. Alas, it's not a 2. Look crestfallen, and say, "I guess that was *two* much to hope for. Maybe it's because of my counterclockwise approach." Turn over the spectator's remaining four cards, still going counterclockwise. None of them match the corresponding hours on a clock.

"But I haven't given up hope. Remember you copied my every move a little while back, with the random cards we both obtained after all of the earlier mixing and shuffling," you remind the spectator. "Perhaps, by doing so, you turned your cards into matches of mine. Then mine, of course, will match yours! The cards I just exposed were yours, let's take a look at mine now." Turn over the next card, in the nine o'clock position.

It is a 3, just like the spectator's first card. Turn over the following one, in the eight o'clock position. It matches the corresponding spectator's card (in the two o'clock position). Continuing, it will be found that all six of your cards match the spectator's. Moreover, when the spectator's card is Red, in most cases yours is Black, and vice versa. The situation will resemble what's depicted in Figure 7.6.

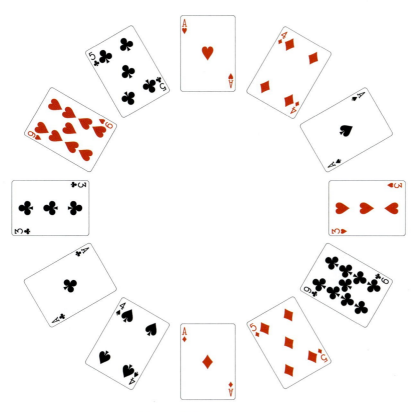

Figure 7.6. Clock match.

The effect appears to be over. An extraordinary sixfold match has occurred with seemingly random cards that were certainly mixed and shuffled in many ways earlier. Now comes the kicker: draw attention to the first six exposed values. "Look: three—in the correct position—then one, four, one, five, nine, ..., does that ring a bell?"

Assuming your audience is well educated and alert, somebody should perk up and exclaim, "It's the start of the decimal expansion of pi." Reply *radiant*ly, "Yes, indeed, note that we have two copies here, your upper semicircle and my lower semicircle. It's inevitable if you think about it: after all, a full circle corresponds to two pi!"

How it works: The first secret underlying the 3.14159 effect is that the original twelve cards dealt off the "shuffled" deck are far from random. They are A♠, A♣, 3♠, 4♠, 5♣, 9♣, 9♥, 5♦, 4♦, 3♥, A♦, and A♥, and care must be taken to maintain this stack at the top of the deck while genuinely shuffling the rest. You can riffle shuffle, to give the illusion of fair mixing, provided that the hand holding the top part of the deck drops those twelve cards last, as a unit. The haphazard suit selection here is an attempt to make the final outcome not look so carefully planned.

The second secret is that the repeated dealing into piles—of various sizes—and the final riffle shuffle and even split into two piles of six cards, turns out to ensure that you and the spectator each end up with a 3, 4, 5, 9, and two Aces. Just why this is the case, we will see in due course.

When the spectator arranges her cards in ascending numerical order, you, of course, do the same with yours, but without drawing attention to that. At this point, nobody suspects that your cards already match. (The colors will complement each other too, but it's not easy to arrange that the Aces come out in the appropriate order later on, unless the spectator asks, "What do I do if I have two cards of the same value?" and you answer accordingly.) When you deal into a small circle, "clockwise from the top," and do the subsequent "cross-diagonal pickups," both actions being copied by the spectator, it's merely a ruse to go from ascending numerical order to digits of π order; in other words, it's the necessary setup for the final large circle deal.

Why it works: When a palindromic twelve-card packet is dealt into two, three, four, or six piles, and then gathered by piles (in order), the re-assembled packet is still palindromic. This follows from the Stay-Stack Principle. The nature of the setup now makes the claimed outcome inevitable.

Source: Published online in August 2008's *Card Colm* "(A) Pi Evolved Set—Harmonic Split Drill" [Mulcahy 08_08], dedicated to the memory of high school teacher and geometer extraordinaire Steve Sigur, of the Paideia School, Atlanta.

Presentational options: When dealing with π, circles are hard to avoid, as we saw above. There are additional circular π effect possibilities, based on a 7- or 8-digit approximation to that important irrational ratio, i.e., 3.1415926(3). As before, cards with those values—eight of them this time—can be stacked at the top of a deck and maintained there through some shuffles. Deal these into a face-down circle, counterclockwise starting at the three o'clock position. Unknown to anyone but you, the cards—note the CHaSeD suit order—are as shown in Figure 7.7.

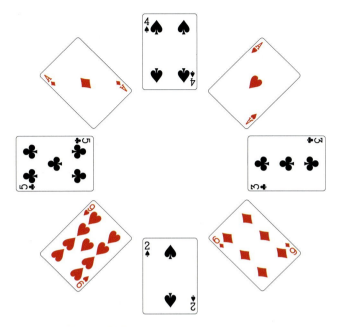

Figure 7.7. π to more decimal places.

There are several properties of this display that can be exploited for other effects. (1) The "compass points" are 2, 3, 4, and 5 (and Black). (2) The sums of pairs of opposite numbers are 6, 7, 8, and 10. (3) The sums of three adjacent numbers (around the circle) are the numbers 8, 6, 10, 15, 16, 17, 11, and 10, so we almost have a sum-rich bracelet (see page 51). (4) The products of three adjacent numbers (around the circle) are the distinct numbers 12, 4, 20, 45, 90, 108, 36, and 18, so this "pi bracelet" is definitely product-rich (see page 52).

The duplicate triple sum of 10 can be avoided by dealing the eight cards into a row instead of a circle, thus removing wrap-around sums from consideration. Knowing the sum or product basically allows one to deduce the identity of the three cards used to compute it.

Oh I See

We wrap up with a no-holds-barred matchmaking application of a Martin Gardner favored mathematical property. Indeed, our discussion of the palindromic principle involved is largely based on Martin's *An Amazing Mathematical Trick with Cards* [Gardner 09].

In 1982, in the final two volumes of his series *OICUFESP* [Adams 82_01, Adams 82_11] ("Oh, I see you've ESP"), magician Howard A. Adams published some related effects, one with an Egyptian presentational theme, based on the following observation.

Ramasee Principle

Start with a packet that runs $X_1, X_2, \ldots, X_n, Y_1, Y_2, \ldots, Y_n$. Have it cut several times. Next, n cards are counted out, reversing their order, to yield two packets of equal size. Then $n - 1$ cards in total are transferred from the top to the bottom of the two packets, any number being transferred in one packet, and the rest being transferred in the other.

It turns out that the resulting top cards of the two packets match in the sense that they are X_j and Y_j for some j between 1 and n.

Have this top pair set aside face down—without being inspected yet—and then have the whole process repeated over and over, scaling everything down by 1 each time. There are now n pairs of face-down cards on the table. It turns out that all of them match.

It's fun and instructive to explore why this works; we leave it for the motivated reader to verify.

The audience has no reason to expect any matches at all, so merely revealing the first match is an attention-getter in itself. Then, as a kicker, you can reveal that actually all discarded pairs match.

It's popular to apply this principle with identical card values for each j between 1 and n—for example, the X_js are all Red and the Y_js are all Black.

It's also standard to package the counting and transferring as a spelling routine. The phrase "Those Last Two Cards Match" works with six pairs. Starting with five pairs instead of the six we suggested, the phrase "Last Two Cards Match" does the trick.[4]

We close with the promised matchmaking routine, that handily solves the problem of male dominance in the usual deck of cards.

7◆   Intrinsically Disordered

How it looks: *At a social event with numerous couples known to you in attendance, produce a large sheet of paper and write down a dozen names as predictions, then set it aside out of sight. Take out a deck of cards and split it in two, having a spectator riffle or rosette the resulting piles together. Split it in two again, and have a second spectator perform the same kind of shuffle.*

---

[4]Max Maven notes, "The first to apply a logical phrase to be spelled, related to the activity, was Larry Becker, who in 1983 came up with a version using the phrase "Will The Cards Match"—published in his 1984 lecture notes, under the title 'Duck Dupe.'"

Reclaim the deck, saying, "I'm sure you've all heard of 'Seven Brides for Seven Brothers.' Let's try a little matchmaking here right now, using these thoroughly mixed cards, but on a slightly less ambitious scale. We'll settle for six matches, using all the guys and gals here. There are eight guys and four gals, so a little open-mindedness is required. The predictions I made just now refer to twelve of you, paired off as six couples according to how I see it."

Turn the deck face up, and go through it, tossing out the royal cards face down as you find them, until you have the required packet of twelve cards. The remainder of the deck is removed from the theater of operation, not being needed any more.

Retrieve the sheet of paper containing your predictions. Say, "Earlier, on the other side of this piece of paper, I made a list of some of the qualities people look for in a relationship or admire in a potential partner." *Place that side of the sheet in plain view, so that all can see the words written on it:* romance, empathy, affection, honesty, wealth, integrity, excitement, patience, confidence, *and, of course,* sexiness.

"Recall the deck was shuffled twice earlier. I'm going to ask you to mix up the cards further with the help of words from this list. Let me demonstrate. For me, honesty is very important: I expect it of any partner." *Holding the packet in one hand, count cards to the table, one for each letter of the word* honesty, *and drop the rest on top.* "Of course, I expect honesty of myself too," *you add, repeating this spelling-and-dropping move.*

Now hand the packet of a dozen royal cards to a third spectator and ask that what you just did be repeated, using any desired quality from the list. The word can be spelled out loud, or the selection can be private. Make sure that the same spelling and dropping is done a second time.

Have a few more spectators do likewise. Take the cards back and count out six to a pile on the table, saying, "Now I need the help of somebody very special. Who fancies themself as a matchmaker?" *Sooner or later, someone will step forward.* "Place one of these piles on top of the other, and cut the packet as often as you like. Then deal out six cards like I just did. Good, we now have two new totally randomized piles."

For the first time, reveal your predictions. The sheet is seen to contain a list of a dozen names, each one matched with one of the royal cards.

"The guys and gals represented by these twelve cards are really mixed up now," *you declare.* "Indeed, some would say that they are intrinsically disordered. Yet I believe you can tease a match out of this chaos." *Draw attention to the bottom of the list, where the phrase* "Those last two cards match" *is written.*

Have the Ramasee Principle invoked, setting aside a pair of face-down cards after each spelling, until two cards remain. Have those turned over and checked against the list: it's a perfect match.

Congratulate the matchmaker for a job well done, saying, "The chances that you matched Kevin to Priscilla were less than ten percent, but you did it. You could go into business with that talent! If you charge enough for your successful hitchings, it won't even matter if you have to return your fee to the couples not hitched up correctly." As you say this, casually turn over another of the face-down pairs, adding, "Look at these two. Checking the list here, it appears that you've matched up Brian with, let's see, Michael! You got that one right, too. I'm impressed, that's two perfect matches. Let's check the rest."

They are all matched up correctly: the other Michael with Adrian, Mohammed with Almaz, Janet with Susan, and, of course, Gerry with Jaime.

How it works: The deck and prediction sheet are prepared in advance. The deck starts with the twelve royal cards in some specific order at the top. The first riffle (or rosette) shuffle distributes these cards in order in the top half the deck, and the second shuffle distributes them more widely, but still in order. Hence, when you casually extract them later, the original order is preserved (or reversed, it doesn't matter which).

The prediction sheet has the list of desired qualities on one side, not to be revealed until halfway through, and a list of the twelve royal cards on the other side (as well as the phrase "Those last two cards match"). When you first produce it and write down some predictions, what you are really doing is writing the names of the people in question beside the card names, one for each one. Furthermore, the names/card correspondence should reflect the desired matchmaking pairings. Perhaps assign a King and Queen to Kevin and Priscilla, two Jacks to Brian and Michael, and so on. Customize the card and people pairs to your audience as you see fit.

Source: Original. February 2013.

Presentational options: There's no need to use all twelve royal cards. Feel free to drop two or four—either of the same or different genders—if it makes it easier to work with the audience in front of you. Adjust the phrase "Those last two cards match" accordingly.

Parting Thoughts

- Can you prove that the Stay-Stack Principle works?

- Consider a packet of twenty cards consisting of two sets of Ace–5, 5–Ace, back to back, namely two identical palindromes one after the other. This is in the family of palindromes that are also cycle; note that nothing changes if half of the cards are COATed. Since this can be visualized as having two symmetric peaks, it seems natural to associate it with a bactrian (two-humped) camel.

  Such a bactrian palindrome remains a palindrome under all of the mixing types considered earlier, as well as under appropriate numbers of underCOATs, such as COATing four cards five times, or five cards four times. It's also true for two COATs of ten cards (no surprise there), or ten COATs of two cards, but we do not recommend the latter as keeping count of the underCOATs is no fun.

  Can you come up with an interesting effect using such a packet?

- Can you prove that the Ramasee Principle works?

Bligreath and Beyond

In this chapter, we examine new options for riffle shuffling prearranged cyclic piles, by eliminating one of the hallmarks of the Gilbreath shuffle studied in Chapter 6: the tedium of generating one of those piles by first dealing out about half of a large packet of cards.

"Flushed with Embarrassment" (8♠) packs two punches, and should get the attention of any audience.

> *Five poker players and a poker novice are located. The novice is handed a packet of cards and invited to cut several times, then cut it into two piles and riffle shuffle those, and finally give it one more cut. Poker hands are dealt out, and each poker fan discards his or her worst cards, which you take. Each now gives his or her best card to the novice. Hand over the remainder of the deck, face up, so that the poker enthusiasts may make free choices to replenish their hands, until each has five cards.*
>
> *The resulting poker hands are compared, and it will be seen that some players have done quite well. However, when the novice turns over the cards she was given, she beats them all, having four Aces and a King.*
>
> *Just when it seems that the show is over, turn the discards you were given earlier face up for all to see. They constitute the winning hand: a straight flush.*

Measure of Imperfection

Before we try to tease new structure out of riffle-shuffled packets, consider a basic example. Start with any Ace–6 in Red on top of Ace–6 in Black, split in the middle so that Reds are on the left and the Blacks are on the right. While there are 12!, or about 480 million, ways to jumble twelve cards overall, if we riffle shuffle these six-card piles together, there are only $\frac{12!}{6!6!} = 924$ possible outcomes (see pages 16 and 128).

As discussed and illustrated on pages 4 and 129, it is helpful to take advantage of the equivalence of riffle shuffling and rosette shuffling. In the latter, the dealt and remaining parts are twirled into rosettes with the fingers before being pushed together and then squared up.

Interweaving cards in a perfectly alternating fashion when table riffle shuffling is highly unlikely; that kind of Faro shuffle is one of the hardest of all moves to master, even for skilled magicians and card handlers (see page 6). Furthermore, exact interweaving when rosette shuffling is not something that occurs often by accident, and it's debatable if it can happen by design.

One way to see how much a shuffle deviates from perfection is to simply deal the resulting packet into two new piles, alternately dealing left then right, and compare those to the original piles.

Before continuing, we note that such dealing into two piles leads to a kind of reversal of Faro shuffling, as mentioned on page 7. If, in addition, each of the resulting piles is dealt to the table to reverse its order, and the two new packets are combined in an appropriate order, then the effect of the Faro is undone. If we skip those additional steps, however, we can still take advantage of some of the fascinating mathematical properties of the Faro discussed by Elmsley, Morris, and others (e.g., see [Minch 91,Minch 94,Morris 98,Diaconis and Graham 11]), although it is tedious when applied to a whole deck. Another popular option consists of up-jogging and stripping out (see page 7, [Diaconis and Graham 11] or [Goldstein 02]).

Below, we show three of the 924 possible outcomes when A–6♦ and A–6♣ are riffle or rosette shuffled together, and then dealt to two piles. These images provide information about the quality of the shuffles from which they arose, ranging from barely mixed at all, to quite well mixed.

Figure 8.1 shows two piles that are almost identical to the original piles; hence, the shuffle hardly mixed the cards at all. Also note that had the two 6s shown there been switched, we would have perfectly reconstructed the initial piles, meaning that the shuffle was perfect in the first place. This kind of flawless interweaving of cards is known as *out-Faro shuffling*. For such shuffled cards, prior to dealing into two piles again, each pair in positions 1–2, 3–4, and so on, consisted of a Red card followed

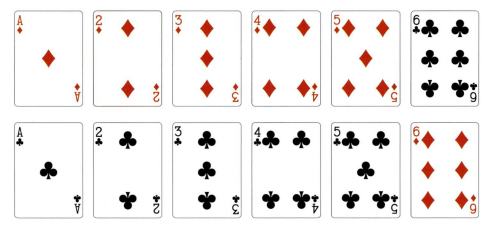

Figure 8.1. One possible result for left and right piles.

by a Black card of the same value. "Out-" refers to outside cards at the start—namely, A♦ and 6♣—still being on the outside after shuffling.

Had such a perfectly reconstructed pile situation arisen with a twist—the left pile being all Black and the right all Red—the shuffle would have been perfect in a different sense. Again, flawlessly interwoven, but this time prior to dealing into two piles again, each pair consisted of a Black followed by a Red card of the same value. This is *in-Faro shuffling*: after the shuffle, the A♦ and 6♣ ended up "inside"—in positions 2 and 11, respectively—rather than in positions 1 and 12 where they started.

The piles in Figure 8.2 came from a packet in which more mixing is evident, although a little structure clearly remains. The piles in Figure 8.3, however, came from a quite well mixed packet.

Vertical thinkers may see a hint in those three images of another measure of a shuffle's deviation from perfection, something we'll say more about soon enough.

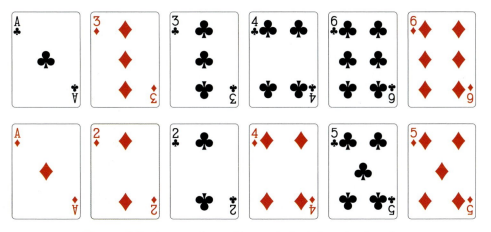

Figure 8.2. A second possible result for left and right piles.

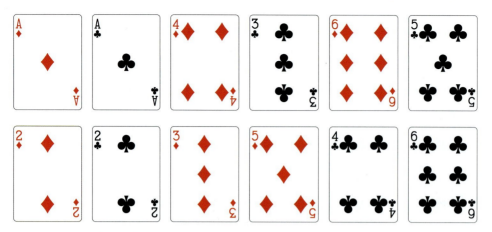

Figure 8.3. A third possible result for left and right piles.

Tweaking Perfection

We just had a look at what can happen when we riffle two piles of the same size, cycling in the same sense. In Chapter 6 we surveyed Gilbreath shuffling, which is the analogous situation where the piles riffled together cycle in the opposite sense. That gave rise to a different kind of perfection.

Recall that in Gilbreath shuffling, we start with a cyclically arranged packet of cards, deal some to the table (thus reversing their order), and riffle or rosette the resulting two parts. Some regularity, if not order itself, is retained. For instance, if the card suits cycle CHaSeD—or any other well-defined order—at the outset, then each group of four pulled off the top of the shuffled packet (i.e., those in positions 1–4, 5–8, and so on), will contain one card of each suit, though not necessarily in the original suit order.

How can one improve on that? People have tried, for decades. In what follows, we look at some other options for riffle shuffling prearranged cyclic packets.

The simplest question, extrapolating from our opening exploration, concerns what happens if, instead of dealing out some of the cards from a cycling packet to shuffle into the rest, we just cut into two piles and riffle or rosette shuffle those together. It's a lot quicker and easier to do than Gilbreath shuffling, so it would be nice if there was a payoff. One case of note is cutting so that the same type of cards start (and finish) both piles.

Should we focus on whether prescribed pairs match (or not), or seek structure in groups of a certain size peeled off the top? First things first.

Ladder Arguments

Imagine a ladder with six rungs, numbered A, B, C, D, E and F from the top down. Ann is in charge of checking the integrity of rung A, Bonnie of rung B, Colm rung C, and so on. Periodically, each of these six people checks to see whether her or his assigned rung appears to be in place, firmly attached at both ends to the appropriate part of the uprights. Broken or bumped rungs are frowned upon.

The ladder takes an unfortunate tumble down the stairs, almost certainly sustaining nontrivial damage in the process. It's likely that some rungs have become detached at one or both sides of the ladder. However, somebody correctly observes that each rung half miraculously ends up attached somewhere, namely to one of the original rung slots and, moreover, none of these rung halves (on either side) is now below one on the same side that started off lower down than it did.

You can't see the results of this, and ask, "Please let me know if you see an intact rung where you expect your rung to be." Bonnie, Declan, and Evin report positively; there are unbroken horizontal rungs where rungs B, D, and E should be. Rungs that have flipped around, their ends switched, may not have been noticed, however. You conclude, "In that case, I'm confident that those are actually your original rungs in position, essentially undamaged."

Brilliant logic, eh? It certainly is, when applied to card shuffling. You'll need at least six or seven friends to try this out. If you don't have that many, it's time to call your imaginary friends into action.

Let's return to two piles, one arranged Ace–6 in Red on the left, and another Ace–6 pile in Black on the right. Have those piles riffle or rosette shuffled together by a spectator while you turn away. Have another audience member take the first two cards off the top, a new person take the second two, and so on.

When all six pairs have been taken, turn back, asking those persons now holding pairs matching in value to say so. If the first, third, and sixth persons admit to this, you can confidently announce that the first person has a pair of Aces, the second has a pair of 3s, and the sixth has a pair of 6s. What's going on here?

Think of the shuffling in question as the equivalent of dropping the ladder down the stairs. It may indeed break some pairs ("rungs"), but those that are unbroken at the end are exactly where they were at the outset. It seems inevitable for ladder rungs, but why is it true in the riffle shuffling context? All things considered, it's really for a similar reason.

Consider the two six-card piles shuffled above, together with the fact that the third person ends up with a matching pair. If that pair has a value greater than 3, that means that the six cards comprising the two sets of Ace–3 are in the hands of the first two people, contradicting the fact that they have two cards each. If that pair has a value less than 3, that leaves at most two of the four cards comprising the two sets of Ace–2 in the hands of the first two people.

Go back and inspect Figures 8.1, 8.2, and 8.3 again. The first shows all pairs intact, although one has been reversed; the second shows two intact and reversed; and the third shows none intact. Most significantly, in no case is there a misplaced matching pair.

The statistics of riffle (or rosette) shuffling may be asymmetric—leading to more broken links early on and fewer later. A similar statement is true for the physics of a ladder, either in the context suggested or if somebody stood high up on it and some weak rungs smashed due to the person falling—we would expect more broken rungs near the top. However, forget the likely "top down" statistical nature of riffle or rosette shuffling in practice, and the directional bias of gravity.

Such shuffling doesn't distinguish one end of the resulting packet from the other. The "internal coherence" condition, as seen through the lens of the Solitaire Principle (see page 129), works just as well if we express everything from the bottom up rather than the top down.

There are internal tensions working from *both* ends of the two piles in these kinds of shuffles, and the result is a "conservation of displacement" balancing act of sorts. Here it is stated more formally.

Intact Rung Principle
Suppose two piles consisting of cards numbered 1–n from the top down are riffle or rosette shuffled together, and cards are then peeled off two at a time. If the jth pair matches, then both cards have value j.

The proof in the general case mimics the argument provided above for a specific example. If the jth pair has value greater than j, this results in at least $2j$ cards (the two sets of 1–j in order) being among the cards in the first $j - 1$ pairs, which is impossible. If the jth pair has value less than j, this results in fewer than $2(j - 1)$ cards being available for the first $j - 1$ pairs, which is likewise impossible.

Magic inventor and prolific author Karl Fulves[1] treats this in "Self Correcting Set Ups"—in the context of two packets running Ace–King—in his rare magic-insider book *Riffle Shuffle Set-Ups* [Fulves 68, pages 20–21]. If thirteen audience members take pairs of cards from the shuffled packet, and the second, tenth, twelfth, and last ones indicate that they have matching pairs, you can be sure that those have values 2, 10, Queen,

---

[1]Pronounced *FULL-ves.*

and King. Fulves also has suggestions for applications to other packet orders.

We think it's a good idea to have larger packets (not to mention more friends), so that you can try this out on a scale where it's more likely to work. So if you haven't been getting out much lately, it's never too late to give it a shot.[2] In practice, small piles (such as the Ace–6 ones originally suggested above) result in few if any of the desired matches.

The repeated cuts and shuffles in the next effect seem to eliminate the possibility of any setup. An audience of one is adequate.

8♣ Matchmaker Instincts

How it looks overall: *Say that you're going to perform an experiment in which the cards will be randomized with much help from the spectator and then dealt off into pairs while you turn away. You'll use your matchmaking instincts to predict any perfect matches that show up.*

The deck is cut and riffle shuffled by a spectator, then cut again, following which you separate the Red cards from the Black cards. The spectator riffles the resulting two piles together.

Turn away, and ask the spectator to take off pairs of cards from the top, one by one. You are told if their values match. Every time they do, you can name the card values.

How it looks in detail: The deck is actually first cut by the spectator, then cut into two equal piles by you, then riffle shuffled by the spectator, who cuts it again. You run through the faces of the cards, tossing out the Red cards to a face-down pile on the table. The spectator riffles the Reds into the Blacks.

The rest is as already described. It's important that for each pair taken off, the spectator reports to you whether it's a match. You may not get the first match right, but you soon hit your stride and perform flawlessly from then on.

How it works: The deck is prepared by separately ordering the Red and Black cards in two stacks that repeat according to a specific value order. Assume that Conway's *The Five Tenacious Boys Nicely Joke to Hated Servant Girls Sick For Absent Kings* cycling (see page 13) is used, freely mixing Clubs and Spades, and Hearts and Diamonds. Omit Jokers. Apart from suits, the two resulting half-decks are identical. Place one on top of the other, and cut somewhere.

---

[2]Remember, there are no strangers, only friends you haven't met.

At showtime, have the deck cut by a spectator. As you take the cards back, peek at the bottom one and use that to guide you as to where to split the deck; you seek a card of the same value but opposite color half a deck away. Put the resulting two piles on the table, and have the spectator riffle shuffle one more time.

Unknown to the audience, the two piles just shuffled were in identical order, and so the Intact Rung Principle holds. Hence, as pairs are taken off, once you know which ones are identical, you know what the values must be, provided you can get a foot in the door.

The first time, you simply guess. In all likelihood you are wrong, but when the spectator corrects you, pay close attention to where you are in the Conway cycle. For instance, if the first matching pair consists of 10s, the next three pairs do not match, but the following one does, then you know it must consist of 8s (remembering "... *Tenacious* Boys Nicely Joke To *Hated*...," and skipping Jokers). This logic continues, cycling through the phrase over and over and carefully keeping track.

Source: Original. March 2013.

Holly Days

The ladder rung considerations above concern pairs taken from packets obtained by riffling together two identically cycled piles. What about the case where the cycles in the two piles are of length three, and we peel the cards off three at a time? We call these triplets.

Let's switch from cycling values to cycling suits. Suppose we have two packets that run Clubs, Hearts, Spades, repeatedly, from top to bottom, with no Diamonds used at all. Riffle or rosette shuffle these piles together, and peel off triplets. Ignoring the perfect ones in which all three suits are present, we find that the following rules hold.

Bookends Principle

1. *Triplets in which Clubs repeat start with a Club.*
2. *Hearts will never be repeated.*
3. *Triplets in which Spades repeat end with a Spade.*
4. *Triplets with two Clubs alternate with those with two Spades.*
5. *The Diamonds are still nowhere to be seen.*
6. *There is no Rule 6.*

This Karl Fulves discovery arose from some summer 1959 experiments with magician buddy Steve Schimm, as reported in "The 3-Card Cycle"

[Fulves 68, pages 17–18]. Fulves expressed everything in terms of cycling Aces, 2s, and 3s, and gave several applications to card magic, including a Martin Gardner effect called "A Better Swindle" [Fulves 68, page 88], the latter being a modification of a betting game based on the Fulves result that Gardner had presented in one of his *Scientific American* columns [Gardner 83, Chapter 19].[3]

A tête-à-tête argument along the lines of those in Chapter 6 (e.g., see page 134) is easy to construct, using 012012012... 012 to represent the suit cycling.

Gilbreeath Shuuffling

Here's a twist on the General Gilbreath Principle inspired by another Karl Fulves creation for a related suit cycle of length three published as "ESP + Math" in his more accessible *More Self-Working Card Tricks* [Fulves 84]. Ironically Fulves didn't actually use ESP cards (which cycle around five at a time), but he has a version under the same title that does [Fulves 68, page 8]. He clearly knew how the principle worked in general.

In a sense, what we consider next is akin to dealing from a cycling packet in preparation for a Gilbreath (riffle or rosette) shuffle, but then inserting a second copy of one of the "exposed" cards, so that both piles start with the same card before the shuffle. That's what inspired the name *Gilbreeath Shuuffling*[4] given to the online August 2012 *Card Colm*, in which some of what follows first appeared [Mulcahy 12_08].

Fulves has the reader assemble two packets of twelve cycling cards side by side. From the top down, the first consists of four rounds of Spades, Hearts, Diamonds, and the second one runs Spades, Diamonds, Hearts four times over. No Clubs are used, and the card values play no role.

If we now riffle shuffle these together, and peel off cards in triplets (i.e., three at a time), then a curious pattern emerges. Ignoring those perfect triplets—three consecutive cards in which each of Spades, Hearts, and Diamonds are present (in some order)—a cascading sequence of deductions can be made, as seen here.

> ### Gilbreeath Shuuffling Principle for Triplets ▶◀
>
> 1. *The first triplet with a repeated suit features two Spades; moreover, one of them is the first card.*
>
> 2. *Whichever suit is not represented above appears twice in the next triplet with a repeated suit, and one of them is the first card there.*

---

[3]We recommend looking up this Gardner column, as his general treatment of the Fulves principle is notable.

[4]Pronounced *gil-BREETH SHOE-fling* to distinguish it from *GIL-brith* shuffling.

3. *This logic repeats as long as triplets with repeated suits show up.*

We are only speaking of the eight consecutive triplets consisting of the cards in positions 1–3, 4–6, ..., 22–24. The same claims hold if we start with two packets of size eighteen, or any other multiple of three, cycling in the same way.

Let Spades, Hearts, and Diamonds be represented by 0, 1, 2, respectively, so that the packet on the left, from top to bottom, is 012012012012, and the packet on the right is 021021021021.

The kind of tête-à-tête argument given on page 134 in the Gilbreath context works here too. The result is the following key observation.

> The effects of riffle (or rosette) shuffling these two packets and then pulling off triplets are
>
> 1. to line up the left packet in reverse beside the right one to get 210210210210 ‖ 021021021021, and
>
> 2. to successively extract three adjacent cards from the "middle" starting with at least one of the two initially adjacent 0s.

Ignoring the undeniably perfect cases in which all three cards come from one of the packets, we therefore get as our first interesting triplet either two 0s and a 1, with a 0 on top, or two 0s and a 2, with a 0 on top.

In the first case, the next triplet that involves cards from both packets is extracted from the middle of 2102102102 ‖ 21021021021, and hence consists of two 2s (and either a 0 or a 1). In the second case, the next triplet is extracted from the middle of 21021021021 ‖ 1021021021, and so consists of two 1s (and either a 2 or a 0). Now the pattern that gives rise to the claimed prediction principle should be clear.

Let's extend this to groups of four cards each time. Arrange two packets of sixteen cards, one cycling Spades, Hearts, Clubs, and Diamonds, from the top down, and the other cycling Spades, Diamonds, Clubs, and Hearts. Note that, apart from the top card, the suits in each packet are in reversed order. Let 0, 1, 2, and 3 represent Spades, Hearts, Clubs, and Diamonds, respectively (this is not the same correspondence we had above), and try a little tête-à-tête. From top to bottom, we have 0123012301230123 on the left, and 0321032103210321 on the right. This time, we observe the following.

> The effects of riffle (or rosette) shuffling these two packets and pulling off quadlets (four cards at a time) are
>
> 1. to line up the left packet in reverse order beside the right one to get 3210321032103210 ‖ 0321032103210321, and

2. to successively extract four adjacent cards from the "mid-dle" starting with at least one of the two initially adjacent 0s.

We claim that at most one suit is repeated in each set of cards in positions 1–4, 5–8, and so on. Moreover, the following rules apply.

Gilbreeath Shuuffling Principle for Quadlets

1. *The first quadlet with a repeated suit features two Spades; moreover, one of them is the first card.*

2. *Whichever suit is not represented above appears twice in the next quadlet with a repeated suit, and one of them is the first card there.*

3. *This logic repeats as long as quadlets with repeated suits show up.*

Indeed, ignoring the perfect quadlets for which all four cards come from one of the packets, we then get as our first quadlet two 0s, a 2, and a 1; or two 0s, a 1, and a 3; or two 0s, a 3, and a 2. In all three cases, a 0 is on top, and the next nonperfect quadlet contains two cards of the suit not represented in the previous quadlet.

We leave it to interested readers to verify that the remainder of the prediction claims hold.

The principle extends to five or indeed any larger number. Fulves has versions for various cycle lengths [Fulves 68].

When rosette shuffling two piles of equal size, we suggest trying to ensure that the top cards of each pile end up together. It seems to cut down on the number of perfect triplets or quadlets.

8♥   The Guessing Game

How it looks: *Two packets of twelve cards are placed on the table. These could be dealt from a full deck, either one pile at a time, or alternating left to right. These are rosette shuffled by an audience member, following which you deal them into three piles, one for yourself, and the others for two spectators.*

Have each person in turn try to predict the top card of his or her pile before turning it face up. You do much better than either of the spectators at this guessing game.

How it works: Following Fulves [Fulves 68], the piles initially on the table should each consist of twelve cards, the first cycling Spades, Hearts, and

Diamonds, the second Spades, Diamonds, and Hearts. Plan accordingly if you wish to deal these from a full deck.

Now have the two piles rosette shuffled, and then deal into three piles from left to right. This distributes the structured triplets among the three of you in an organized way. Retain the first pile for yourself and give the others to two spectators. Request that each person deal to the table to verify that each has eight cards, doing likewise yourself; this is just a ruse to reverse the order of each pile.[5]

Next, suggest that a guessing game be played. Have each spectator guess the suit of the top card in his or her respective pile before turning that card over. Of course, you already know that one of the three cards is a Spade, so if the first person's card turns out to be something different, you can even gamble on guessing that the next person's is a Spade. In any case, you end up by correctly guessing that yours is a Spade, before you turn it over to confirm.

Have those three cards set aside face down, quietly noting which suit you did not just see, and proceed to the guessing and uncovering of the suits of the next three exposed face-down cards. After a few successful rounds say, "Perhaps you think I need to see your card faces to guess mine correctly. This time, let's all guess before we exposes any cards!" You still come out ahead most of the time, despite the misdirection of the words you just uttered. In the hopefully rare cases in which all three cards are of different suits (so that your reasonable guesses are wrong), modestly say, "Nobody's perfect all the time."

Chances are, you'll still do much better than the spectators, who are totally in the dark and may not notice that there are no Clubs in play. (Feel free to guess that suit now and then, when you have nothing to go on, in an attempt to further throw everyone off the scent.)

Source: Based on "ESP + Math" from the book *More Self-Working Card Tricks* [Fulves 84]. Published online in the August 2012 *Card Colm* "Gilbreeath Shuuffling" [Mulcahy 12_08]. The repeated letters of that title[6] are a playful reference to the fact that it is as if we Gilbreath shuffle a cycling packet that had one letter repeated, the dealing being stopped between these two letters.

Presentational options: Your chances of success increase if the given packets are larger: for example, each could contain eighteen cards (or even twenty-four if you don't mind using cards from two decks).

A version using all four suits and involving three spectators is easy to do. Have thirty-two suitable arranged cards (CHaSeD will do nicely)

---

[5]As the fella says, "The first shall be last, and the last shall be first."

[6]Recall that this is pronounced *gil-BREETH SHOE-fling*.

at the outset, split into two piles of sixteen that are then rosette shuffled together. Deal into four piles of four, from left to right, and have those redealt to reverse the card order. Now, if you claim the first such pile for yourself, you can correctly predict the suits of most of those cards provided that you first get to see the corresponding cards in the other three piles. This too can be done with larger packets, if you wish.

There is a connection between what we have just considered and the General Gilbreath Principle. In the case of cycles of length four, the latter is equivalent to starting with something like 2 1 0 3 2 1 0...3 2 1 0 in pile A, representing cards counted out (hence reversed) from a packet, leaving the 3 0 1 2 3 0 1 2 3 ... 0 1 2 3 in pile B. The effect of riffle (or rosette) shuffling these together and then pulling off four cards at a time is to reverse pile B and place it beside pile A, yielding 3210...321032103 ∥ 2103210...3210, and then extract four cards at a time, starting in the "middle" with at least one of the 3 or 2 that abut the dividers, along with several adjacent cards. Each time, we get one of each of the four card types. A similar connection exists between "ESP + Math" and the Gilbreath Principle for packets consisting of repeated cycles of length three.

When all is said and done, what we have explored above is the case of inserting one additional "repeated" card near the middle, beside one of the same type, in the usual Gilbreath context, and then proceeding to riffle or rosette shuffle. Although it changes the nature of one's prediction, predict one certainly can, for at least for a quarter (or a third) of the shuffled cards, namely the top or bottom card of each group removed, ignoring the cases in which all types are represented in the cards pulled off.

What about the Fulves effect for cycles of length two? There is nothing new there; in fact, we're back in familiar Gilbreath terrain. When we riffle (or rosette) shuffle together two packets of the same type, each alternating between the same two types, we have an "out of sync" basic Gilbreath shuffle, which was already covered in Chapter 6 (page 132).

From this perspective, it is now clear that "ESP + Math" and its extensions generalize the binary ($s = 2$) Gilbreath Principle in a way that is different in spirit from the General (arbitrary s) Gilbreath Principle. Many other shuffle variations may be found (in a manner of speaking) in the hard-to-find *Riffle Shuffle Set-Ups* [Fulves 68].

From Gilbreath to Bligreath

The relationship between Gilbreath shuffling and the variation on it that
we are about to introduce—and the reason we've named the latter as we
have—is reflected in the fact that the top parts of the one of the shuffled
packets are in opposite orders when the two shuffles are compared.

As seen in Chapter 6, Gilbreath shuffling a packet of cards that consists of repeated cycles of length nine, such as $G\ I\ L\ B\ R\ E\ A\ T\ H$, first
involves dealing out some cards into a pile C, reversing their order. If we
denote what's left by pile D (the original bottom), and assume that the
last card dealt from the top to C is of type B, then the shuffling can be
modeled with this tête-à-tête display:

$$G\ I\ L\ B\ \dots\ T\ H;\ G\ I\ L\ B \text{ (top of C)} \parallel \text{(top of D) } R\ E\ A\ T\ H;\ \dots\ T\ H.$$

where the dividing bars $\parallel$ mark the break between the dealt and undealt
cards. Riffle shuffling mixes cards from the tops of each of those piles,
and can be viewed as selecting nine adjacent cards from the mid-region of
this display, starting with those on one or both sides of the dividing bars.
The result will be a packet for which the cards in positions 1–9, 10–18,
and so on, each contain one of the nine types being considered.

In practice, dealing out about half of what you are working with takes
time, and may test an audience's patience, especially if you are playing
with a full deck (or think you are). What alternatives are there?

Here is the most obvious one. If the original packet has a good proportion of face-up cards randomly positioned within it, while maintaining
suit cycling as above, then it will look really jumbled. You can even spread
it, and show both sides to onlookers, without revealing the suit regularity.
Now cut off about half, flip those cards over, and riffle or rosette shuffle
the resulting parts together. The usual property holds: each four taken
from the top will consist of one of each suit. In performance, assuming a
color-alternating suit cycle order, such as CHaSeD, is used at the outset,
you can start by revealing (an even number of) pairs of different colors
and then switch to quadlets of different suits.

There is another option for face-down packets: forget Gilbreath shuffling. That is to say, forget the dealing out. Suppose, instead, that we
simply cut the original packet beneath a card of type B, to form a pile C
above the break, without any reversal of card order, the remainder once
more being denoted by pile D. Since the cards in C here are in the opposite order to those in C above, it's as if we cut the following packet at the
dividing bars $\parallel$, and proceeded as before, "from the inside out":

$$B\ L\ I\ G;\ H\ T\ \dots\ B\ L\ I\ G;\ \text{(top of C)} \parallel \text{(top of D) } R\ E\ A\ T\ H;\ \dots\ T\ H.$$

Since *Gilbreath shuffling* denoted the act of dealing cards from a prearranged cyclic packet to reverse their order, and riffling the results together,

then an examination of the middle region of the last display should make it clear why we suggest *Bligreath shuffling* to refer to simply cutting such a packet prior to riffling.

A little tête-à-tête shows that the top nine cards of the resulting Bligreath-shuffled packet arise from selecting that many adjacent cards from the mid-region above, starting with those on one or both sides of the dividing bars.

As in the Gilbreath case, there are ten possibilities, depending on how many of the cards (say, k, a number between 0 and 9) come from pile C. The other $9 - k$ cards must come from pile D, of course. However the $9 = k + (9 - k)$ cards obtained here were almost certainly not together in the original packet, and so can't be expected to contain one card of each type. For instance, if $k = 2$, then we end with the types *IGREATHGI* (in some order).

Proceeding under that $k = 2$ assumption, what could the next nine cards in the shuffled packet be? Deleting the nine just indicated, we see that effectively we are now selecting nine adjacent cards from in and around the bars here:

$$B\ L\ I\ G\ \dots\ E\ R\ B\ L\ \text{(top of C)} \parallel \text{(top of D)}\ G\ I\ L\ B\ R\ E\ A\ T\ H; \dots\ T\ H.$$

The two easy cases are those in which we get all nine cards from one or other of the piles, yielding a perfect set (all nine types present and correct). Perfect sets show up in some of the other eight possibilities, obtained by sliding a window of length nine over the above display, and staying within sight of the bar. Do you see any patterns in the others? Remember that in our first set of nine obtained earlier, G and I are repeated, and L and B are missing.

Despite the shortcomings of the Bligreath shuffle just encountered, some useful bookkeeping can be done. Various special cases, for cycles of small length, have been explored through the decades by Fulves and others. We have already pointed out that the case of cycles of length two is, at worst, just a step away from basic Gilbreath shuffling.

In "The Bligreath Principle" (the online August 2009 *Card Colm* [Mulcahy 09_08]), we discussed observations on the topic about which binary magic creator Leo Boudreau had written in a web magic forum in 2005.

The Triskadetres Principle

Magician John Hostler recently shared an exciting discovery of his from the summer of 2012 called the *Triskadequadra*[7] *Principle* [Hostler 12]. He

---

[7]Pronounced *TRISK-a-de-QUA-dra*.

only developed it for four cycling suits, such as CHaSeD, but some version of the key ideas work for cycles of any length.

It turns out that his results bear a strong resemblance to those discussed in "Gilbreeath Shuuffling," and the same mathematical analysis explains both principles.

Let's start smaller, with the cycles of length three, for instance—a packet running runs Clubs, Hearts, and Spades repeatedly, from top to bottom, with no Diamonds used.[8] We represent this as

$$(\text{top}) \ 0 \ 1 \ 2; 0 \ \ldots \ ; \ 0 \ 1 \ 2; \ 0 \ 1 \ 2; \ \ldots \ ; \ 0 \ 1 \ 2; \ (\text{bottom}).$$

Now let's cut this packet somewhere to form pile A (including the original top) and pile B (including the original bottom) in such a way that pile B has a different starting card suit from pile A. Since A starts with a 0, this means that we either cut between a 0 and a 1, or between a 1 and a 2, and there is no loss of generality in assuming that the first case holds. So we now have

$$(\text{top}) \ 0 \ 1 \ 2; 0 \ \ldots \ ; \ 0 \ 1 \ 2; \ 0 \ \| \ 1 \ 2; \ \ldots \ ; \ 0 \ 1 \ 2; \ (\text{bottom}).$$

The usual tête-à-tête argument applies: riffle shuffling A and B together and taking off the top three cards of the result is like selecting three adjacent cards from the mid-region of the next display, starting with those on one or both sides of the dividing bars:

$$0; \ 2 \ 1 \ldots \ 0; \ 2 \ 1 \ 0; \ (\text{top of A}) \ \| \ (\text{top of B}) \ 1 \ 2; \ \ldots \ ; \ 0 \ 1 \ 2.$$

Depending on how many of the cards come from pile A, there are four possibilities for that first triplet, namely, 2 1 0, 1 0 1, 0 1 2, or 1 2 0 (in some order). Three of these form perfect sets, in which all three types (suits) are present. In the case of the 1 0 1 triplet, upon its removal we are then left with

$$0; \ 2 \ 1 \ldots \ 0; \ 2 \ (\text{top of A}) \ \| \ (\text{top of B})1 \ 2 \ \ldots \ ; \ 0 \ 1 \ 2.$$

Returning to the three perfect triplets, in two cases it turns out that removing them leads to the same display and bar placement as above (only shorter). In the remaining perfect case, where 0 came from A and 1 2 from B, removal of this triplet leaves us with

$$0; \ 2 \ 1 \ldots \ 0; \ 2 \ 1 \ (\text{top of A}) \ \| \ (\text{top of B}) \ 0 \ 1 \ 2 \ \ldots \ ; \ 0 \ 1 \ 2.$$

Removing three cards in the usual way from this display results in either a perfect triplet or 2 1 2 (in some order). Note that this includes two cards of the very type not present in the earlier triplet 1 0 1.

What do we learn from all of this, when peeling off triplets from the shuffled packet, namely, the cards in positions 1–3, 4–6, 7–9, and so on?

---

[8]It seems that diamonds are forever ... missing!

Triskadetres[9] Principle   ▶◀

1. *The first nonperfect triplet encountered features two cards of type* 1.

2. *The next nonperfect triplet encountered features two of whichever type was missing above.*

3. *This logic repeats as long as triplets with repeated suits show up.*

This translates into concrete predictions for suits given our initial packet of cards cycling Clubs, Hearts, and Spades. The repeated suit represented in the first nonperfect triplet must be a Heart, and if there is a Club in that triplet too, then we can be sure that the next nonperfect triplet contains two Spades. If those repeated suits are also accompanied by a Club, there are definitely two Hearts the next time there's suit repetition in a triplet.

If we had cut the packet so that pile B started with a card of type 2, the above discussion would apply, appropriately modified. But what if we had cut the packet so that both piles A and B started with Clubs? After the shuffle, simply cut a single card from one end of the packet to the other. That will convert it to the desired state, as if we had done it properly in the first place.

Suppose the packet was cut and shuffled without checking whether the top cards matched in type. Is there a simple way to scan the card faces and tell whether an adjustment needs to be made, as there was in Chapter 6 for basic Gilbreath shuffling?

The Triskadequadra Principle

What we just explored for triplets has extensions to packets built from cycles of any length. It is especially interesting and fruitful for cycles of four suits, and it was in that form that we first saw it, courtesy of magician John Hostler's manuscript [Hostler 12]. This contains some great applications that are quite distinct from our offering below. What follows is a reworking of his findings using our methods.

Start with $4m$ cards consisting of m repeated ordered sets of four items, for instance, twenty cards consisting of five sets of cycling cards in CHaSeD order. Cut as often as you wish.

Now cut the cards into two packets of any size and riffle shuffle these together. If you like, you can cut them again at this stage. Hostler insists at the outset that the top cards of each subpacket have different colors,

---

[9]Pronounced *TRISK-a-de-TRES*.

but we throw caution to the wind, emboldened by our successful earlier explorations in the cruel jungle of shuffleland.

Scan the card faces. Because the card colors alternated in the original packet, we know from Chapter 6 that we'll never find a run of more than two cards of the same color together.

Consider quadlets from the top, namely cards in positions 1–4, 5–8, 9–12, and so on.

Triskadequadra Principle

Case 1 (good): *Every quadlet is color balanced; in other words, it consists of two Reds and two Blacks. If, in addition, every quadlet is perfect, with all suits represented, then break out the champagne.*

Otherwise, in the first nonperfect quadlet, exactly three suits are represented. If the repeated suit is a Club, then the next nonperfect quadlet will have Spades repeated, the next nonperfect quadlet after that will have Clubs repeated, and so on.

A similar statement holds with Clubs and Spades switched if the repeated suit of the first nonperfect quadlet is a Spade. Also, in both cases, every quadlet will have a Heart and a Diamond.

Similar comments apply if the repeated suit is Red, and moving two cards from one end of the packet to the other toggles between these two possible states.

Case 2 (not good): *At least one quadlet has an odd number of Reds. This is an off-balance packet that needs adjusting. There must be two cards of the same color beside each other. Find them and inform them that this beautiful relationship is over. Cut the packet between these, and reassemble it the other way. Now we are back to Case 1.*

A possible alternative to fanning the card faces to check the exact state, in preparation for a corrective cut if necessary, is to just turn over the top four to eight cards. If you spot two cards of the same suit in a single quadlet, then you know the situation and innocently adjust by cutting the appropriate number to the bottom.[10]

A post-shuffle packet—perhaps with a final cut thrown in—is in one of four possible states that can be cycled through by moving one, two, or three cards from one end to the other. Two of those states are highly desirable, and you can get to one of those from the other two. Hostler's preshuffle color check guarantees getting one of the desirable states.

---

[10] If you don't spot two such cards, dig deeper before jumping to any conclusions.

Focus on a particular suit, say, Diamonds. Scan the card faces seeking two adjacent Black cards, separate the packet between these and reassemble it the other way. Now each quadlet has a Diamond (and a Heart) in it. A similar step can be taken, with a color switch, to ensure that each quadlet has a Club and Spade in it.

Of course, by doing such a separation you also restore the packet to the binary post-Gilbreath state, which considers only the initial R/B state. Hence, no quadlet will be either RRBB or BBRR: the colors in each quadlet will either alternate, or be BRRB or RBBR. Furthermore, considering just the ten cards of a particular color as they occur within the twenty cards, we'll never get more than two in a row of the same suit.

Why does it all work? Naturally, we suggest a little tête-à-tête. Arrange a packet of cards cycling CHaSeD = 0123 from the top down. Cut it to form two piles, A and B, as before, with the lower one (B) having a Red card on top, say, a Heart. Put pile B to the right of pile A. From top to bottom, we have 01230123 ... 01230 in pile A, and 1230123 ... 0123 in pile B. Now, we can assert the following:

The effect of riffle (or rosette) shuffling these two piles and then pulling off four cards at a time, is to (1) line up the left pile in reverse beside the right one to get 03210 ... 32103210 ∥ 1230123 ... 0123, and (2) successively extract four adjacent cards from the "middle" (denoted by ∥ here), starting by including at least one of the initially adjacent 0 and 1 pairs. Ignoring the perfect cases in which all four cards come from one of the piles—and hence represent all suits, we therefore get as our first groups of four cards 2101 or 1012 (i.e., a Club, two Hearts, and a Spade), or another perfect set. In the first two cases, we are left with 03210 ... 32103 ∥ 230123?0123 or 03210?321032 ∥ 30123 ... 0123, and the next (nonperfect) group of four is 0323 or 3230 (i.e., a Club, a Spade, and two Diamonds, hence containing two cards of the suit not represented in the previous group of four) or is another perfect set. Continuing, it can be checked that these two situations alternate.

If, instead, we cut the thirty-two-card packet so that the lower pile has a Diamond on top, then a similar analysis is valid, with Diamonds and Hearts switched. This is "two cards away" from the above situation. Also, if we cut so that the lower packet has a Black card on top, we are "one or three cards away" from the above situation.

Here's a magnetic suits type of application. Take off cards in pairs, out of view, so that the audience has no idea how they are being sourced. Claim that the first pair will be one of each color. Correct. Repeat. Right again. If you have just seen two Spades, you can safely predict that the next cards will be a Club and a Red one. You then know the suit of the

Red card in the fourth pair, as every quadlet has one Diamond and one Heart. There are many guessing games one can conjure up.

Roundies versus Sharpies

We've highlighted just some aspects of the results of riffle shuffling a packet of four-suit cycling cards, having simply split it in two somewhere prior to the shuffle. There is more waiting to be discovered. We have assumed an initial color-alternating situation, but without any additional work, we can deduce some results for other suit cycle possibilities.

For instance, suppose the suits cycle CDHS (alphabetical order), giving rise to color block alternation (i.e., two Reds and two Blacks over and over), with a single Black card at the top and bottom. This is not unlike what was considered in "Doubled-Up Alternation" in Chapter 6, except that a specific suit cycling facilitates easier color tracking. This is a less obvious alternating pattern, namely of "roundies" (Clubs and Hearts) with "sharpies" (Diamonds and Spades) (see page 14).

Splitting such a packet so that the two piles start with a Club and Diamond then riffle shuffling—and turning a blind eye to the perfect quadlets, as usual—we find that the remaining quadlets all consist of pairs of sharpies along with one of each type of roundy. Ignoring the card order within the quadlets, these actually alternate between DDCHs and SSCHs, there being an equal number of each type.

Had we moved two cards from one end of the packet to the other, right after the shuffle, we would have obtained a packet for which the last statements above would hold (with the words "roundy" and "sharpy" switched). Of course, had we split the packet differently—or randomly—in the first place, then the usual post-shuffle type of adjustment could be done, if needed, to restore the packet to either of the two above states.

In addition to the suit-guessing games now made possible, there are color patterns in the shuffled packet that have actually been known for a long time. Only the blocks-of-two color pattern in the initial packet is needed for these results, not the full suit cycling we're assuming. Elementary proofs are easily provided in these cases, independent of the above, as already suggested back in Chapter 6.

For instance Fulves, in his "Backwards Bet" [Fulves 01], points out that riffling two piles that cycle Black, Red, Red, Black, over and over, leads to quadlets consisting of either two mixed-color pairs, or a same-color pair followed by another same-color pair of opposite color. In other words, all of the quadlets are color balanced, and if the first pair is all Red, then the next one is all Black, and vice versa. Four consecutive cards of the same color can occur, but these will always straddle two quadlets, in positions such as 3–6 or 7–10.

Returning to the Triskadequadra principle, the original manuscript [Host-ler 12] is a great source of suit-based applications. We now present an effect that has nothing to do with suits, showing the breadth of application that is possible when "suitable" adjustments are made. We believe it's the strongest magical offering in this chapter.

8♠   Flushed with Embarrassment

How it looks overall: *Five poker players and a poker novice are located. The novice is handed a packet of cards and invited to cut several times, then cut it into two piles and riffle shuffle those, and finally give it one more cut. Poker hands are dealt out, and each poker fan discards his or her worst cards, which you take. Each now gives his or her best card to the novice. Hand over the remainder of the deck, face up, so that the poker enthusiasts may make free choices to replenish their hands, until each has five cards.*

The resulting poker hands are compared, and it will be seen that some players have done quite well. However, when the novice turns over the cards she was given, she beats them all, having four Aces and a King.

Just when it seems that the show is over, turn the discards you were given earlier face up for all to see. They constitute the winning hand: a straight flush.

How it looks in detail: Find five poker fans, along with somebody new to the game. Explain that you are going to deal out some hands of cards and give the poker enthusiasts a chance to improve their hands by discarding some cards and freely picking replacements.

"We'll start with just enough cards for the first round of poker hands," you say, as you thumb off or silently count out cards from the top of the deck to form a pile in the other hand. The rest of the deck is set aside for now. "We should shuffle, to be fair," you continue, and then address the audience member who is not familiar with poker, "Would you, please?" Hand her the packet and have her cut it completely a few times; then ask her to cut it into two roughly equal piles. Direct her as to how to rosette shuffle these together. Finally, have her cut the packet once more.

"You can't ask for fairer than that, can you?" you ask, taking the cards back and fanning the card faces, briefly showing the audience. "Completely mixed up," you say, lying through your teeth.

Now deal the cards from left to right into piles in front of the poker fans, one card at a time, until you have dealt four hands of five cards.

Feign puzzlement, and say, "My mistake—there are five of you, not four; sorry." Gather up the cards and deal again, this time into five hands. Comment, "It seems we don't have enough cards here, only four each. I was never very good at numbers. We'll fix that in a moment. First I want each of you to look at the cards you already have. There are probably some good ones and some bad ones. I'll make you a deal. Get rid of the single worst card you have, and just give it to me. Please keep all cards face down as you discard them." Briefly look at the cards you are given, and comment, "Nothing too high there—oh, well," and place them face down on the table in front of you.

Indicating the poker novice, say, "Since she's new to the game, let's give her a break. Please give her the best card in your hand. Don't worry, you'll have a chance to replace it soon." Have those five cards set in front of the person without any poker experience, also face down. She is not to look at them yet. Turn back to the poker players.

"Since the five of you now have only two cards apiece, you may, in turn, go through the rest of the deck here, and find three cards to complement what you have so far. This isn't the way real poker works, I know, but it's fun to do once in a while. Knock yourself out!"

This should generate a buzz of excitement. Hand the remainder of the deck, face up, to the first poker fan, on your left, and in due course to the next one, and so on, until all five of them finally have five cards.

Recap the initial shuffling of the cards by a spectator, the discarding of low and then high cards, and the complete freedom for each person to pick three cards to add to the two received earlier. Ask the poker fans, "How did each of you fare? It's time to show all!"

With any luck, there will be proud displays of full houses, flushes, straights, maybe even a four of a kind. Everyone has at least a pair.

"Amazing! I guess it does help to have free selections for three of your five cards. Let's see how the novice did. Remember, she had no input at all into her poker hand." The cards in front of her are turned over to reveal four Aces and a King. Everyone should be very impressed.

"Incredible," you say. "You saw her shuffle the cards; she even decided where to cut the packet. What are the chances that four of you end up with an Ace? That's what must have happened, since you each gave her your best card."

Here comes the kicker. "I'm not a poker player myself," you add casually. "It's a good thing. As mentioned earlier, my cards are all of low value." Turn your cards over; they turn out to be the 2, 3, 4, 5, and 6 of Clubs. "Oh, dear—this is rather embarrassing. Isn't that a straight flush?"

Your humble low-valued cards, taken together, do indeed surpass each of the other six poker hands on display. Victory is yours.

How it works: The cards you thumb off or count out (the latter action reversing their order) at the start form a stack that you have carefully constructed in advance and planted at the top of the deck. Twenty cards are used, in four interwoven sets of five.

List A: 2♣, 3♣, 4♣, 5♣, 6♣;
List B: 7♣, 7♠, Q♣, Q♥, Q♠;
List C: A♣, A♥, A♠, A♦, K♥;
List D: 7♦, 8♦, 9♦, 10♦, J♦.

The precise order within each list is not important. Note that the A list has the cards you want, all Clubs, with 6 being the highest value. The C list is the the poker hand destined for the poker novice, the lowest card there being a King. These hands are illustrated in Figure 8.4.

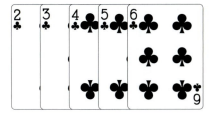 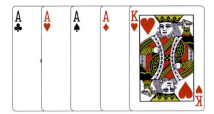

Figure 8.4. Lists A and C: inevitable winners.

The other two lists consist of cards of "middling" value, between 7 and Queen, as illustrated in Figure 8.5.

Figure 8.5. Lists B and D: doomed to be split up.

Arrange these cards cyclically as ABCDABCDABCDABCDABCD, by which we mean: start with any card from A, then any card from B, then one from C, and then one from D, then repeat in that pattern four more times. It doesn't really matter if you cut off or count out these cards from the deck at the outset. Also, they may be cut any number of times, for instance, just before or after you hand them to the poker novice. She cuts somewhere near the middle, and rosette shuffles the two piles she gets together, under your guidance.

Everything learned for packets of four cycling suits can be adapted to work here; one just has to make some mental adjustments. When you scan the card faces, you are really looking for an appropriate place to cut, so that when you reassemble the two parts the other way, the packet is then restored to the desired state. Simply cut it between two adjacent cards neither of which is on lists A or C.[11] If no such cut is possible, then none is necessary.

Unsuspected by all, the result is five stacked quadlets with a special property we discuss below. Ideally, you'd like to dole these out four at a time; however, that's not how poker hands are dealt. First deal left to right into four piles of five, and then under the pretext of having made a mistake, collect those piles in the same order and recombine. Dealing once more from left to right into five piles of four accomplishes your original goal, while seemingly following normal left-to-right dealing conventions.

After the restoration adjustment suggested above, every quadlet from the top down is balanced: it definitely contains one card from list A, one card from list C, and two from the other two lists, B or D. Hence, nonperfect quadlets contain cards from exactly three of these lists, and if a quadlet has two cards from list B, then the next nonperfect quadlet will have two from list D, and so on.

The upshot is that each of the five poker fans now has a card from the C list, which is the four Aces and one King, and moreover, all other cards in play at this stage have lower value. More importantly, each of the poker fans also has something from the A list, and these Clubs are the lowest-valued cards in hand. As a result, the A list cards will be set aside face down near you, early on, seemingly of no consequence, and then the C list cards will be donated face down to the poker novice, none of these faces being revealed until the end.

That leaves each poker aficionado with two cards from the B and D lists. The best-case scenario is that some of them have a pair of 7s or a pair of Queens. Once they are let loose with the rest of the deck, they'll have fun building what they think are strong hands, but it's all in vain. The game has been rigged so that no straight flushes are possible. You will be victorious!

Well, not quite: if one of the poker fans ends up with a certain straight flush in Diamonds, you're sunk. One way around this is to remove 6♦ from the theater of operation at the outset.

Source: Original. Conceived November 2012 in a surgery waiting room. Stack refinement input from mathematician Neil Calkin followed in De-

---

[11] Thanks to De Morgan's laws, this is equivalent to the clumsier "both of which are from lists B or D (considered together)"

cember. Published online as the February 2013 *Card Colm* "Flushed with Embarrassment" [Mulcahy 13_02a].

Presentational options: The bogus and somewhat contrived double dealing can be avoided, as John Hostler points out, by simply handing out blocks of four cards to the five poker fans after the shuffle and cut (by the novice) and your own correction cut.

This effect can also be performed as a Gilbreath application; in that case, the novice deals about half of the twenty-card packet before rosette shuffling. There is extra certainty about the outcome: each of the distributions of four-card hands is sure to contain one card from each of the four lists. This may allow for more fun, as there is much flexibility in how lists B and D are chosen.

A Bonnie Bunch of Roses

Is there any way to shuffle three or more piles of prearranged cards and expect some order to remain? Certainly, if the three piles are perfectly interspliced with surgical precision. That's the triple Faro (or weave) shuffle, whose mathematical properties are documented in the wonderful book *Magic Tricks, Card Shuffling and Dynamic Memories* by mathematician S. Brent Morris [Morris 98]. That author also provides suggestions for how to master the difficult physical skill of actually doing it, which readers with only two hands may find mind-boggling.[12]

The idea of riffle shuffling multiple piles is not new. Many have considered riffling several piles two at a time (see "The Double Riffle Shuffle" [Fulves 68, pages 44–46]). For instance, given three piles, A, B, and C, we can shuffle A and B together to form a new pile D, and then shuffle C and D. If the original three piles were the same size, then the second shuffle here would have some physical limitations. It seems more symmetric to shuffle four piles, as we can first do them in pairs, and then shuffle the results together. We may even worry about whether it makes a difference how we do the initial pairing up.

Rosette shuffling three or more piles is certainly faster than doing several riffles—all you have to do is twirl each pile first and then mush them all together—and in many ways it seems more equitable. The case of three piles is is illustrated in Figure 8.6.

The Solitaire Principle from Chapter 6 can be extended.

---

[12]Indeed, Fulves and others had earlier dismissed this as even being a possibility in practice.

Figure 8.6. Rosette shuffling three piles.

General Solitaire Principle

Suppose k piles of face-up cards running 1–n, representing k "suits," are rosette shuffled together. If the cards are now dealt one at a time to form k face-up piles separated by suit, then the original k piles will be reformed.

Indeed, this observation essentially characterizes the internal coherence of rosette shuffling, and shows the equivalence of multiple rosette shuffles to multiple riffle shuffles of pairs of piles, no matter in what order the piles are paired off.

In the case of four suits running Ace–King, this just says that you can deal—into four face-up piles, one pile for each suit—the result of rosette shuffling those four packets and you will encounter the cards of a given suit in order. You will start with some Aces, then possibly a 2 or two before dealing another Ace. When you encounter the 3♠, the 2♠ is already face up on the table, even though the A♦ may not have been dealt yet. More generally, if working with five or more over-sized suits, when the time comes to deal the Seventeen of Strawberries, the Sixteen of Strawberries will already be there on top of the appropriate pile, just waiting for this moment.

It has long been our belief that there must be an extension to three or more piles of the way to illustrate the Basic Gilbreath Principle (for an alternating deck) by rosette shuffling two piles of cards. For instance, surely there is a structure one can impose on a packet of cards so that if it is somehow converted to three piles, and those piles are rosette shuffled, something of note can be said about the result.

There is, of course, the all-important question of how the three piles might be derived from the starting packet. We surmised that perhaps one could deal out about a third of the cards into one pile, hence reversing their order, and then ... what? We got confused, in part because dealing out a second pile results in two reversed piles and one pile in the original order. The lack of symmetry was bothersome.

There was also the question of what structure the starting packet should have. Perhaps ternary? We tried cycling Clubs, Hearts, and Spades, and then dealing about two-thirds of the cards into two piles as above, and rosette shuffling those with the remainder. Although we frequently got a packet in which many of the triplets pulled off the top were perfect (namely, had one card of each suit), the pattern of the other triplets wasn't obvious.

We had this nagging feeling that three wasn't the right cycle length to consider, for the following reason: what if two of the three piles we pushed together were much bigger than the third? Then, in practice, the first cards in the shuffled packet would have come from only two of the piles, seeming to suggest that we were back to binary considerations. Perhaps cycles of length six were called for? Or maybe the way to go was simply to cut the packet into three piles and rosette those.

We went back to binary arrangements and also failed to get anything of note. With three piles being combined, it's certainly possible to end up with three cards of the same color together. We tried rosette shuffling three piles that each alternate Black/Red and all have the same color on top. We did it again for different top color combinations. Was it the case that every sexlet pulled off the top would be color balanced, containing three Blacks and three Reds?

We gave up, and wrote to Ron Graham in desperation. At first he didn't see what was going on either, but in August 2012, he shared the following.

> Arrange a packet in cyclic order $0, 1, 2, 0, 1, 2, \ldots$. Now split into three stacks that have as their top cards, 0, 0, and 1. Rosette shuffle them together (so that in the final deck, the relative order of the three stacks is preserved). Start removing sets of three consecutive cards at a time. Then none of these three cards sets will have all values the same.

Graham went on to note that (1) in the case of two stacks rosette shuffled together, the corresponding result (where the two stacks have 0 and 1 on top) just says that each removed pair will have one 0 and one 1, and (2) the three-stack result also holds if the top cards are 0, 1, and 1 (or any set that has exactly two different values).

What's really being asserted here is that three cards of the same type may end up together, one coming from each stack, but they won't all be in any one of the triplets 1–3, 4–6, 7–10, and so on. In due course, Graham pointed out that these are all special cases of the following general result for s stacks.

Non-Alliance Principle

Start with a packet arranged cyclically as $0, 1, \ldots, s-1, 0, 1, \ldots,$ $s - 1$, and then break it into s stacks. Rosette shuffle those together, and then remove sets of s cards at a time from the top. If the top card of the ith stack is a_i, and the sum of the a_i is not $0 \pmod{s}$, then none of the sets of s cards are all of the same value.

For $s = 2$, this is equivalent to something we have known for a long time. It just says that after breaking an alternating packet into two piles whose top cards don't match and rosette shuffling, then the cards in positions 1–2 won't match either, and likewise neither will those in positions 3–4, 5–6, and so on. This is just the Basic Gilbreath Principle once more!

What is it really saying for three piles? As essentially stated above, starting with a packet that cycles Clubs, Hearts, and Spades and splitting it into three piles that have as their top cards Clubs, Clubs, and Hearts, then no triplet from the rosette-shuffled packet will consist of three cards of the same suit. The same conclusion also holds whenever the top cards of the three piles account for exactly two of the three suits in play. It's not promising anything when the top cards of the three piles are all the same, or account for one of each suit. An interesting question is whether these cases can be adjusted, to bring them into line. Is there a restoration principle here, as we've seen before in similar situations?

It's also saying nothing at all of note in the case of rosette shuffling just two piles (think of the third one as being infinitesimally small!). It's hardly any surprise that we won't end up with three cards of the same suit in a row, given that we're merging only two piles, in each of which the suits were all separated.

For the sake of completeness, let's look at the case $s = 4$. Naturally, our starting point is a packet in CHaSeD order. Let's split it into four piles, with top cards all of different suits, noting that in this case the sum of the four a_i values is $0 + 1 + 2 + 3 = 2 \pmod{4}$. Hence, when we rosette these piles together, we can be sure that none of the quadlets pulled off the top will have four cards of the same suit. Perhaps there is more structure to be found in the shuffled packets that can arise.

Hammock Arguments

Now, imagine three (or four) packets side by side on a table, organized by suits, each one running Ace–King from the top. Instead of a ladder of rungs, think of a generalized ladder supporting a tier of hammocks.

Rosette these packets together, and remove triplets (or quadlets). If the sixth set matches in value, there is a simple explanation. They are all Sixes! Here is the appropriate generalization of The Intact Rung Principle.

> ### Intact Hammock Principle
> *Suppose k piles consisting of cards numbered 1–n, from the top down, are rosette shuffled together, and cards are then peeled off k at a time. If the jth set of k cards match across the board, in value, then they all have value j.*

Karl Fulves also has this in his book *Riffle Shuffle Set-Ups* [Fulves 68, page 45].[13] The proof we gave for the case $k = 2$ extends easily.

Is there any entertainment to be derived from all of this? Our final effect is the best we have to offer.

8♦ Wholesome Threesomes

How it looks overall: *Produce two decks of cards, a red-backed one that you keep for yourself, and a blue-backed one that you offer to a spectator. Announce that you first wish to sync the two decks by extracting the diamonds from each. "Perhaps you can assist me. Scan the card faces of your deck, and remove all of the diamonds. We won't need those; just set them aside face up on the table here, please. I'll do the same with my deck."*

Look at the faces of your deck and pull out the cards in question, leaving them in an untidy face-up pile. The spectator does likewise. The Diamonds of each packet emerge in a different order.

"Phew—that's the hard part over with," you comment. "It always makes me nervous." Now have the two decks compared; it will be seen that the remaining cards of each are in exactly the same order. "Good: it looks like they are indeed perfectly in sync."

Now demonstrate the breaking up of your deck into three face-down piles, and have the spectator mimic your actions. Each of you then rosette shuffles your piles. You bet that if each of you now puts your packet of thirty-nine cards behind your back, and brings forth three cards at a time, there will be a significant difference. You will be able to produce three

---

cards of the predictably same suit much more often than the spectator will.

How it looks in detail: A key point not mentioned above is that when each of you brings forth your triplets, you ask each other in turn, "Are these cards all of the same suit?" When you are asking the spectator, and he answers "Yes" for a triplet of your choosing, you have an advantage when it comes to identifying that suit; by the Intact Hammock Principle you know what the common suit must be. However, when you are being asked by the spectator about three cards of his choosing, your answers of "Yes" (in the cases of a triple suit match) do him no good at all.

How it works: Before meeting your audience, arrange the Clubs, Hearts, and Spades of each deck cycling in exactly the same way, paying close attention to card values. Then go back and insert the thirteen missing Diamonds randomly in each deck. Replace the decks in their cases, ready for showtime.

The opening patter and suggested actions are just window-dressing— it's all about how the two packets of thirty-nine cards are broken up into three piles. You are careful to ensure that the top cards of your three piles represent exactly two of the three suits in use. The spectator's piles, determined randomly, may or may not enjoy this property.

Once your rosette-shuffled packet is behind your back, all you need to do each time is extract triplets from the top or bottom. We suggest alternating from which end you take these cards, so that there is less risk that those produced by the spectator—possibly at random but perhaps also from one particular end of the packet—bear too strong a resemblance to yours.

Source: Original. December 2012.

Presentational options: You may not want to draw attention to the fact that the decks started out in the same order, apart from the Diamonds.

Parting Thoughts

- Prove that the recent (Ron Graham) Non-Alliance Principle holds.

- Is there any situation in which a restoration principle akin to the one on page 132 holds when rosette shuffling multiple piles?

- Prove that the old (Karl Fulves) Intact Hammock Principle holds.

- How many intact rungs (or hammocks) can we expect, on average, when rosetting two or more identically valued piles of size n?

- We've considered various extensions of the Basic Gilbreath Principle, including the case of riffling or rosetting more than two piles of cards. Another option that has potential is riffle or rosette shuffling more than once. Is it possible to double riffle shuffle[14] a cyclic packet of cards and still see some semblance of order in the output?

Surely the simplest case is to start with an alternating Red/Black pattern. There are two riffle (or rosette) shuffles to be done. We must decide whether to deal out cards to form one of the piles to be shuffled, or simply cut the packet in two. If we opt for the latter, presumably we should make the usual adjustment afterwards, namely, cutting between cards of the same color and then reassembling the packet the other way to restore it to a desired state before going on.

We need to make similar decisions (or perhaps the same ones?) about how to break the shuffled packet into two before we do the second shuffle. No matter which decisions are made, we have something of great interest after two shuffles. Three (and maybe four) cards of the same color can be adjacent here. Are there adjustments that should be made, so as to restore the packet to one with more desirable properties?

In many cases, we seem to get a packet that has a high proportion of "perfect pairs"—that is to say, the cards in positions 1–2, 3–4, and so on, are often of opposite color. But the pattern is by no means perfect.

Quadlets—that is to say the cards in positions 1–4, 5–8, and so on—also seem to be color balanced (with two of each) a lot of the time, but it certainly can't be guaranteed.

Maybe some structure—even simple matters of color distribution—could be observed in the results if more structure was imposed in the initial packet. What if we start with the cards in CHaSeD order, perhaps also adding a length 8-cycle aspect, such as cycling CHaSeD (even), CHaSeD (odd), over and over? This might facilitate better tracking, for instance, when deciding where to cut to make adjustments after one or both shuffles.

---

[14]Driffle shuffle?

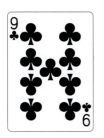

Flipping Miracles

In our chapter closer, "Any Mathemagician" (9♦), we finally get some joy out of a kind of double-dealing with a largely unprepared deck.

> *Shuffle the deck over and over, and then place it on the table. Have a spectator take off roughly the top quarter, then name a mathemagician and a descriptive word. The spectator uses the mathemagician's name to count out and reverse some of the cards in his packet, and the word to transfer some cards from top to bottom. This is repeated. Finally, the spectator presses down hard on the top card as you request that he turn it into a specific card, say the Jack of Diamonds.*
>
> *Despite the fact that you never touched the cards after the spectator selected his packet, and you had no control over the choices of the two words used to mix up the cards, the top card turns out to be the one you requested.*

On the way there, we combine various concepts from earlier in these pages with older ideas, some of which date back to the 1940s.

Bob Hummer's CATO

One of the two or three factors that led to the Low-Down Triple and Quadruple Dealing (overCOATing) principles of Chapter A was playing around with a Bob Hummer discovery that he published in 1946. Most of this mathematically-minded magician's amazing output was assembled by magic inventor and publisher Karl Fulves in *Bob Hummer's Collected Secrets* [Fulves 81]. In the introduction to that book, Martin Gardner writes, "I believe that if Hummer had obtained an education in mathematics he might have become a great mathematician or physicist," before going on to explain why.

The Hummer move we now discuss is known as CATO[1] for **C**ut **A**nd **T**urn **O**ver. According to Max Maven, this acronym was introduced by magician Charles M. Hudson in 1979 to represent "Cutting And Turning Over." It obviously inspired our COAT designation for the reversed transferring move of Chapter A.[2]

CATO refers to cutting any number of cards from the top to bottom of a packet, without altering their order, alternating with turning exactly two cards over together, namely, flipping as a unit the two cards currently at the top of the packet.[3] It's crucial to note that the first letter of this acronym stands for Cut, not Count—the latter would reverse the card order, as COATing does.

A quick in-hand way to effect the cutting is to use one thumb to push off any number of cards as a unit into the other hand, without changing their order and without needing to consciously count them, before tucking them behind the remainder in the first hand.

For instance, if we start with any Ace–10, some face up, some face down, and cut three cards to the bottom, turn the top two over, cut four cards, turn over the top two, cut five, and turn over the top two (one of which had already been turned over earlier), then we end up with $5, -3, -4, 6, 7, -9, -8, 10$, Ace, 2, the negative signs denoting cards that now face opposite to the way that they originally did.

Had the packet been face down to begin with, then Figure 9.1 shows what it would look like before and after this sequence of CATOs, if it were flipped over and fanned left to right. As a result, the second image actually shows the faces of the face-down cards in the packet; the others are really the face-up cards, but their faces cannot be seen here. In the order displayed, they are the 3♠, 4♥, 9♠, and 8♥.

The cutting and turning over actions can be performed—by you or a spectator—while the packet is held out of sight, hence apparently randomizing it. It's not immediately clear if any feature of the outcome is predictable, and presumably things are even more chaotic if the starting packet mixes face-up and face-down cards. In Figure 9.1 there appear to be even-sized clusters of cards facing the same way, but that is just

[1]It's pronounced *KATE-oh*.

[2]We do not recommend using CATO for Counting out (individually) And Turning Over (those, as a unit), before replacing them on top; apart from anything else, this move puts each card back where it started, with the top ones flipped over in position.

[3]We're tempted to render it as CAAATTOT to emphasize that it means "Cut Any Amount And Turn Two Over Together." Indeed, Max Maven notes that back in 1974, Hudson had referred to it in print as CATTO for "Cut And Turn Two Over" to emphasize how many cards were flipped each time.

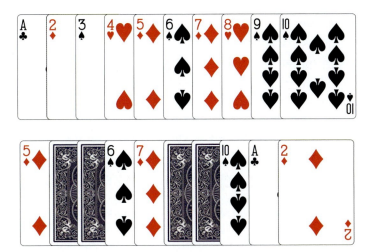

Figure 9.1. A face-down Ace–10 fanned "face up" before and after CATOing.

a coincidence. It might be tempting to conclude that CATO does not preserve any structure.

However, Hummer's 1946 classic "The Lonely Card" (see [Fulves 81, page 7] or [Simon 64, page 167]), which we now sketch, suggests otherwise. It exploits a hidden parity preservation. In a nutshell, when the parity is no more, having ceased to be, you're able to isolate a selected card.

A spectator shuffles a deck and hands you about a third of it under the table. You claim to be mixing the cards, then hand them back. The spectator subjects the packet to some randomly repeated actions (e.g., cutting, flipping pairs of cards), perhaps also with the cards out of sight. She then notes the identity of the new top card, flips it over relative to how she found it, and replaces it on the top before cutting the packet once more. You take the cards back, do more under-the-table manipulations, and bring them forward to reveal that they all face one way—except for the spectator's card.

How is this possible? The first secret is that you must work with an even number of cards; we suggest between twelve and twenty. The second is that when you are handed the cards, you pass them from hand to hand out of view and flip over those in the even positions $(2, 4, \ldots)$. If you've been given an odd number of cards, you'll soon discover this, in which case simply drop the last one on the floor. In this way, the cards handed back to the spectator consist of pairs of face-down and face-up cards, a condition that is not altered at all by the subsequent cutting and flipping. The third secret is that when you get the packet back, you once again flip alternate cards, out of view, which orients the packet all one way save for the spectator's card, which is thus unmasked without fail. This all works because of the following observations of Hummer.

Hummer's CATO Principle

Start with an even-sized packet of 2k cards.

1. *If the packet alternates face-up and face-down cards, or vice versa, then that kind of alternating structure is preserved after any number of CATOs, namely, random cuts mixed with the flipping over of the top two cards as a unit. (Actually, if an odd number of cards is cut overall, then the cards end up alternating opposite to the way they started.)*

2. *If the cards all face the same way, then after any number of CATOs, the odd $(1, 3, \ldots, 2k-1)$ and even $(2, 4, \ldots, 2k)$ positions contain the same number of face-up cards.*

It's worth inspecting Figure 9.1—and the information given on the identities of the card faces that are not visible—in light of the second observation. Some structure can indeed be found, once we know where to look.

Returning to Hummer's "The Lonely Card," note that when the spectator above notes the top card after all of the CATOs, flips it over, and cuts it into the packet, this causes the chosen card to be "out of step" relative to the rest.

We hope that we have whetted your appetite on this topic, because there is so much more to explore. Martin Gardner introduced Bob Hummer's magic to the wider public in his landmark *Mathematics, Magic and Mystery* [Gardner 56, page 17], which has a version of the above effect along with many other parity-based Hummer classics. The February 2006 *Card Colm* "Many Fold Synergies" [Mulcahy 06_02] has more of those, and also some recent twists on Hummer's ideas. Over the decades, Gardner presented many fresh and entertaining versions, sometimes posed as puzzles (e.g., about water and wine mixing, or coins). See the June and August 2010 *Card Colm*s for some of Martin's own writing on the subject [Mulcahy 10_06, Mulcahy 10_08], or the book chapter "Mathemagical Miracles" [Mulcahy 11] or the *Huffington Post* blog "Flipping Miracles (or Bar Bets to Amaze Your Friends)" [Mulcahy 12_12a].

The definitive modern reference on Hummer's CATO is the Euler prize-winning book *Magical Mathematics* [Diaconis and Graham 11], whose first chapter includes proofs of the two complementary Hummer parts of the principle that we have highlighted above, along with some very clever recent incarnations of them. There, it is also mentioned that—*ignoring card orientations*—any possible arrangement of $2k$ cards is attainable from any starting point if appropriate CATOs are done.[4] *Bob Hummer's Collected Secrets* [Fulves 81] is also highly recommended.

---

[4]Ignoring card orientations, the flips are transpositions of the top two cards, and since cuts are also thrown in, all flips of initially adjacent cards are certainly attainable. Furthermore, all possible arrangements can be attained using only transpositions, as mentioned on page 18.

George Sands' "Lucky 13" and Extensions

Another strand in the discovery of the Low-Down Triple and Quadruple Dealing principles was playing around with a principle due to magician George Sands. In August 1975, a Sands creation called "Lucky 13" was published in *The Pallbearers Review* [Fulves 75], and Martin Gardner offered his own take on it under the title "Negative Psi" in *Martin Gardner Presents* [Gardner 93, page 140], noting that "Tokyo amateur magician Mitsunoba Matsuyama independently discovered the same principle."

Let's start with the mathematics that makes these effects possible. With hindsight, it's really based on the following fact about reduced residue systems from elementary number theory [Rosen 00, page 216].

> ### Reduced Residues Modulo a Prime
> *Let n be prime. If k and a are both between 1 and $n-1$, and hence have no factor (other than 1) in common with n, then ka has no factors in common with n, and in particular can't be a multiple of n.*

As originally applied to packets of cards, this was only used in the special case where $n = 13$.

Fixing k, and taking $a = 1, 2, 3, \ldots, (n-1)$ in turn, we deduce that none of the numbers $k, 2k, 3k, \ldots, (n-1)k$—some of which are bigger than n—have any interesting factors in common with n. It follows that $k, 2k, 3k, \ldots, (n-1)k$ are all different mod n, and so they all have different remainders when divided by n (see page 16), giving back the numbers from 1 to $n-1$ again in some order. To see why that holds, suppose $s < t$ are two numbers strictly between 1 and n with $sk = tk \mod n$. Then $(t-s)k = 0 \mod n$, i.e., n divides $(t-s)k$, which is impossible because n is prime and both $t-s$ and k are strictly between 1 and n [Rosen 00, page 97].

For example, if $n = 13$ and $k = 4$, then none of 4, 8, 12, 16, ..., 48 is a multiple of 13, which isn't so surprising. Hence, none of 5, 9, 13, 17, ..., 49 is 1 more than a multiple of 13. Moreover, mod 13, the list 4, 8, 12, 16, ..., 48 comes out to be 4, 8, 12, 3, 7, 11, 2, 6, 10, 1, 5, 9, which is indeed 1–12 reordered. Those sequences show up—in different contexts—in the next example.

Effectively, what Sands took advantage of in "Lucky 13" was how this impacts a packet of thirteen face-down cards if a selected card is controlled to the bottom and a random number, such as four, is used as follows. Turn the fourth card face up—of course, it's not the selected one—and then cut the top four cards to the bottom, thereby putting the flipped card at the bottom. Turn the new fourth card face up, and then cut the new top four cards to the bottom. Repeat, over and over. Following along with any Ace–King, we find that the 4, 8, Queen, 3, 7, Jack, 2, 6, 10, Ace, 5, and 9 are flipped face up, in that order, before the King—representing the chosen card—finally suffers a similar fate. The punchline is that the original bottom card is the last one to show its face, and that holds if four above is replaced by any other number between one and twelve.

Now forget the flipping above, and focus on what cards show up on top. Starting with the Ace, the top card in turn becomes 5, 9, King, 4 (the remainder when 17 is divided by 13), 8 (the remainder when 21 is divided by 13), and so on. After the Queen, 3, 7, Jack, 2, 6, and 10 have also come to the top, in that order, the Ace (1 being the remainder when 53 is divided by 13) finally does so. All told, the original top card returns to the top for the first time after thirteen cuts of four cards, and every other card in the packet has visited the top once in the interim. If we have the continually-changing top card flipped over each time in between the cuts, we obtain another interesting way to track progress. There are two natural options to consider, depending on which is done first, the flipping or the cutting. Assume, for now, that we start with a face-up packet Ace–King.

Option 1: If we start by turning the Ace face down, then we will in turn flip the 5, 9, King, ..., and finally the 10, in sequence, having not yet encountered a face-down card on top after cutting. One final cut of four cards brings the Ace back to the top, with the result that the packet is restored to its original order, only with all cards now face down instead of face up.

Option 2: If we start by cutting the Ace (still face up) and the three cards beneath it, then flipping over the new top card (the 5), and continue as before, then after the 10 shows up on top and is flipped face down, only the Ace remains face up, in position 5 in the packet. (One more cut-four-and-flip produces the same result as the end of Option 1.)

As already mentioned, there is nothing special about the choice of four for the number of cards cut each time, any number k between 2 and 12 inclusive works for a packet of thirteen cards.

Again, we recommend cutting by using a thumb to push off exactly k cards as a unit into the other hand before tucking them behind the remainder. Then flip over the new top card. Repeat as desired. For a packet of size n, we tend to insist that $2 \leq k \leq n - 2$, as spectators often find it confusing cutting one (or all but one) card to the bottom in conjunction with turning over just the top card. What we saw above in the case $n = 13$ and $k = 4$ works in general.

Sands' Lucky Principle

Start with a packet of n cards, where n is prime. Fix some k between 1 and n, and cut k cards to the bottom, over and over. Different cards appear on the top, and after n cuts of k cards, the original top card returns to the top for the first time.[5] All told, each card in the packet has visited the top once.

While we've only stated this for the case where n is prime, everything works just as claimed provided that k and n share no interesting factors. When two numbers have no common factors larger than 1, we say they are *relatively prime*, or *coprime*. For instance, 6 and 11 are coprime, and so are 3 and 10. Furthermore, 4 and 25 are coprime, as are 63 and 8, but while 25 and 63 are also coprime, 4 and 8 are not.

A number of effects in magic inventors Peter Duffie and Robin Roberston's *Card Conspiracy Vol. 2* [Duffie and Robertson 03b, Chapter 18] make use of this extension of Sands' Lucky Principle.

Since 9 and 4 are coprime, then none of $4, 8, 12, 16, 20, 24, 28, 32$ is a multiple of 9, and so none of $5, 9, 13, 17, 21, 25, 29, 33$ is 1 more than a multiple of 9. Take a packet of nine cards, say, any face-up Ace–9, and cut four cards from top to bottom, over and over. The top card becomes $5, 9, 4, 8, 3, 7, 2, 6$ in turn, before the Ace resurfaces for the first time.

Two implementations of Sands' Lucky Principle are worth exploring. In one, a spectator is handed a shuffled packet containing a prime number of cards, and you request that a number k be called out. Pick up a different packet, and demonstrate some counting, moving, and turning over of cards, all determined by k. Once the spectator understands the motions, have her perform the same actions on the packet given to her earlier. At the conclusion, only one card remains face-down, and it matches what is written on a prediction sheet that has been lying in full view all along.

A packet of $n = 11$ (or 13 or 17) cards is used. It is genuinely shuffled, and the bottom card is peeked at as the cards are handed to the spectator. Have the packet dealt out and counted, which, of course, puts the known card at the top. While this is happening, write down the prediction. Inquire how many cards there are, then request that a number k strictly between 1 and $n-1$ be called out. Of course k and n share no factors, and so the smallest number a for which ak is a multiple of n is $a = n$. The usual sequence of moves ensures that the initial top card is always the last one remaining face-down, after exactly $n-1$ rounds of cutting k cards and flipping the top card. It's the one you peeked at before and then wrote the name of as a prediction.

---

[5]At this stage the whole packet has of course been restored to its original state.

You can't afford to have the spectator go over the limit: one too many cuts puts the *force* card (see page 2) at the top, and if that card is then flipped, the whole effect is spoiled. To avoid this possibility, we suggest that after about half of the cutting and flipping is completed, you invite the spectator to fan through the cards; you point out, "More and more cards are now face up. Keep going until just one is left face down." This way, hopefully, she will fan the cards often and stop at the right point. Note that the force card here ends up at position $k + 1$ in the packet, something that can be taken advantage of if you use your imagination.

Another implementation sees us start with a face-up packet of prime size. Then the last remaining face-up card, after all the transfers and flips are done, can be predicted. However, in that case, this card starts face-up on top at the outset, for all to see. We prefer an approach that doesn't draw attention to card faces until the grand finale.

9♣ I'll Be Lucking Out for You

How it looks: *The spectator shuffles the deck. You then scatter several face-down piles about the table. Have the spectator select one, and ask her to call out a number k "between one and six" (but don't accept one or six). Using one of the other piles, illustrate what you want her to do to her selected pile in due course.*

Start by shuffling your pile, turning it face up, showing around the card face now on the top, and then flipping that card face down again, noting that it's your "lucky card." Next, demonstrate the cutting of k cards and the turning over of the new top card. Point out that more and more cards get turned face down as a result. Stop after the third or fourth turning over.

Once the spectator understands the procedure, turn your back and ask that the same actions be performed on her own freely chosen pile, being careful to show her lucky card to the rest of the audience at the outset before turning it face down. Indicate that after five or six top cards have been turned face down, the packet should be fanned now and then, to ensure that the process is stopped as soon as the last card is turned face down. Finally, request that this pile be dealt face down from left to right alternately to form two smaller piles, one of those being placed on top of the other.

Finally, turn around to face everybody again. Have the cards spread face down on the table. Pick up the pile you used earlier for the demonstration, fan the cards, and locate your lucky card, saying, "Here it is." Hover it over the spectator's spread until you get good vibrations. Ask what the spectator's lucky card was, and say, "I wonder if my lucky card has by any chance found your lucky card?" Use your lucky card to flip over the spectator's card, proving that this is indeed the case.

How it works: The first point is that the piles you scatter should contain seven, eleven, or thirteen cards each; we recommend a mixture so it looks haphazard and unplanned. These are easy to assemble by dropping clumps of three or four cards at a time. No matter what number k is called out, it will share no factors with the selected pile size n, for some possibly unknown n, since the latter is, by design, a prime number.

At the conclusion of the spectator going through the cutting and flipping moves with her selected pile, all of the cards will be face-down because $n - 1$ rounds of cutting and turning over the top card have been done, bearing in mind that her lucky card was turned over at the outset. Hence, her lucky card is in position k, ready for a final trip to the top that it is doomed never to make. The purpose of the left-to-right dealing into smaller face-down piles at the end is to hide this last fact. Once the spectator does that final dealing, it's easy to figure out where her lucky card is, even though you are turned away and you probably don't know what n is. It all depends on whether k is odd or even.

For instance, if k is odd, say $k = 5$, then since $\frac{k}{2}$ rounds up to 3, her lucky card is third from the bottom in the left pile dealt. Have the right pile placed on top of the left pile, so that the spectator's card is still third from the bottom. Your lucky card will easily find hers after you turn around once more, and the cards are spread. On the other hand, if k is even, say $k = 4$, then since $\frac{k}{2} = 2$, her lucky it will be in position 2 in the right pile dealt. This time, have the right pile added to the left one; once again, your lucky card can quickly locate the spectator's one.

Why it works: This has been cooked up as an application of Sands' Lucky Principle.

Source: Original. February 2006.

The next effect requires some fast thinking on your feet, and appropriate adjustments must be made to the way in which you then proceed. Basically, you need to arrange it so that two packets, each consisting of about a dozen cards, have sizes that are each coprime to a third smaller number, which is called out by the spectator early on in the game. This one is repeatable if you can convincingly gloss over the slight differences in procedure that may be necessary on subsequent performances. It makes use of some old standard ideas, such as forced twins and double discoveries.

9♥   Coprime Twins

How it looks: *A spectator shuffles the deck thoroughly. Take it back from her, ostensibly to remove jokers and seek your "lucky card"—setting the latter on the table face down—then cut and place the deck face down on the table. Say, "I'd like somebody to shout out a magic number between one and ten, for later use." The deck is then cut into quarters (approximately), and you each take one of these packets, the other two being discarded, as they are no longer needed. Each of you counts your packet and announces the result. Switch packets, and hold them face down. You each use the magic number called out earlier to transfer that many cards from top to bottom of your respective packets, one at a time, then you each turn over your new top card. Repeat, until each of you is left with just one face-down card in your packet. Say, "We each have one special card hiding its face in our random packets of different sizes. Those cards were determined by the lucky number called out earlier. Wouldn't it be interesting if our cards were somehow related?" Wave your lucky card over the two face-down cards and turn them over simultaneously to reveal that they are twins: the same color and value.*

How it works: After receiving the shuffled deck back from the spectator, fan through the faces, secretly looking for adjacent twin cards. We claim this is possible with high probability (see "Parting Thoughts" later in this chapter). It doesn't hurt to set up several adjacent twins ahead of time; it's likely that some will remain together despite the shuffling. If you find such a pair, then cut the deck so that one is on top and the other is on the bottom, and toss some other card as an aside, face down, claiming it's your lucky card. If you don't find adjacent twins, try to locate twins that are separated by only one card, tossing out the card in between as your lucky card. You can cover all of that suspicious fumbling by muttering, "I nearly forgot, I need to make sure there are no jokers here."

Request that a number k strictly between one and ten be called out. Have the deck cut in half, and take the bottom half for yourself. Have the spectator cut the top half in half again, retaining the top portion (the original top quarter of the deck). He should have eleven to thirteen cards. Cut the bottom half in half yourself, and hold on to the bottom quarter, hopefully eleven or twelve cards.

At this point one of the twins is at the top of the spectator's packet, and the other is at the bottom of yours. You want them at the top of both packets. Ask the spectator to count his packet out loud, and watch to see if he reverses the card order by so doing. If he does, count your own, also out loud, but without changing the card order, and then switch packets before both of you count into piles "to check those counts." If the spectator doesn't alter the order of his cards, calmly count yours to a pile on the table, hence getting the bottom card on top, and then switch packets.

Before proceeding, it is essential that the number of cards in each packet be different, and each coprime to k. For instance, if $k = 5$ or 7, then packets of size eleven, twelve, or thirteen are fine, but if $k = 2, 3, 4, 6, 8,$ or 9, we must avoid packets of size twelve. Ideally, one of you has eleven cards and the other has thirteen—you do not need to announce that, but you can, if you wish. Emphasize the importance of having packets of different sizes. The eleven-thirteen setup, if it does not occur naturally, can be achieved as follows. Note that the suggested adjustments can be made before or after the packet switch; do them whenever it is the case that the spectator's packet is no larger than yours. If you both have twelve cards, say, "This will be more interesting if we have different numbers of cards," as you pluck one from the middle of the spectator's packet, and stuff it deep into yours. If the spectator has twelve cards, and you have thirteen, ask him to reach into his packet and take out a card that will serve as his lucky card. This card is set to one side on the table. If the spectator has eleven cards, and you have twelve, have him (1) pick a random card from the unused half-deck and (2) insert it into your packet, "for good luck." If both packets contain eleven cards, have two additional cards from the unused deck inserted into your packet. Finally, if both packets contain thirteen cards, simply have the spectator ditch two from the middle of his pile (using the "unequal sized packets" justification, as necessary).

Recap the number k selected earlier, and show how to transfer k cards from top to bottom, one at a time, before turning over the new top card. Repeat a few times, and then have the spectator do the same with his packet. After five or six transfers and flips, fan your cards to show that more and more are now face up. Clarify that you must each continue until only one card is face down in each packet.

When this state is reached, each of you places your one remaining face-down card on the table. Wave your lucky card over them as you recap the arbitrary nature of the shuffling, cutting, and selection of a random number used for transferring cards. Suitably built up, you should reach the dramatic climax just as you turn both cards over to reveal that they are indeed twins.

Why it works: This too has been engineered to work thanks to Sands' Lucky Principle.

Source: Original, but hardly new. May 2003.

COATing and Flipping One

So far, we've explored cutting cards, in various quantities, alternating with flipping the top one or two cards. Alternately cutting an arbitrary number of cards and flipping the top two cards is Hummer's CATO move, the highlights of which were surveyed earlier in the case of even-sized packets. Alternately cutting a fixed number k of cards from a packet of size n and flipping the top card is Sands' move, which we explained in the cases where k and n are coprime.

What about COATing instead of cutting? We leave it to curious readers to investigate alternately COATing some number of cards (not necessarily the same number each time) and flipping the top two. (See "Parting Thoughts" at the end of this chapter.) Next, we focus on COATing alternating with flipping just the top card. The final effect in Chapter A, "Ace Combination" (A♦), was a tentative step in this direction, in the case of overCOATing. There, we found that only four cards ever showed up on top, the easily identifiable top visitors.

First we look at the specific case of COATing (variable) even numbers of cards in an odd-sized packet, alternating with flipping the top card. The kind of COATing in question was already explored on page 118.

> **COAT Any Even Number and Flip One Over Principle** ⋈
> *Start with a face-down packet of $2k + 1$ cards. Alternate the reverse transferring (COATing) of any even number of cards with the flipping over of the top one. Provided that the number of COATed cards is adjusted upward or downward as needed to avoid flipping face down a card previously flipped face up, then after exactly $k + 1$ rounds of this move, the cards in the odd-numbered positions are face up.*

This is a simple consequence of the First Odd Drop Principle (see page 118), since repeatedly COATing an even number of cards results in the cards that started in the odd-numbered positions moving around among those slots, and likewise for the cards that started in the even-numbered positions. The only top visitors are the $k + 1$ odd-positioned cards, and these are the ones that get flipped over, one by one.

Let's use this to produce a royal flush unexpectedly.

9♠   Royal Flush at the Eleventh Hour

How it looks overall: *Eleven face-down cards from a shuffled deck are handed to a spectator, who is requested to reverse transfer (COAT) any even number of cards to the bottom, repeatedly, alternating that move*

with the flipping over of the top card. As soon as it becomes impossible to continue without turning a face-up card face down again, the spectator stops. It transpires that five cards are still face down, and they constitute a royal flush.

How it looks in detail: The deck is shuffled, and eleven cards are dealt out face down to a pile, which is given to a spectator, the remainder of the deck being set aside. To start with, request that she count off two, four, or six cards into her other hand, thus reversing their order, before tucking them underneath the rest. Have her do this several times, using different even numbers each time, to mix up the cards.

"You're ready to flip, aren't you?" you ask. "That's understandable, merely transferring cards isn't that exciting. Let's spice it up a little." Request that the top card be flipped face up. Have the spectator do more counting and transferring as before, always reversing an even number of cards, but this time interspersed with the *optional* flipping of the newly exposed top card. Claim that it's considered unlucky to flip a face-up card face down again, so that if ever that seems inevitable, more reverse transferring of cards should be done as needed, to avoid that happening.

After six cards have been flipped face up, the spectator is stuck. No matter how many cards she attempts to count out and transfer—two, four, six, eight, or ten—it always leads to a card on top that is already face-up. Take the cards back and comment, "You know what this means? The cards still face down are beyond reach." Shuffle the packet in hand and turn it over, fanning the newly exposed card faces. "I guess there was a reason: those cards would give you a stunning poker hand." There are five faces on view, forming a royal flush.

How it works: Alternate five cards for a royal flush with six generic cards, so that the good cards are in the even positions in an eleven-card packet, such as those shown face up in Figure 9.2. Have this stock on top of the deck, so that some shuffling is possible without disturbing those eleven cards.

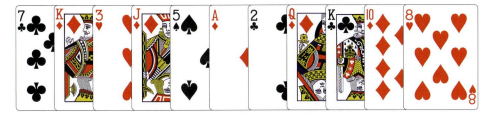

Figure 9.2. A royal flush hiding in broad daylight.

It is inevitable that no matter what choices the spectator makes regarding transferring even numbers of cards—sometimes alternating that with turning over the ever-changing top card—after six cards have been flipped, no more moves following your directions are possible. Five cards remain face down; they are, by the COAT Any Even Number and Flip One Over Principle, the royal flush. Your last-minute shuffle is merely a halfhearted attempt to hide the fact that the packet ended up alternating face-up and face-down cards.

Source: Original. December 2012.

Presentational options: Any cards can be forced in a similar way. For instance, you could work with thirteen cards instead, planting the key cards from "Little Fibs" (4♣) in the even-numbered positions. Then, as in that effect, you could ask for the sum of the values of any two of them before revealing what each card is.

What about COATing and flipping one card for even-sized packets? In the case of minimal underCOATing, it's time to put on your dancing shoes.

Riverdance—The Principle ▶◀

Start with a packet of size 2d. *Consider minimal underCOATing—thought of as reversing the order of the top two cards before moving them together to the bottom—alternating with flipping the top card. After exactly* 2d *rounds of this, the packet is once more in the original order, but with all of the cards now facing in the opposite direction (i.e., flipped). Doing this a further* 2d *times restores the packet to its original state.*

Momentarily turning a blind eye to the name of this principle, let's see why it holds. Suppose for the sake of concreteness that $2d = 12$. The packet 1–12 assumes the order $3, 4, 5, 6, \ldots, 11, 12, 2, 1$ after COATing two cards to the bottom. After the new top card is flipped, it's $-3, 4, 5, 6, 7, 8, 9, 10, 11, 12, 2, 1$. The next round of COATing and flipping transforms it to $-5, 6, 7, 8, 9, 10, 11, 12, 2, 1, 4, -3$, the round after that to $-7, 8, 9, 10, 11, 12, 2, 1, 4, -3, 6, -5$, and so on. First, five of the odd-numbered cards are flipped, then 2, resulting in $-2, 1, 4, -3, 6, -5, 8, -7, 10, -9, 12, -11$. It can be checked that if we continue to do six more rounds of COATing and flipping, then the other five even-numbered cards are flipped, before the original top card is too. After twelve rounds we have $-1, -2, -3, \ldots, -11, -12$. Clearly another twelve rounds restores the packet to its initial order.

Here's a whimsical application from May 2003 for lovers of structured dancing. Address your audience, "You've heard of *Riverdance—The Show*, the Irish music and dance spectacular? One of the more dramatic sequences of moves can be demonstrated with cards. It's not as tuneful or toe-tapping as Bill Whelan's Grammy-winning score, granted, but it's a lot cheaper to stage and there are no long queues for tickets." Hold a packet of cards at the ready.

"How's your dancing? Mine is terrible, but I want to show you, using these cards, how it can be done." Drop some face-up cards to the table, and draw attention to the top two, a King and a Queen.

Continue, "The original *Riverdance* dancing stars, Irish-Americans Michael Flatley and Jean Butler, were known for their extraordinary onstage chemistry, not to mention their elegant and fantastically chore-ographed footwork." Get ready to demonstrate with the cards.

"In one of the more popular routines, Michael would start at the front of a line of twelve forward-facing dancers in which the men alternated with the women, Jean being right behind him. He'd go first, stepping to the right, then she'd step to the right in front of him, and they'd proudly dance backwards to the rear of the line, still facing forward. The next couple at the head of the line would do it a little differently. The man would first turn around to face the woman, before stepping to the right, then she'd step to the right too, in front of him, and back to back they'd dance to the rear of the line, without turning around. The next couple would repeat that move, and so on."

Continue, "When the last couple had done this, Jean would now be at the front of the line, with Michael behind her, both of them still facing forward. Next, she'd turn around to face him and step aside, he'd step in front of her, and back to back they'd dance to the back of the line. From then on, each couple would consist of a forward-facing woman in front of a backward-facing man, and she'd also turn around and step to the right, he'd step in front of her, and they'd both dance to the rear of the line, their backs to the audience. When that phase was complete, Jean would be at the front of the line, her back to the audience, facing Michael who'd be right behind her. She'd turn around and step to the right, he'd step in front of her, and they'd dance to the back of the line as usual. All of the other dancers would have their backs to the audience, but by repeating the kinds of actions already outlined, eventually everyone would be facing forward, in their original positions, with Michael and Jean at the front again. Brilliant stuff! That's how I remember it. Did I mention my memory isn't what it used to be? I forget."

Demonstrate using the following cards to simulate the dance. Start with this face-up twelve-card packet: K♣ (representing Michael Flatley), Q♥ (representing Jean Butler), and then five other couples, 6♣, 6♥, 5♣, 5♥, 4♣, 4♥, 3♣, 3♥, 2♣, 2♥.

From FLOATS to COATs with a Difference

We move on to consider more sophisticated interactions between card reversing and cutting than those discussed earlier. These, in turn, suggest further avenues to explore, as well as new magical possibilities.

Start with a packet of n cards in the left hand. Fix $k > \frac{n}{2}$ and compute $n - k$ (it's less than $\frac{n}{2}$). Push off k cards into the right hand as a unit. In practice, it's easier to focus on the $n - k$ cards retained in the left hand, since that's a smaller number, perhaps sliding them off from underneath with the left fingers to achieve the desired packet split. Flip over the larger set of cards in the right hand as a unit and drop them *on top* of the cards in the left hand. Finally, cut $n - k$ cards from top to bottom, either as a unit or individually. We can't resist denoting this by FLOATS for **F**lip **L**ots **O**ver **A**nd **T**ransfer **S**ome. (Here, "lots" and "some" determine each other, between them accounting for the whole packet.)

To get a sense of what is going on, and perhaps spot some patterns, let's try some examples. First, let $n = 12$ and $k = 7$. Follow along with any Ace–Queen packet, initially all facing the same way, doing the moves yourself. Flipping over (as a unit) the top seven cards from a packet that runs 1–12, and replacing them on top, before cutting five of them from top to bottom, yields $\{-2, -1, 8, 9, 10, 11, 12, -7, -6, -5, -4, -3\}$, where the negative signs indicate cards that face opposite to the way they started.

Next, try doing it for $n = 13$ and $k = 10$. For a packet that runs 1–13, flipping over (as a unit) the top ten cards and replacing them on top, before cutting three of them from top to bottom, yields the rearranged packet $\{-7, -6, -5, -4, -3, -2, -1, 11, 12, 13, -10, -9, -8\}$.

From those two examples, we can already see a hint of the fact that the packets naturally break up into three subpackets, which can be thought of as lopsided tops, middles, and bottoms, each of which moves around intact, subject to occasional total subpacket reversals.

Let's keep going in the second example, and apply a second FLOATS. This transforms $\{-7, -6, -5, -4, -3, -2, -1, 11, 12, 13, -10, -9, -8\}$ to $\{1, 2, 3, 4, 5, 6, 7, -10, -9, -8, -13, -12, -11\}$. Note how the top seven cards are back on top again, and the bottom three are on the bottom in reverse order. A third FLOATS move transforms the last arrangement to $\{-7, -6, -5, -4, -3, -2, -1, -13, -12, -11, 8, 9, 10\}$.

Even better, it can be checked that a fourth application restores the entire packet to its original order, so that for these values of n and k, this move has period four. This is no fluke: under the same conditions as for the COATs of Chapter A—actually here we need to insist that k exceeds $\frac{n}{2}$—it always does. This suggests a link with overCOATing, the latter being viewed in the TOAFUH incarnation from page 162.[6] FLOATS and

---

[6]Recall that our TOAFUH offering originated as a better way to market FAT (**F**lipping **A**nd **T**ransferring).

COATs can't be the same thing in disguise, however, because it's easily checked that different top visitors are associated with each.

This period-four FLOATS move was discovered in the context above,[7] but from a magic perspective, there's the disadvantage that too many card faces are seen by the card handler. Bearing in mind that flipping chunks of cards as a unit was introduced in Chapter 7 as a low-impact alternative to dealing into piles, let's go back to dealing out to reverse card order, as a "less-revealing" alternative to flipping. We have come full circle, returning to that most essential ingredient of COATing: reversing card order while all of them remain face down.

Count out k cards (representing "lots") from a packet of size n held in the left hand, perhaps using the right thumb to pull off cards one by one to form a growing pile in the right hand, naturally reversing the card order. Next, replace them as a unit *on top* of those remaining in the left hand—we have to be very careful to break with habits ingrained since Chapter A. Finally, transfer (i.e., cut) $n - k$ cards (representing "some") from top to bottom. From now on, in this context, the post-reversal replacement step will always be understood when we have cards reversed or counted out.

It's different from overCOATing in the following way. For $n = 12$ and $k = 7$, the latter can be thought of as reversing the top seven cards and cutting them to the bottom. Replacing them on top first makes little sense, and just seems like an unnecessary burden, which is why it wasn't required or mentioned in the ordinary COATing context. In contrast, we are now proposing reversing those same seven cards, *in situ*, as it were, then cutting only five of them to the bottom.

Let's use COAT(ML) to refer to such "COATing with a difference"— namely, the **C**ounting **O**ut **A**nd **T**ransferring (**M**ore and **L**ess cards, respectively) move. Of course, "more" (as in the "lots" of FLOATS) and "less" (the "some" of FLOATS) determine each other, representing complementary amounts of cards from the packet of size n. (There is a reason we exclude the possibility that k is half of n for even-sized packets.)

Tracking this in general is a little tricky, but it can be shown that dealing out (and hence reversing) k cards from the packet $\{1, 2, \ldots, n\}$, replacing them on top of the rest, and then cutting $n - k$ of those from top to bottom, yields the rearranged packet: $\{2k - n, 2k - n - 1, \ldots, 2, 1, k + 1, k + 2, \ldots, n - 1, n, k, k - 1, \ldots, 2k - n + 2, 2k - n + 1\}$. Just why this holds will become clear shortly.

Let's repeatedly COAT(ML) a thirteen-card packet in the case $k = 9$. It's helpful to follow along with a face-down packet that starts A–K♣. Reversing nine cards, replacing them on top, and then cutting four of

---

[7]While seated in the upper saloon of a No. 9 bus in Dublin in March 2013.

them leads to $\{5, 4, 3, 2, A, 10, J, Q, K, 9, 8, 7, 6\}$. Figure 9.3 shows this packet fanned face up. Note that COAT(ML)ing more than half a packet does not preserve the top and bottom halves (as overCOATing does, see page A♥).

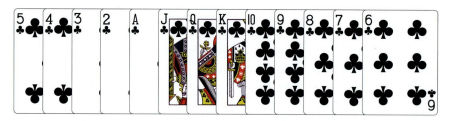

Figure 9.3. After COAT(ML)ing nine cards from A–K♣.

Again reversing (in place) the top nine cards of the face-down packet $\{5, 4, 3, 2, A, 10, J, Q, K, 9, 8, 7, 6\}$ and then cutting four of them, and fanning the result face up, we obtain what is seen in Figure 9.4.

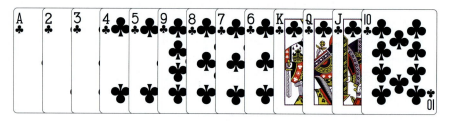

Figure 9.4. After a second COAT(ML) of nine cards from A–K♣.

Note that the original top card of the packet A-K♣ is on top again. This holds in general, as we now note.

Top to Top Principle   ▶◀

Provided that $k > \frac{n}{2}$, the original top card of any packet of size n ends up back on top after two rounds of reversing the top k cards followed by cutting $n - k$ of those to the bottom.

(If the reversal is accomplished by flipping over cards as a unit, this is FLOATSing, and if the cards are dealt or counted out to reverse their order, it's COAT(ML)ing. We proceed under the latter assumption.)

The full justification for this and the next two claims follows in due course. We ignore the case $k = n$, since there are no cards to cut. Surprisingly, it doesn't work if $k = \frac{n}{2}$, which is why we avoid that possibility.

As the last example suggests, the original top card may not be the only one that is back where it started after two COAT(ML)s. Actually, it's possible for all of the cards to return to their original positions.

Save up to 100% Principle   ▶◀

If k cards from n cards are reversed and replaced on top of the rest, and then $n - k$ cards are cut from top to bottom, and this process is repeated, then provided that $k > \frac{n}{2}$, the original $2k - n$ top cards (and possibly others too) are back on top in order.

Of note is the special case of reversing $n-1$ cards and replacing them on top before cutting one card from top to bottom. Doing this twice actually restores the entire packet to its original order.

At the other extreme, there are times, such as when $n = 13$ and $k = 7$, when only the top card is restored to its original position. That's a representative example, since $2k - n = 1$ if and only if $2k - 1 = n$ if and only if $k = \frac{n+1}{2}$, corresponding to what we might call *minimal COAT(ML)ing* for odd-sized packets.

Returning to the packet that started A–K♣ (face down), a third application of reversing nine cards, *in situ*, and then cutting four, yields the image in Figure 9.5 when the packet is fanned face up.

Figure 9.5. After a third COAT(ML) of nine cards from A–K♣.

There are no prizes for guessing what a fourth application produces: the original packet again. Can you see why that holds for the cards depicted? The upshot is a second Low-Down Quadruple Dealing Principle.

Four COAT(ML)s Principle   ▶◀

Four reversals of the top k cards in a packet of size n, in place, alternated with four cuts to the bottom of the top $n - k$ cards, restores every card in the packet to its original position, assuming $k > \frac{n}{2}$.

You'll never notice the effect of four COAT(ML)s when k is more than half of n—it's as if you never COAT(ML)ed at all. As noted before, the period of the move is two, not four, if $k = n - 1$, just as in the case of overCOATs, though it's less obvious in the COAT(ML) situation. (Is there a card effect of interest to be derived from that last observation?)

To get a firm understanding of all of this, let's do some bookkeeping. As in the case of overCOATing, there are three portions of the packet to keep track of. These move around intact, subject at most to some internal reversals. Since they are quite different from the top, middle, and bottom portions from Chapter A, we don't use those names. We also denote them by different letters.

Note that since $k > \frac{n}{2}$, we have $k - (n - k) = 2k - n > 0$.

> ### Quomodo secatur, semper est Gallia divisa in partes tres[8]
> *Writing* $n = k + (n - k) = (n - k) + (k - (n - k)) + (n - k) = (n - k) + (2k - n) + (n - k)$, *we see that a packet of* n *cards naturally breaks into three pieces* X, Y, Z, *of sizes* $2k - n$, $n - k$, $n - k$, *respectively.*
>
> *Starting with the packet* $\{1, 2, \ldots, n\}$, *we thus get*
>
> $$X = \{1, 2, \ldots, 2k - n - 1, 2k - n\},$$
> $$Y = \{2k - n + 1, 2k - n + 2, \ldots, k - 1, k\},$$
> $$Z = \{k + 1, k + 2, \ldots, n - 1, n\}.$$

Now we are in a position to analyze COAT(ML)ing when $k > \frac{n}{2}$, as the basic count out (and replace) then transfer move can be represented by the notation $X, Y, Z \to \overline{X}, Z, \overline{Y}$, where the bar, as always, indicates a complete subpacket reversal.

Given this decomposition, it's easy to verify that two such COAT(ML)s have the effects claimed earlier. First, they transform the original X, Y, Z to $\overline{X}, Z, \overline{Y}$, and then to $X, \overline{Y}, \overline{Z}$, thus restoring the top $2k - n$ cards to their original positions. Second, the reason up to 100% of the packet may be preserved is that not only can $2k - n$ be large, resulting in big savings, but in the extreme case $k = n - 1$, we find that X contains $n - 2$ cards and Y and Z each consist of one card. Hence, the final state $X, \overline{Y}, \overline{Z}$ after two COAT(ML)s is in fact the same as the initial state, that is to say, you save 100%![9]

The period four property can be verified in a similar way. Here's a more subtle point.

> ### Another Special 4-Cycle Principle   ▶◀
> *If* $k > \frac{n}{2}$, *then under a sequence of four COAT(ML)s, the initial bottom card in position* n *orbits through positions* k *(the last card in* Y *),* $k + 1$ *(the first card in* Z *), and* $2k - n + 1$

---

[8]Our final Roman empire reference, this one courtesy of Tony Phillips of SUNY Stony Brook (see http://www.condenaststore.com/-sp/CROSSED-PATHS -Caesar-Meets-de-Gaulle-New-Yorker-Cartoon-Prints_i8575328_.htm).

[9]Please note that your actual savings may vary from the amounts shown.

(the first card in Y), in turn, before returning to the bottom (the last card in Z).

Equivalently, the bottom position is visited by the cards that started in positions $2k - n + 1$ (the first card in Y), $k + 1$ (the first card in Z), and k (the last card in Y), in turn, before the original bottom card returns there.

Note that those four bottom visitors are the bookends of Y and Z. Also note that the inhabitants of X are kept away from the bottom (and the middle) by the arguably more experienced travelers in Y and Z.

For example, if $n = 12$ and $k = 7$, the bottom card cycles around positions 7, 8, and 3 before going back to the bottom (position 12), and the bottom visitors are those that started in positions 3, 8, 7, and 12.

There is an appropriate statement about the cycle decomposition of the permutation corresponding to any COAT(ML), parallel to that in the Special 4-Cycle Principle on page 41.

Plain TACOs (with Many Possible Toppings)

The Top to Top and Period Four principles seen above actually hold under much broader conditions than we have revealed, thus giving rise to more flexible magic possibilities. Before we get to that, note that we transitioned from COATing k cards from a packet of size n, namely, reversing the top k cards and cutting them all to the bottom, to the contrasting COAT(ML)ing, in which we reverse the same k cards, but only cut $n - k$ of them to the bottom. Both of these moves have period four if $k > \frac{n}{2}$. It's certainly reasonable to ask what happens if we cut first, and then reverse.

Suppose, for starters, that we cut $n - k$ cards from top to bottom in a packet of size n, then reverse the top k cards. A little experimentation reveals that doing this three times takes the original top card to the bottom. We first nudged readers to look into this back at the end of Chapter A, starting on page 44, doing the reversing in the usual counting out fashion (e.g., into a waiting hand) before replacing them on top of the rest. Later, in the "Parting Thoughts" of that chapter, we referred to it as TACO(LM)ing, for **T**ransferring **A**nd **C**ounting **O**ut (**L**ess and **M**ore cards, respectively). Since it has an associated Top to Bottom Principle, as mentioned there, it's natural to think of it as a kind of dual of COATing. Again, "less" $(n - k)$ and "more" (k) determine each other, and, as in the case of overCOATing, we're assuming that k is at least half of n.

At this point, it's inevitable to consider a plain TACO: **T**ransferring (cutting) k cards from top to bottom **A**nd then **C**ounting **O**ut (reversing) the same number of cards at the top. When $k \geq \frac{n}{2}$, it turns out that two of these restores the original top card to the top, and four of them have the same impact as none at all.

If you're feeling slightly confused by the proliferation of similar, yet subtly different moves just surveyed, welcome to the club. Table 9.1 may help; it summarizes some key information about the four main period-four moves we have highlighted. The TACO claims are featured in "Parting Thoughts" later in this chapter, as things to confirm.

COAT	TACO(LM)
Reverse $r = k$, then cut $c = k$.	Cut $c = n - k$, then reverse $r = k$.
If $k \geq \frac{n}{2}$, then $X, Y, Z \to Z, \overline{Y}, \overline{X}$	If $k \geq \frac{n}{2}$, then $X, Y, Z \to \overline{Z}, \overline{Y}, X$
and Bottom to Top with three moves.	and Top to Bottom with three moves.
COAT(ML)	TACO
Reverse $r = k$, then cut $c = n - k$.	Cut $c = k$, then reverse $r = k$.
If $k > \frac{n}{2}$, then $X, Y, Z \to \overline{X}, Z, \overline{Y}$	If $k > \frac{n}{2}$, then $X, Y, Z \to \overline{X}, \overline{Z}, Y$
and Top to Top with two moves.	and Top to Top with two moves.

Table 9.1. A summary of sorts.

But wait, there's more! The COATs and COAT(ML)s turn out to be "extreme cases" on a spectrum of transferring and cutting possibilities, all of which have period four. A similar statement holds for the TACO(LM)s and TACOs. This was stumbled on in the spring of 2013, ten years after the Low-Down Triple and Quadruple Dealing principles of COATing first came to our attention.

For a packet of size n—let's fix $n = 13$—we need to consider the effects, for various values of r and c, of (1) reversing r and then cutting c cards, and (2) cutting c and then reversing r cards. When $r = c \geq 7$, we are COATing or TACOing, and when $r = k \geq 7$ and $c = n - k$, we are TACO(LM)ing or COAT(ML)ing, but there are so many other possibilities to explore.

Table 9.2 documents some of the impact of "reversing r cards (on top) then cutting c cards (from top to bottom)" on a packet running 1–13, for many values of r and c. Specifically, we list those cards that are back in place after two and four applications of this move.[10] The claimed results can be verified by tracking a face-up Ace–King packet under the indicated moves. Note that when 1–13 is listed in the fourth column, we have a move of period four; there are many cases with $r > c$ where this happens.

Several patterns and new principles are suggested, and related magic begs to be invented. The $r > c$ cases are equivalent to this modification (or

[10]It's also worth tracking the position of each card after one and three applications, respectively; the former in case there might be interesting fixed points or other regularities, and the latter for hidden Bottom to Top or Top to Bottom principles.

r	c	After two moves	After four moves	Name
7	5	1–2	1–2, 4–7, 9–12	
7	6	1	1–13	COAT(ML)
7	7	7	1–13	COAT
7	8	6–7	1–4, 6–7, 9–12	
8	4	1–4, 7, 11	1–4, 6–8, 10–11	
8	5	1–3, 6, 11	1–13	COAT(ML)
8	6	1–2, 5, 8, 11	1–13	
8	7	1, 4, 7–8, 11	1–13	
8	8	3, 6–8, 11	1–13	COAT
8	9	5–8, 11	1–13	
9	3	1–6	1–6, 8–9, 11–12	
9	4	1–5	1–13	COAT(ML)
9	5	1–4, 9	1–13	
9	6	1–3, 8–9	1–13	
9	7	1–2, 7–9	1–13	
9	8	1, 6–9	1–13	
9	9	5–9	1–13	COAT
9	10	4–9	1–2, 4–9, 11-12	
10	2	1–8, 10, 12	1–8, 10, 12	
10	3	1–7, 9, 12	1–13	COAT(ML)
10	4	1–6, 8, 10, 12	1–13	
10	5	1–5, 7, 9–10, 12	1–13	
10	6	1–4, 6, 8–10, 12	1–13	
10	7	1–3, 5, 7–10, 12	1–13	
10	8	1–2, 4, 6–10, 12	1–13	
10	9	1, 3, 5–10, 12	1–13	
10	10	2, 4–10, 12	1–13	COAT
10	11	1, 3–10	1–13	
11	1	1–10	1–10	
11	2	1–9	1–13	COAT(ML)
11	3	1–8, 11	1–13	
11	4	1–7, 10–11	1–13	
⋮	⋮	⋮	⋮	
11	10	1, 4–11	1–13	
11	11	3–11	1–13	COAT

Table 9.2. What's fixed after reversing r, then cutting c, from 1–13.

generalization) of COATing: each time, COAT r cards and then transfer (cut) a few (specifically, $r - c$) cards back from bottom to top.

For a fixed r, as listed in the first column, certainly all rows between the COAT(ML) and COAT extremes have period four.

Generalized COATs Principle   ⋈

If $r > \frac{n}{2}$, then a packet of n cards ends up restored to its original order after four rounds of reversing the top r cards followed by cutting c of those to the bottom, provided $n - r \leq c \leq r$.

The original bottom card shows up on the top after three such moves.

The original top card shows up on the top again after two such moves, provided $n - r \leq c < r$.

Something obvious that is not revealed by Table 9.2 is that if the reversing precedes the cutting, and $c = r - 1$ (or equivalently $r - c = 1$), then the original top card comes back to the top after one such move. The corresponding condition when the cutting precedes the reversing is worth noting, and it has a simple but fun magical application.

One Too Many Complementary Principle   ⋈

Provided $c + r = n + 1$, the original top card of any packet of size n ends up back on top after a single round of cutting the top c cards to the bottom followed by reversing the r top cards.

Ask a spectator to shuffle the deck and hand you about a quarter to a third of it, which you then hold behind your back briefly, claiming that you're looking for your lucky card. Give the packet to the spectator and have more shuffling done. Have the top card looked at and remembered.

Produce a list such as the one depicted in Table 9.3, containing the names of several mathemagicians along with matching descriptive terms. Have any row selected, and have the name in it used to transfer cards

Mathemagician	Description
Benjamin	Amazing
Elmsley	Stunning
Diaconis	Awesome
Gardner	Dazzling
Graham	Wonderful
Green	Staggering
Hummer	Brilliant
Maven	Astounding
Sands	Delightful
Steinmeyer	Nifty
Trost	Formidable
Wohl	Astonishing

Table 9.3. Mathemagician and superlative pairings.

from top to bottom, one for each letter; then have the description there used to reverse cards at the top, one for each letter. The words in each row have been paired in such a way that between them they always have fifteen letters; the typesetting makes this constant length less obvious. The spectator's card will end up back on top provided that the packet you give her contains fourteen cards: when the cards are behind your back you secretly count them, discarding (or pocketing) any surplus ones. At the conclusion of the double spelling, take the cards back, and under the guise of additional shuffling briefly shuffle the top card to the bottom so that you can peek at it before losing it in the packet once more. You are all set for a mind reading conclusion.

Table 9.4 documents some of the impact of "cutting c cards (from top to bottom) then reversing r cards (on top)" on a packet running 1– 13, for many values of c and r. In a sense it's the dual of Table 9.2. Again, we list those cards that are back in place after two and four applications of this move, and we still recommend tracking the position of each card after one and three applications for the same reasons as before. For a fixed r—listed in the second column here—certainly all the rows between the TACO(LM) and TACO extremes have period four.

Again, there are new patterns and principles to think about, and hopefully new magic effects to be devised. Once more, there are many cases where $r > c$ and the period is still four.

> ### Generalized TACOs Principle   ▶◀
> *If $r > \frac{n}{2}$, then a packet of n cards ends up restored to its original order after four rounds of cutting c cards from top to bottom followed by reversing the top r cards, provided $n - r \leq c \leq r$.*
>
> *The original top card shows up on the bottom after three such moves.*
>
> *The original top card shows up on the top again after two such moves, provided $n - r \leq c < r$.*

In a sense, this wrap-up of our explorations of low-down triple dealing and generalizations (and origins)—spread across three chapters—sees us complete a circle.[11]

---

[11] Or at least a closed curve; self-intersections cannot be ruled out.

c	r	After two moves	After four moves	Name
5	7	3, 6–8	1–13	
6	7	7	1–13	TACO(LM)
7	7	1	1–13	TACO
8	7	1–2	1–2, 4–7, 9–12	
9	7	1–3	1–3, 6, 10–11	
10	7	1–4	1–4	
4	8	5–8, 11	1–3, 5–8, 11–12	
5	8	3, 6–8	1–13	TACO(LM)
6	8	1, 4, 7–8, 11	1–13	
7	8	1–2, 5, 8, 11	1–13	
8	8	1–3, 6, 11	1–13	TACO
9	8	1–4, 7, 11	1–4, 6–8, 10–12	
10	8	1–5, 8, 11	1–5, 8, 11	
3	9	4–9	1–2, 4–9, 11–12	
4	9	5–9	1–13	TACO(LM)
5	9	1, 6–9	1–13	
6	9	1–2, 7–9	1–13	
7	9	1–3, 8–9	1–13	
8	9	1–4, 9	1–13	
9	9	1–5	1–13	TACO
10	9	1–6	1–6, 8–9, 11–12	
2	10	1, 3, 5–10, 12	1, 3–10, 12	
3	10	2, 4–10, 12	1–13	TACO(LM)
4	10	1, 3, 5–10,	1–13	
5	10	1–2, 4, 6–10, 12	1–13	
6	10	1–3, 5, 7–10, 12	1–13	
7	10	1–4, 6, 8–10, 12	1–13	
8	10	1–5, 7, 9–10, 12	1–13	
9	10	1–6, 8, 10, 12	1–13	
10	10	1–7, 9, 12	1–13	TACO
11	10	1–8, 10, 12	1–8, 10, 12	
1	11	2–11	2–11	
2	11	3–11	1–13	TACO(LM)
3	11	1, 4–9	1–13	
4	11	1–2, 5–11	1–13	
⋮	⋮	⋮	⋮	
10	11	1–8, 11	1–13	
11	11	2, 4–10, 12	1–13	TACO

Table 9.4. What's fixed after cutting c, then reversing r, from 1–13.

Our final effect shows that despite our seemingly dismissive words at the start of Chapter A, low-down double dealing types aren't so bad after all. It's just one of many ways to exploit the generalized principles we've surveyed above. Now that we know that this works, justifying just why can be done using much more elementary means than is suggested by the deliberations of the last several pages.

9♦ Any Mathemagician

How it looks overall: *Shuffle the deck over and over, and then place it on the table. Have a spectator take off roughly the top quarter, then name a mathemagician and a descriptive word. The spectator uses the mathemagician's name to count out and reverse some of the cards in his packet, and the word to transfer some cards from top to bottom. This is repeated. Finally, the spectator presses down hard on the top card as you request that he turn it into a specific card, say the Jack of Diamonds.*

Despite the fact that you never touched the cards after the spectator selected his packet, and you had no control over the choices of the two words used to mix up the cards, the top card turns out to be the one you requested.

How it looks in detail: Shuffle the deck over and over, and then place it on the table. Addressing a spectator, say, "I'd like you to try your hand at doing some mathemagic. You'll need about a quarter of the deck." Ask the spectator to lift up roughly that much of the deck from the top, setting the rest aside.

Next, ask him to shout out the name of his favorite mathemagician, dead or alive. Then have a descriptive word such as "magic" or "amazing" be selected. Direct the spectator carefully in (1) the use of the mathemagician's name in the spelling out (counting out) of cards, one at a time, from one hand to the other, one card from the packet being transferred for each letter, thus reversing their order, before replacing those on top of the remainder, and (2) the cutting from top to bottom of one card for each letter in the descriptive word selected. Finally, have the spectator repeat those two steps.

Announce, "These are random cards, and neither of us knows exactly how many you have. I haven't once touched them, and you randomized them further using two words of your own choosing to move cards around. Here comes the fun part. I want you to channel the powers of the mathemagician you named, by pressing down hard on the top card and turning it into, I don't know, let's say the Jack of Diamonds." Alternatively, you could produce a prediction slip, and say, "If you do this right, I bet it will match what I predicted earlier."

Once the spectator has pressed down hard on the top card, have him turn it over. It is miraculously found to be just as you requested (or predicted). With any luck, those present will burst into spontaneous and sustained applause.

Conclude by saying, "Congratulations! Just don't let that applause go to your head. In this business, you just have to get used to it."

How it works: Needless to say, the card that shows up on top of the spectator's packet at the end is the one that started out at the top of the deck. Let's suppose it's the J♦, and that it's been planted there ahead of time, and maintained there in the usual way throughout some convincing looking shuffling.

The spectator lifts off the top twelve to fourteen cards, approximately. Let's assume that he then names "Diaconis" in response to your prompts, following which he selects the word "magic."

You direct the spectator to thumb off eight cards from one hand to the other, one at a time to reverse their order, as D-I-A-C-O-N-I-S is spelled out, before replacing those eight cards on top of the remainder. Next, have him cut five cards from the top of the packet to the bottom, one for each letter in the word "magic."

Have the spectator repeat both steps, and all is set for a successful conclusion.

Why it works: It's a straightforward application of the Top to Top Principle. The descriptive word should be shorter than the mathemagician's name; hence you must adapt the words offered depending on who is named. For "Gardner" or "Graham," keep it short: magic, cool, and nifty all. For "Hammingway" there are additional options, including elusive, witty, and talented.

You can certainly make your life easier by offering a list of longer names like Boudreau or Goldstein. We don't recommend allowing spectators to use any name with five or fewer letters—sorry Max!—as these don't always work when using packets of size thirteen or larger. (If the spectator appears to cut off too many cards at the start, have a few removed, from the bottom.)

Source: Original, from May 2013.

Presentational options: Do whatever it takes to know the top card of the deck at the outset; if you'd rather not plant a specific card there at the outset as we've suggested, then wild shuffling followed by peeking at the bottom card, and running off so that it ends up on top, should be adequate.

If you are feeling exceptionally brave, you can adapt this effect to also ensure that any card the spectator desires is at the top at the conclusion.

See "Any Card (and Any Magician)" (A♥) for some hints; there are quite a few details to perfect before trying this out live.

Parting Thoughts

- Prove that the Sands' Lucky Principle still works for nonprime values of n provided that k is coprime to n.

- Mathematician Neil Calkin provides a rough analysis of an issue raised in "Coprime Twins" (9♥) on page 218: for a shuffled deck, what are the chances that there are adjacent twins, namely, a pair of cards of the same value and color (such as 3♥ and 3♦) beside each other? (The motivated non-mathematical reader is encouraged to explore the probability and statistics terminology used by Calkin, which we will not define here.)

  First, let's focus on getting two adjacent cards of the same value, regardless of color. The probability that the kth card matches the $(k + 1)$st card in value is $\frac{1}{17}$, as there are fifty-one cards, only three of which match. Hence if X_k is a random variable, equal to 1 if they match or 0 otherwise, and $X = X_1 + ... + X_{51}$, then X is the number of matching pairs (counting a matching triple as 2 and a matching quadruple as 3). Thus, there are no matching pairs if $X = 0$. The expected value of X is $\frac{51}{17} = 3$. Hence the average number of matching pairs is exactly 3.

  If we approximate X by a sum of independent Bernoulli trials, we'd get a distribution close to a Poisson distribution with parameter $\lambda = 3$. The probability that such a random variable is 0 is about $\frac{1}{e^3}$, which is just slightly smaller than 0.05, so with probability about 95%, we find a match. The exact probability of no matches makes a nice exercise in inclusion-exclusion.

  When seeking two adjacent cards of the same value and color, there's only one possible match, not three, and the above argument holds with 3 replaced by 1 throughout. So with probability about $1 - \frac{1}{e}$, or 63%, we find adjacent twins in a randomized deck.

- Is there an unCOAT(ML)ing analog of the unCOATing on page 43?

- How many cards are restored to their original position when minimal COAT(ML)ing an even-sized packet twice? Here, $k = \frac{n}{2} + 1$. Try it with a packet of size twelve, for which k must be seven. Is the result typical?

- Prove that the Four COAT(ML)s Principle holds.

- Adapt "Three Scoop Miracle" (A♣) from COATs to COAT(ML)s.

- What special 4-cycle appears in the cycle decomposition of the permutation corresponding to a COAT(ML)? Are there any other 4-cycles of particular interest?

- Discover and prove an analog of the First Odd Drop Principle from page 118 for COAT(ML)s.

- Now suppose that $k < \frac{n}{2}$ and we reverse k cards at the top of a packet of size n before cutting the resulting top $n - k$ cards to the bottom. How many such moves—which we could denote by COAT(LM)—are needed to restore the packet to its original order?

- When $k \geq \frac{n}{2}$, we know from Chapter A that COATing corresponds to $T, M, B \to B, \overline{M}, \overline{T}$, for appropriate subpackets T, M, B of a packet, where the bar indicates a complete subpacket reversal, and TACO(LM)ing corresponds to $T, M, B \to \overline{B}, \overline{M}, T$, for the same subpackets T, M, B.

  On the other hand, when $k > \frac{n}{2}$, we saw that COAT(ML)ing corresponds to $X, Y, Z \to \overline{X}, Z, \overline{Y}$, for different subpackets X, Y, Z. Under the same assumption, show that a packet of size n naturally breaks up into subpackets X, Y, Z such that a plain TACO corresponds to $X, Y, Z \to \overline{X}, \overline{Z}, Y$.

  Now use that to state and prove various TACO principles.

  Is there a "continuum" of such operations $X, Y, Z \to \ldots$ for the rows between the COAT(ML) and COAT extremes in Table 9.4?

- Are there other possibilities here worth investigating, namely partitions of a packet into three pieces, along with associated rearrangements—similar to the four just surveyed—giving rise to card moves of period four?

  If such variations exist, how are they implemented with cards and do they give rise to any interesting magic effects?

  Hint: mustn't there be period-four moves such that two applications of them have associated Bottom to Bottom principles?

- Show why the Generalized COATs and TACOs principles hold.

- In Chapter 5 we extended much of the material in Chapter A to the situation where two different numbers s and t of cards were alternately COATed from a packet of size n. Provided that s and t are on average at least half of n, things work out well. Can any of the corresponding COAT(ML)ing or TACO(LM)ing results, or their later generalizations, be extended in a similar fashion?

- Consider the following FOAC(2,1)^{12} (**F**lip **O**ver **A**nd **C**ut) move applied to a packet of size n: flip over the top two cards as a unit, and then cut one card to the bottom. How many times a must this be done to the packet to restore it to its original state (order *and* orientation)? Certainly a is even, since each application of the move keeps the original top card on top, while toggling its orientation between the two possible states. For $n = 2, 3, 4, 5, \ldots, 13$, experimentation with cards shows that $a = 2, 4, 6, 8, \ldots, 24$. Show that for $n \geq 2$, if $n = k + 1$, then $a = 2k$.

  Can any magic effects be derived from this move? Are there other aspects of interest, such as parity preservation?

- Consider FOAC(3,2) applied to a packet of size n: flip over the top three cards as a unit, and then cut two cards to the bottom. How many times a must this be done to restore the packet to its original state? Again a is even, for the same reason as in the FOAC(2,1) case above. We must have $n \geq 3$, and for the odd values $n = 2k+1$, it turns out that $a = 2k$. Can you see why?

  It's a different story for the other half of the possible values of n, i.e., the evens. For $n = 4, 6, 8, 10, 12, 14, 16$, experimentation reveals that $a = 4, 12, 24, 40, 60, 84, 112$. Do you see a pattern here? Seek and you shall find! Can you prove that what you discover must hold? Over 50 values of a are listed at *The On-Line Encyclopedia of Integer Sequences (OEIS)* at https://oeis.org/A225232.

  Is there a magic effect of value lurking in the wings, perhaps for odd-sized packets, using $\frac{a}{2}$ applications of the move? Similar questions are worth asking for the other variations discussed below.

- Consider FOAC(4,3) applied to a packet of size n: flip over the top four cards as a unit, and then cut three cards to the bottom. How many times a must this be done to restore the packet to its original state? For $n = 4, 5, \ldots, 16$, direct experimentation shows that $a = 2, 4, 4, 4, 12, 12, 6, 24, 24, 8, 40, 40, 10$. Do you see a pattern?

  There's a pattern for a third of the cases, when the packet size is one more than a multiple of three. For $n = 4, 7, 10, 13, 16$, we have

---

^{12}Pronounced *folk.*

$a = 2, 4, 6, 8, 10$. We must have $n \geq 4$, and for $n = 3k + 1$ it seems that $a = 2k$. Can this be proved? The other two-thirds of the cases are less obvious. See *OEIS* at https://oeis.org/A221564 for additional values of a.

- Try to repeat the above for FOAC(5,4) applied to a packet of size n: flip over the top five cards as a unit, and then cut four cards to the bottom.

- What about for FOAC(5,3), i.e., flipping over the top five cards as a unit, and then cutting three cards to the bottom?

- What about COATing in general alternating with flipping both top cards, or the top three cards? COATing k cards is of course related to flipping over k cards as a unit, and since cards maintain their original orientation when COATing, the periods of associated moves can be expected to be more manageable. (Is it safe to assume that it cuts the period in half?)

Red, Black, Silver, and Gold

Each earlier chapter has a distinct mathemagical theme; this one has two. It also has a colorful Swedish undercurrent. We start with a separation algorithm that is more than half a century old—though it's still not that well-known—give a few applications of it in action, and then move on to another ingenious principle of similar vintage and two terrific effects that arise from it and its generalization. This is all made possible thanks to four very creative people from various places on the magic-mathematics[1] spectrum: Lennart Green, Ron Wohl, Persi Diaconis, and Ron Graham.

There is no question that the last two effects are the real show stoppers. Of the two original ones that precede them, "Top Twenty Hit" (10♥) should impress the average audience.

> *Hand out a deck for shuffling, then count twenty cards into a face-down pile. Pick these up and look through them face up, commenting on how well mixed they are.*
>
> *Deal out left to right to form five piles of cards. Say, "That's only four cards each, let's try that again." Gather up the piles in any order and deal left to right once more, this time into four piles. Hand out three of those to spectators, keeping the last one for yourself.*
>
> *Guess who has the best poker hand?*

Among the deck setups sometimes required earlier, we never asked for anything as contrived as having all the Red cards on top of all the

---
[1] Not to mention medical and chemical.

Black cards, or vice versa. Yet, there are numerous popular card effects, none more impressive than Paul Curry's "Out of This World" from 1942 [Curry 01], in which a miracle happens in the eyes of the audience because the deck secretly starts out separated by color.[2] Such a division may be maintained through some honest-looking shuffles to give the impression that the cards are well mixed.

Similar effects can be pulled off with a genuinely shuffled deck, thanks to a brilliant in-hand separation move due to Swedish magician Lennart Green. We now present that with his kind permission.

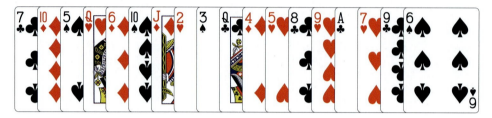

Figure 10.1. Random mix of cards.

Suppose we have a face-up packet of cards, fanned as in Figure 10.1, with a good irregular distribution of Red and Black cards. We'll learn how to process this in an apparently innocent way to end up with the Reds separated from the Blacks, as shown in Figure 10.2.

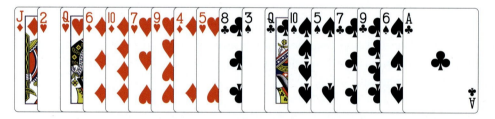

Figure 10.2. Separated by color.

This requires two rounds of the sneaky move introduced next.

Green's One And Two (GOAT) Separation

Before we describe Green's idea, we note that its application is by no means restricted to separating a packet (or full deck) into Reds and Blacks. It works just as well for odds and evens, or primes and composites (see page 14), provided you can see the card faces as you proceed. Although he did not publish it until relatively recently [Green 84, Green 89, Green 95, Green 00], Lennart recalls discovering it around 1960.

---

[2]See the June 2011 *Card Colm* "Out Of This Whirled (The Odds Are Not Even)" for a separation effect inspired by this.

Green's **O**ne **A**nd **T**wo (GOAT) separation algorithm applied to a full and randomized deck to yield total color separation generally requires three passes of the cards, face up—that is, you go through the face-up deck three times. If that sounds daunting, relax. As we will soon see, it can be done right in front of an audience without arousing too much suspicion. It also works well for isolating smaller collections of cards, such as the four Aces, any desired poker hand, or all of the Spades, with fewer passes. We describe it as performed in-hand, assuming that the audience does not see any card faces during the second or third passes.

We describe it as performed in-hand, usually without the audience seeing any of the card faces. Later, we'll discuss a version in which clumps of cards are openly dropped face up on the table to form an ever-increasing pile with better separation than the original packet.

The first goal is to be able to perform the steps below speedily and without hesitation. Turn the packet over in your left hand, and fan the card faces to the right with your thumb, so that the rightmost card is the original bottom card. For the sake of argument, we assume that the card is Black.

Green's One And Two (GOAT) Separation Principle
Any face-up packet of cards can be color separated as follows. Use your left thumb to push over to your right hand as a unit all of the Black cards (one color) at the face. Now push over the next complete runs of Black and Red cards (two colors), as viewed from left to right, from your left hand, as a unit. Your right hand takes these on top of the cards already there, so that the growing pile in that hand has Reds on top of Blacks.

Push over as a unit the next complete run of Reds (one), then the next complete runs of Reds and Blacks (two). Continue in this way, alternating "one, two" silently in your head, until there are no cards left in your left hand. Always push over into your waiting right hand the largest possible one- and two-color clumps of cards. If possible, sneak in a single cut at the end, from face to back, to help things along.

The final cut just suggested is intended to apply to situations in which the packet ends up with clumps of the same color at both ends. Then, it is prudent to cut those from the face to the back.

It will be helpful to work through some examples, starting with small packets. Suppose we have the six-card packet 5♦, 8♥, 9♠, 4♣, A♥, 2♣, fanned left to right as shown in the first image in Figure 10.3.

First, your right hand takes the 2♣ (representing all of the cards of one color at the face), as shown in the second image in Figure 10.3.

Figure 10.3. First steps.

Second, your right hand takes the next three cards (representing the two colors seen at the face), as shown in the first image in Figure 10.4. Those go on top of what's already in your right hand. Finally, that hand takes the last two cards (representing all of the cards of one color seen at the face), as shown in the second image in Figure 10.4. The cards are now separated by color: Blacks followed by Reds.

Figure 10.4. Final steps.

Something like this works for any packet of size six (or less), with an occasional helping hand, as our next example shows. Starting with the same cards, but in order 8♥, 2♣, A♥, 4♣, 9♠, 5♦, also fanned face up, we first take the lone Red card at the top (one), then the following two Blacks and one Red, as a unit (two), as shown in Figure 10.5.

Next, your right hand takes the 2♣ (one), as shown in the third image in Figure 10.5. There is little choice now but to add the last card (representing two as best it can) to those in your right hand. Since the cards at each end of the fan have the same color, transfer all of the cards to your left hand and use your right hand to cut the 8♥ from the face to the back, as shown in the final image in Figure 10.5. The cards are now separated by color: Reds followed by Blacks this time.

If we are allowed to do two such passes and last minute adjustments as above if needs be, then two GOAT separation moves works for packets of up to eighteen cards. Let's start with the cards shown earlier in Figure 10.1, the goal being to end up with the arrangement shown in

Figure 10.5. Another example.

Figure 10.2. The first image in Figure 10.6 shows the initial state if such a packet is fanned in one hand. In practice, a packet this large is usually held in a tighter fan, so that only a few cards at the face are clearly identifiable.

Nudging off a few cards with your left thumb reveals the appropriate first move, as shown in the second image in Figure 10.6: the taking of the Black Nine and Six as a unit into your right hand (representing one). Then take the Black Ace and Red Seven as a unit (representing two), as shown in the third image in Figure 10.6.

Figure 10.6. With eighteen cards: first steps of the first pass.

Figure 10.7 shows the next three steps: the taking of the Red Nine (one), then the Red Four and Five along with the Black Eight (two), followed by the Black Three and Queen (one).

Figure 10.7. With eighteen cards: the next three steps of the first pass.

Figure 10.8 shows the subsequent taking of the Black Ten and the Red Jack and Two (two), followed by the Red Queen and Six (one). Only three cards remain in the left hand.

Figure 10.8. With eighteen cards: more steps of the first pass.

As shown in the first image of Figure 10.9, the Red Ten and Black Five (two) are now taken as a unit. Finally the Black Seven (one) is taken.

Figure 10.9. With eighteen cards: final steps of the first pass.

A quick fanning of the cards (see the center image of Figure 10.9) shows that both ends are now Black, so you cut the two Black cards at the face to the back (see the last image). This gets us closer to our destination, while being only a slight modification of the one-two mantra. The first pass is complete.

The second pass here is much quicker, as Figure 10.10 reveals. The five Red cards now at the face of the packet are taken first (one), then the next clump of four Reds and four Blacks together (two), and finally the remaining five Blacks (one). The separation is complete, as seen in the final spread, this time without any last minute face-to-back adjustment.

Figure 10.10. With eighteen cards: the second pass.

With proper presentation, perhaps making misleading comments to audience members about the card value distribution but drawing no attention to the card colors, the first such pass can be done quite openly. Most observers will not notice that you are pulling off a substantial and useful rearrangement of the deck right under their noses.

As the last example suggests, fewer steps are involved in later passes, for which we recommend not revealing any card faces. For instance, the audience could be engaged in a distracting one-sided conversation on the second pass, with comments such as, "In a moment, I'm going to ask one of you to choose a card. I'm just checking where the Aces are, because I have a feeling one will be chosen. Did I mention I'm also speed-memorizing the entire deck?"

The GOAT separation algorithm applied to a face-up fifty-two-card deck generally requires three passes, as we see below. Do the third pass with the cards close to your chest—perhaps under the guise of "one last check for that Ace"—and you are all set: either the Reds are all on top of the Blacks or vice versa. In practice—and *with* practice, lots of practice—this goes faster and smoother than seems likely at first. Also, you can train your mind to do it while jabbering away to the audience, so that it goes unsuspected that you are up to anything.

Why are three passes enough to result in total color separation for a full deck? Assume, in the worst-case scenario, that the deck alternates Red/Black all the way.[3] It's easy to see that the first pass results in monochromatic runs of length three, except perhaps at the two ends, but then the suggested adjustment more or less takes care of that. By the same logic, the second pass results in monochromatic runs of length nine, and the third pass separates the deck into two maximal monochromatic runs of length twenty-six as desired. Don't forget to cut at the end of each pass if it helps the cause. (Actually there is a slight flaw in the above reasoning. Can you see what it is? Can it be fixed?) In general, the deck is likely to start out with some color clustering, so that each pass results in even larger monochromatic runs.

---

[3]One person's "worst-case" scenario is another's "best-case" scenario. If the deck alternates colors perfectly, you should be headed for a different effect.

There are other ways to partition a deck of cards, and three (or fewer) GOATs work just as well on those. For instance, the deck could be split into the Spades and Diamonds versus the other suits, or into even-valued card versus odd-valued ones, the latter yielding piles of twenty-four and twenty-eight cards, respectively. You could also divide the deck into prime values (2, 3, 5, 7, J, K) versus composite values (4, 6, 8, 9, 10, Q), throwing the Aces into whichever category upsets you least.

Less visual partitions are unlikely to arouse suspicion, even permitting you to let the audience watch the separation as it unfolds. It is not difficult to train your eyes and brain to implement GOATs with such a separation as the goal. For the first and most time-consuming pass, instead of doing it in-hand, you can drop the "one" or "two" clumps on the table to form an ever-increasing pile, chatting about how well shuffled the deck is.

10♣ Double Location

How it looks: *The deck is shown to be well mixed and is split between two spectators, who are encouraged to shuffle further, before looking at and memorizing their top cards. These two cards are exchanged and shuffled into the rest of the retained packets. Finally, the two piles are recombined, and the reconstituted deck is cut several times. Take it back and fan it publicly, then rapidly pull out two cards and place them face down on the table. Have the selections named, and the tabled cards turned over. They are indeed the spectators' cards.*

How it works: This would be very easy to do if one person got all Red cards and the other all Black cards. The selected cards would break these runs in the reassembled-and-cut deck you later look through. However, this would necessitate you being the only person who got to see the card faces, as the color separation would be far too obvious. We suggest using a prime and composite division, lumping in the Aces with one of those sets, thus reducing the REDS (Risk of Embarrassment, Detection, and Shame) factor should anyone glance at faces (yours or those of the cards). This is achieved via three GOATs, either ahead of time or in front of your audience. Use a pinky break (see page 11) to maintain the separation before the deck is handed out in two packets.

Source: This is essentially "Double Location" from *Scarne on Card Tricks* [Scarne 50] with minor modifications, and prime/composite separation in place of even/odd. Scarne[4] credits the idea of proceeding from a color-separated deck to a (single) location to Martin Gardner.

---

[4]Pronounced *SCAR-nee.*

♣   ♥   ♠   ♦

Contrary to the impression we may have given, there is no need for the deck separation to use nearly equal halves. For instance, three GOATs will always isolate any ten to eighteen cards—try it for the thirteen Clubs in a well-shuffled deck. Two GOATs isolate any four to nine cards of your choice—try that for the worthy poker hand 10♦, J♦, Q♦, K♦, A♦ from another well-shuffled deck.

But what about a single GOAT application? Think of any card, and then remove it from a shuffled deck. A single pass of the deck will unite the three cards of value equal to the one removed.

Start over, with a newly shuffled deck. It's not unusual to find two adjacent cards with the same value (see the "Parting Thoughts" in Chapter 9); a single pass will result in all four cards of that value appearing together. Then, it's not hard to cut them to the top or bottom of the deck, if desired.

For instance, let's apply one GOAT pass to get three Aces together. First, we classify each card as either an Ace or a non-Ace. Hold the deck face up in the left hand, and push off to the right hand, as a unit, all cards up to but not including the first Ace. The first image in Figure 10.11 shows this in action, with the A♥ being the first Ace spotted.

Figure 10.11. Uniting three Aces—first steps.

Now locate the second Ace in the face-up deck—here it's A♦—and push off all of the cards above it, including the A♥ (representing two

Figure 10.12. Uniting three Aces—second steps.

clumps, the first Ace and the non-Aces), to your right hand, as shown in the second image in Figure 10.11 and the first image in Figure 10.12.

This results in the first Ace being at the face of the cards in the right hand, and the second Ace being at the face of the cards remaining in the left hand. Next push off just that second Ace (one) to your right hand so that the two Red Aces are together, as shown in the second image in Figure 10.12.

Continue, pushing off more cards as a clump, until a third Ace is located; here it's the A♣. Push to the right hand all of those cards, including that third Ace (two), as shown in the first image of Figure 10.13. This unites those three Aces, as the second image shows.

Figure 10.13. Uniting three Aces—final steps.

If desired, a very fast second pass will allow you to get the fourth Ace together with the other three, as will several "less honest" methods, that we leave to your imagination.

10♥   Top Twenty Hit

How it looks: *Hand out a deck for shuffling, then count twenty cards into a face-down pile. Pick these up and look through them face up, commenting on how well mixed they are.*

Deal out left to right to form five piles of cards. Say, "That's only four cards each, let's try that again." Gather up the piles in any order and deal left to right once more, this time into four piles. Hand out three of those to spectators, keeping the last one for yourself.

Guess who has the best poker hand?

How it works: It's important not to mention poker until the second deal is in progress, after the "mistaken" first deal. We already saw in "Poker with Any Ten Cards" (3♦) that ten random cards are 98% likely to contain at least one matching pair. With twenty cards, representing almost half the deck, it's 100% certain that there are multiple pairs and extremely likely that there is at least one three of a kind.

Birthday Card Triple Match Principle ⋈

*In any twenty randomly selected cards, there is a 96% chance
of at least one value occurring three times.*

As you scan the card faces while babbling about how well mixed they
are, locate the highest value three of a kind you can. If there are no threes
of a kind, settle on the two pairs of highest value. If possible, select three
or four such cards so that one GOAT application will unite them; apply
GOATs twice if necessary. Finally, cut so that those three or four cards
are at the bottom of the packet.

For instance, suppose you are presented with the twenty cards in Figure 10.14.

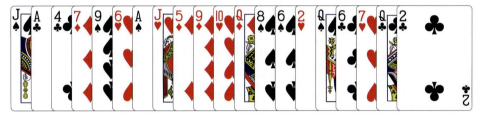

Figure 10.14. Twenty random cards.

A quick glance reveals that this contains a bounty of delights: three
6s, three Queens, and five pairs. As explained above, a single GOAT
application will get the Queens together. If you can ensure that you end
up with those, you will almost surely win, since they beat any other three
of a kind or two pairs that can arise (see Table 3.1). It's possible, but
not too likely, that one of the other hands is a full house, a straight, or a
flush, so victory should be yours.

When dealing, first "mistakenly" deal five piles from left to right, fast,
and when each pile ends up with four cards, apologize. Say, "I guess for
poker hands we need five cards each." This is no accident, the bottom
cards of the last three or four piles are the three or four desired cards.
Now collect the piles in a haphazard order, and recombine as one packet.
Deal again, left to right, this time into four piles of five cards. Give the
first three piles to the three spectators, retaining the last one for yourself.
You get the desired cards (the three Queens above).

Source: Original. February 2013.

♣   ♥   ♠   ♦

From Red & Black to Silver & Gold

Next, we switch gears and present twists on two terrific effects learned from *Mathematical Magic* [Diaconis and Graham 11], based on ingenious and subtly related "universal cycle" principles. In the first, you know which cards several spectators have selected as soon as you know which of them (if any) holds Red ones.

De Bruijn Bracelets

We start by explaining the principle underlying a mathemagical tour de force with permission of Persi Diaconis and Ron Graham. We loosely follow the description they provide of related tricks, and the associated background, in their chapter "Products of Universal Cycles" in the Martin Gardner tribute book *A Lifetime of Puzzles* [Diaconis and Graham 08], and Chapters 2, 3, and 4 of their subsequent award-winning book *Mathematical Magic* [Diaconis and Graham 11].

Imagine a sequence of sixteen cards, viewed as a circle or bracelet by joining the beginning and end. Is there a highly customized and memorized arrangement of Red and Black values, which repeated cuts do nothing to alter, such that knowing the colors of the top four cards allows you to deduce exactly what part of the sequence has been sampled? If so, then assuming your memory is up to the task (or you use a cheat sheet), you can name all four cards. Hence, if such a packet of cards is cut repeatedly, before four people remove a card from the top, you can identify all of the selected cards as soon as you know the color distribution. If you knew that the cards in question were two Reds followed by a Black and then a Red, for instance, there would be only one possibility for the identities of the cards in question.

What is needed to make this possible is what is known as a (binary) de Bruijn sequence of length sixteen. There are only $2^4 = 16$ different arrangements of four Reds and Blacks in a row, so we certainly can't find such a sequence with more than sixteen terms, as a longer one would have to contain at least some repetitions of four in a row. The good news is that length sixteen can be attained.

Similarly, the longest possible sequence with the property that we could identify k cards in a row from their color distribution has length 2^k. This too is attainable, for all k. We now state this in card terms.

> **De Bruijn Bracelet Principle**
> *For all k we can find a sequence of 2^k Reds and Blacks such that when the ends are connected to form a bracelet, each possible pattern of k consecutive color values—imagine sliding a window of width k around it—occurs exactly once.*

Actually we can usually find a lot more than one such sequence. Graph theory is helpful here [Diaconis and Graham 11, page 20], and a careful analysis reveals that there are $2^{2^{k-1}-k}$ distinct such maximal length de Bruijn sequences (or necklaces). For our applications it doesn't matter in which direction we travel around these circles, so we are really interested only in de Bruijn bracelets. (Necklaces are direction sensitive, bracelets are not.)

Small examples are easily constructed by hand. Let's use 0 and 1 instead of Red and Black. The only de Bruijn sequence of length four, corresponding to $k = 2$, is 0011. When $k = 3$, we find that 00011101 is one of two ($= 2^{2^{3-1}-3}$) options, but since the other de Bruijn necklace here is the same sequence in reverse, we consider there to be only one de Bruijn bracelet with eight beads.

There are sixteen ($= 2^{2^{4-1}-4}$) de Bruijn sequences or necklaces of length $16 = 2^4$, corresponding to $k = 4$. (Counting the distinct bracelets is another matter!) Here's one of them, with gaps inserted to aid in readability:

$$0000\ 1111\ 0110\ 0101.$$

Fans of 1970's Swedish pop music, who are also alphabetically inclined, may find it easier to remember that last one converted to AAAA BBBB ABBA ABAB,, which can, in turn, be morphed to RRRR BBBB RBBR RBRB for Red and for Black cards.

Plugging $k = 5$ into the above formula shows that there are 2048 binary de Bruijn sequences, or necklaces, of length 32. Here is one of them:

$$0000\ 0111\ 1101\ 1010\ 1110\ 0101\ 0011\ 0001.$$

The longest possible runs of 0s and 1s—namely, 00000 and 11111 in this case—need not be adjacent in such sequences, as the equally valid example 0000 0101 0010 0011 1110 1110 0110 1011 confirms.

Here's a simple way to construct one de Bruijn necklace of length 2^k: start with $000\ldots01$—that is to say, $(k-1)$ 0s followed by a 1—and continue it by using the rule $x_{k+i} = x_i + x_{i+1} \mod 2$. By this we mean that the next bead x_{k+1} is obtained by adding the first and second, the one after that x_{k+2} by adding the second and third, and so on, additions begin done mod 2. Hence, the sum of two consecutive beads mod 2 shows up k terms later. This procedure produces different "windows of width k" as we progress, and the sequence so generated repeats on cue after an appropriate number of steps, to yield a de Bruijn necklace as desired.[5]

---

[5]This is not the only way to generate de Bruijn sequences: a greedy algorithm approach is suggested in [Diaconis and Graham 11, page 23].

For instance, if $k = 3$, this yields 00101110, giving a new de Bruijn necklace. However, upon reflection (and rotation), it yields the same bracelet with eight beads that we had earlier. If $k = 6$, it yields a de Bruijn bracelet of length sixty-four that starts 0000 0100 0011 0001 0100 1111 . . . .

Note that if we don't seek maximal length when $k = 3$, then 0010110 can be snipped here and there to yield 000111, which has the property that the six "three beads in a row" obtained by sliding a window of width three around it (looping back to the beginning at the end) are all distinct, although we don't get either 101 or 010.

10♠   What's Black and Red and Red All Over?

How it looks: *A red-backed deck is taken out and cut several times before being passed to an audience member, who is invited to cut it further and then take off the top card. A second person now takes the deck and removes the new top card. Two more people do this, until four people each have one card. Take the deck back, shuffle it, and set it aside. Have the four people to look at their cards, noting suits and values. Have the cards displayed in a face-down row, in the order in which they were selected.*

Say, "Let's identify the Red cards first. Who had a Red card?" Pause while corresponding audience members identify themselves. Joke, "Judging by what I can see here," waving at the row of face-down red-backed cards, "I was sure you all had Red cards!" Pause again and then continue, "So, which Red card should I name first?" As soon as somebody indicates a particular card, name it, and have it turned over to confirm that your identification is correct. Continue with the remaining cards, in any order. You should score four out of four.

How it works: The seemingly light-hearted gag about Red cards above is in fact a crucial bit of fishing: Without making it too obvious that you really need to know, you find out which of the selected cards have Red faces; from knowledge of the Red/Black distribution alone, you can, with sufficient preparation, tell what all four cards are.

The red-backed "deck" referred to at the start above is actually a packet of forty-eight cards carefully assembled from parts of three identical regular red-backed decks. The "four shortened" deck you use consists of the same sixteen cards from three normal decks, arranged the same way and stacked on top of one another, so that the sixteen-card cycles of RRRR BBBB RBBR RBRB repeat.

Which sixteen cards? It's up to you. Experiment, starting with something easy to remember—but not too "obvious," bearing in mind that four in a row will end up exposed to view—perhaps alternating Diamonds & Hearts and Clubs & Spades within the subsequences of Reds and Blacks.

More sophisticated selections lead to additional magic possibilities. Once you have committed a particular sequence of sixteen cards to memory, assemble three such packets, taken from three identical red-backed decks, and stack them to form a forty-eight-card "deck," and proceed as above.

Source: This is a simplified version of an effect that can be found in Persi Diaconis and Ron Graham's wonderful book *Mathematical Magic* [Diaconis and Graham 11] (as we see below, they work with a length-five thirty-two-term sequence.) The above version was published online at MAA.org in the December 2008 *Card Colm* "What's Black and Red and Red All Over?" [Mulcahy 08_12].

Presentational options: Instead of basing the effect on the distribution of Red and Black cards, one could rework it so that it's based on evens and odds, or primes and composites, or one specific suit, for instance, Clubs, versus the rest. In the last suggested case, you can say, "Let's start with Clubs, who's got those?"—an innocent enough sounding comment that serves its information-fishing purpose well. Once the spectators with Clubs are known, then everything is known. You can identify those Clubs followed by the other cards in any desired (suit) order. Being able to say things such as, "Let's move on to Diamonds. Yes, that would be you, madam, you must have the King of Diamonds," gives an appropriate (and totally correct) impression that you are all-knowing. For such an effect, each cycle of sixteen cards must, of course, consist of eight Clubs and eight non-Clubs.

More ambitious readers may wish to work with $32 = 2^5$ cards, and five audience members, or a single spectator who grabs five cards in a row as a poker hand. Diaconis and Graham provide details of a de Bruijn sequence with matching cheat sheet in this case [Diaconis and Graham 11, pages 21–23]. For many years, Persi Diaconis has been performing such a version to wide adulation—without any cheat sheet—and nobody's ever accused him of not playing with a full deck. Note that since $64 = 2 \times 32$ is closer to 52 than 32 is, you could consider using two Diaconis decks as above—still with five audience members—if you're concerned that a thirty-two-card "deck" will look too thin.

Can we start with a de Bruijn sequence of length $64 = 2^6$ (for six audience members) and then cut it down to one of length 52 while preserving the distinct property of sliding windows of width 6? Yes, and *Mathematical Magic* [Diaconis and Graham 11, Chapter 2] has all of the elegant details. University of Newcastle staffer Christian Perfect and doctoral student David Cushing came up with the following take on this problem. It uses a different value correspondence ($0 =$ King, $1 =$ Ace, ..., $12 =$ Queen) than the one in the associated *Aperiodical* article [Perfect 12] (and the video linked to that).

The goal is that if six cards are drawn from the deck, the resulting 6-bit string corresponding to the sequence of Reds and Blacks gives the value of the sixth card. So 6-bit strings were to be assigned to the cards in a full deck, four for face value and two for suit, the latter in the order Diamonds, Clubs, Hearts, and Spades.

For each of the four suits, there are three 6-bit strings representing nonexistent face values (13–15). They argued, "Any sequence we used would have twelve unused strings at the end. So we had to make one kludge and manually swap bad strings in the usable part of the sequence with good ones at the end."

With the aid of a Python script, they constructed a de Bruijn sequence of length 64, which was then decoded to a sequence of sixty-four cards (some of them virtual). They wrote (in email) that their code

> finds bad cards in the deck of fifty-two and swaps them with good cards in the pile of twelve castoffs at the end, making sure the colour is preserved. It turns out there are eight bad cards in the deck that need swapping. So, we only need to remember eight pairs of binary strings and a simple iterative rule which constructs the sequence in order to perform the trick.

Table 10.1 and the following list of swaps should help in performing such a version.

Binary	Card	Binary	Card
000000	King of Diamonds	000001	King of Clubs
000010	King of Hearts	000011	King of Spades
101100	Jack of Diamonds	101101	Jack of Clubs
101110	Jack of Hearts	101111	Jack of Spades
110000	Queen of Diamonds	110001	Queen of Clubs
110010	Queen of Hearts	110011	Queen of Spades

Table 10.1. Binary values of cards.

Swaps: 110101 → 010101   (010101: Five of Clubs)
         110110 → 110000   (110000: Queen of Diamonds)
         110111 → 010111   (010111: Five of Spades)
         111101 → 011111   (011111: Seven of Spades)
         111001 → 101111   (101111: Jack of Spades)
         111011 → 101011   (101011: Ten of Spades)
         111010 → 000000   (000000: King of Diamonds)
         110100 → 100000   (100000: Eight of Diamonds)

In *Magical Mathematics* [Diaconis and Graham 11, page 56] there is an even better effect in which a regular deck is cut and five people take consecutive cards, whereupon the identity of each card is deduced merely from having all people with cards of the same suit stand together, without revealing what those suits are.

In the final offering here, you know what cards several spectators have selected as soon as you know how they compare to each other in value.

Rank-Rich Bracelets

Let's focus on the relative ranks of three adjacent values in a row or circle of distinct numbers. There are six possibilities, namely the rank permutations LMH (representing low, medium, high), LHM, MLH, MHL, HLM, and HML. For instance, the circle obtained from 1, 2, 3, 4, 5, 6 (considered wrapped around to 1 again) produces LMH four times, then MHL, and finally HLM. As pointed out by Diaconis and Graham in their chapter "Products of Universal Cycles" in the book *A Lifetime of Puzzles* [Diaconis and Graham 08, page 36], the circle obtained from 1, 3, 2, 1, 3, 3, 4 (considered wrapped around to 1 again) produces LMH, HML, MLH, LMH, MHL, and HLM. That is to say, all six possible rankings occur, so that we have a rank-rich bracelet.

It would be perfect if there were rank-rich bracelets of six distinct numbers. There are $\frac{5!}{2} = 60$ different (reversible and spinnable) circles using the numbers 1–6. Of these, consider the one obtained by connecting the ends of 1, 4, 6, 2, 5, 3, as depicted in Figure 10.15.

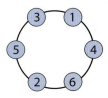

Figure 10.15. A perfect rank-rich bracelet.

This bracelet is highly desirable as it has real magic potential.

Rank-Rich Principle for Triplets
Each possible rank permutation of three objects occurs exactly once in the bracelet obtained by joining the ends of 1, 4, 6, 2, 5, 3, considering the adjacent values: $\{1, 4, 6\}, \{4, 6, 2\}, \ldots, \{3, 1, 4\}$.

Hence, if three adjacent values of these six are chosen by different people, then one need only ask who has the largest and who has the smallest to be able to tell exactly which value each person has.

Interested readers may wish to investigate whether, upon reflection (or rotation), the indicated circle is unique. Incidentally, to remember that circle, we found it helpful to think of the area of a circle of radius 5. (Can you see why? Hint: Think Omega.)

Diaconis and Graham found circles of length $4! = 24$ and $5! = 120$ for which sliding windows of lengths 4 and 5, respectively, cycle through all possible rank permutations of that many objects. They then conjectured, and eventually proved with the assistance of Fan Chung, that this could be done for any k in a suitable circle of length $k!$, although they report that their proof was difficult and nonconstructive [Diaconis and Graham 11, page 50].

The cards in ordinary decks are not naturally numbered up to 24 or 120, however, instead consisting of four suits of thirteen repeated values. Now imagine a deck with five suits, from which we assemble a packet of all five Aces, 2s, 3s and 4s, but only four 5s, having 24 cards total. In a sense, one can pull this off with cards from two ordinary decks.

Rank-Rich Principle for Quadlets
Each possible rank permutation of four objects occurs exactly once in the circle obtained by connecting the ends of $1, 2, 3, 4, 1$, $2, 5, 3, 4, 1, 5, 3, 2, 1, 4, 5, 3, 2, 4, 1, 3, 2, 5, 4$, considering the adjacent values: $\{1, 2, 3, 4\}$, $\{2, 3, 4, 1\}, \ldots, \{4, 1, 2, 3\}$.[6]

It follows that if four adjacent cards from such a packet of twenty-four cards of those values (and varying suits) are chosen by different people, then you need only ask who has the largest, who has the second largest, and who has the third largest, to be able to tell exactly which value each person has. You can also identify the suits, if you've memorized them. Note that you'll need to use cards from two identical decks.

There is a reason Diaconis and Graham needed five copies of each card value above. A recent result of mathematician Robert Johnson establishes an earlier conjecture that $k + 1$ is the smallest number of distinct values that can be assembled into a circular deck of $k!$ cards so that the relative order of each group of k adjacent cards is distinct [Johnson 09].

Much more fascinating material, both mathematical and magical, appears in the chapter "Universal Cycles" of *Magical Mathematics* [Diaconis and Graham 11]; in truth, we have only skimmed the surface.

We close with our version of the Chapter 4 opener from *Magical Mathematics*, a superb cocreation with prolific magic inventor Ronald Wohl, that uses the rank permutation explorations above. Here it is, stripped of all magic coating.

---

[6]While no ties are encountered here—meaning equal numbers within the sets of four considered—that possibility is also entertained by Diaconis and Graham.

10♦   Gold & Silver

How it looks: *Shuffle a deck and deal six cards into a face-down circle, inviting a spectator to pick any one of them. A second person picks one of the two cards beside that card, and finally a third picks either of the two "exposed" cards in the broken bracelet. Have the three cards compared and ask for information: either (1) who has the highest and who has the lowest values, or (2) what the average value is. No matter what response you get, you are soon able to announce the exact identities of all three cards.*

How it works: First note that Diaconis and Graham's rank-rich bracelet, which is obtained by joining the ends of $1, 4, 6, 2, 5, 3$, is also sum-rich for triples in the sense explored back in Chapter 2. Hence, in addition to allowing for three adjacent cards to be identified based only on rank information, this bracelet has the property that the six possible triple sums are distinct. Those sums are $8, 9, 10, 13, 12, 11$, as we go around the circle, based on the central element of each sum, so that $8 = 4 + 1 + 3$ is listed first, followed by $9 = 1 + 3 + 5$, and so on. Notice that these sums are the six consecutive numbers $8, 9, 10, 11, 12, 13$ in a very easy to remember order. Hence, given the mean or the sum, it's not difficult to backtrack and deduce the identity of the three cards that gave rise to that number.

We suggest loading the top of a deck with six cards of the desired values, in order, using any mixed suits you can remember, such as those shown in Figure 10.16 (CHaSeD, if considered in numerical order). Maintain them there throughout several fair-seeming shuffles.

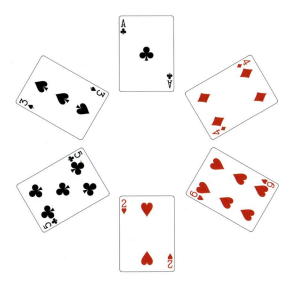

Figure 10.16. Perfect circle of six.

Source: This is adapted from an effect in Diaconis and Graham's book *Mathematical Magic* [Diaconis and Graham 11], and was published December 2011 online at MAA.org in "Magical Mathematics: Recurring Cycles of Ideas of Cycles" [Mulcahy 11_12].

Presentational options: You can opt to ask for "The mode, median, or mean—your choice!" above. (Of course, there can't be a mode, but only you know that.) As long as nobody responds with "two" or "four," all is well; you should be able to say who had which card. If "two" or "four" is called out, then you need to fish a little to determine whether it's a median or mean before successfully concluding the trick.

For the record, the circle of six numbers considered above is product-rich for triples too, so you could just request the product. Also, for what it's worth, only twelve turns up as both a sum and a product, so you may wish to gamble on asking for either.

Here's yet another possibility. From a pre-arranged and cut deck of fifty-two cards, any five adjacent ones are selected by five people. You are able to identify all five cards based only on the knowledge of who holds the top three in rank. Say something like, "Those are your scores, ladies and gentlemen. The highest one gets Gold, the next highest Silver, the third highest Bronze. I'm afraid that's it, the other ones don't get anything." Then hand out mock medals. As soon as you know who gets which one, and assuming you have memorized the initial deck setup, you can then name the exact card (not just the value) that each of the five people has.

Why is there hope of pulling off such an effect? Basically, because, given any five adjacent cards, the number of gold/silver/bronze arrangements of some three of them is $5 \times 4 \times 3 = 60$, which is greater than 52, the number of cards (or possible positions) in a deck.

How does one actually come up with an appropriate deck arrangement in the first place? Very carefully. Good luck.

Parting Thoughts

Many fascinating questions are spawned by the mathematics behind the third and fourth effects above, and a great place to start is with the two Diaconis and Graham sources. Incredibly, they consider long sequences with the property that color *and* rank information for three consecutive values permits the correct deduction of position within the sequence, and hence the identification of those three cards as well as the three following them. (These are the products of universal cycles of [Diaconis and Graham 08] and [Diaconis and Graham 11, Chapter 4].)

Electrical engineer and magic inventor Leo Boudreau has published several fine books of original material inspired by binary mathematics and de Bruijn sequences in particular.

Below, we only suggest fun things to explore relating to GOATs.

- How does Lennart Green's algorithm work if we decide in advance that we always want to end up with the Reds on top of the Blacks, so that when fanned in the usual way, the Reds are to the left of the Blacks? (Above, we were happy either way.)

- Does the GOAT separation algorithm work in essentially the same way if the mantra is two-one instead of one-two?

- Does it matter in the long run if we adjust the adjustment suggested on page 243 by always cutting from the back to the front, if possible, to complete a pass, instead of from front to back? Or by randomly doing one of these two possible adjustments?

- Experiment by applying GOATs to perfectly alternating Red/Black packets of at least a dozen cards in which the values are in order, such as a packet that starts A♥, A♣, 2♥, 2♣, 3♥, 3♣, ... What numerical patterns emerge in the separated parts, and can any mathemagical advantage be taken of these, for instance, if different cards are used?

- Can Green's algorithm be modified to partition a shuffled deck into four suits?

To be sure of having a "Top Twenty Hit" (10♥)—or success at "Poker with any Ten Cards" (3♦)—we must appeal to the odds gods. Let's close by crunching some numbers.

- We need to justify the Birthday Card Triple Match Principle on page 251. There are $\frac{52!}{20!32!} = 1.26 \times 10^{14}$ ways to select twenty cards from a full deck,[7] and multiple value matches are inevitable by the pigeonhole principle. At one extreme, we could have five fours of a kind; though this is very unlikely indeed. At the other extreme, if all values are represented and there are neither fours of a kind nor threes of a kind, this leaves seven cards $(7 = 20 - 13)$ that must all be different from each other, so that we have seven distinct pairs.

  Hence, if there are no fours of a kind or threes of a kind, there must be between seven and ten pairs, inclusive. These four possibilities can be quantified individually, using the multiplication rule, and

---

[7]That's more than the number of cells in the human body!

their cumulative probabilities are accounted for using the addition formula (see pages 16 and 21).

For instance, to get exactly eight pairs, none of them being part of threes or fours of a kind, we must select eight from thirteen values; for each such value select two from four possible suits, which determine the eight pairs; select the other four cards ($4 = 20 - 16$) from the five remaining values ($5 = 13 - 8$); and finally select a suit for each one of those four. This can be done in

$$\frac{13!}{8!5!} \times \left(\frac{4!}{2!2!}\right)^8 \times \frac{5!}{4!1!} \times \left(\frac{4!}{1!3!}\right)^4 = 1{,}287 \times 1{,}679{,}616 \times 5 \times 256$$

$$= 2.77 \times 10^{12} \text{ ways.}$$

Dividing by $\frac{52!}{20!32!} = 1.26 \times 10^{14}$, we find that the probability of getting exactly eight pairs, none of them being part of threes or fours of a kind, is about 0.0219, or 2.19%.

Repeating that kind of calculation for exactly seven, nine, and ten pairs, and adding up the four results, we find that the probability of avoiding fours or threes of a kind is about 3.98%, which is to say, about 96% of the time, we will get at least one three of a kind (or better). It's true!

Slippery Slopes

In the final three chapters of this volume, we move on to what we consider to be the true royalty of mathematical card effects. Your reputation as a mathemagician should be elevated to a whole new level once you've mastered a few of these.

"No Drama Queen" (J♦) is impressive, but tricky to pull off in the usual setting. It adapts well as a phone (or email, or instant message) effect. The much easier "Erdős Numbers" (J♠) is just as good for the average audience, as long as nobody demands a repeat performance.

> *Select two volunteers from the audience. Shuffle the deck and hand it to the first volunteer. Now turn away and gradually give the following directions: "Deal the top five cards into a pile, and set the rest of the deck aside. Mix up those five cards further and display them in a face-up row. Note how random they are. Turn each of them face down."*
>
> *Turn around and ask the second volunteer to turn two of the cards face up once more. With little hesitation, you identify the three face-down cards one by one. They are exposed to reveal that you are correct.*

We start with something simpler that works for "any ten cards"—as opposed to "any five."

J♣ Ten Soldiers

How it looks: *Select two volunteers from the audience. Shuffle the deck and hand it to the first volunteer. Now turn away and gradually give the following directions: "Deal the top ten cards into a pile, and set aside the rest of the deck. Mix up those ten cards further and display them in a face-up row. Note how random they are. After a suitable period of reflection, turn each of them face down."*

Turn around, and ask the second volunteer to silently turn five or six of the cards face up once more. These may be as shown in Figure J.1, interspersed with the rest face down.

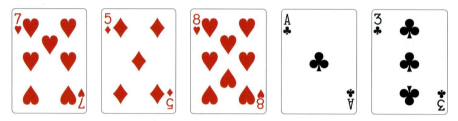

Figure J.1. The face-up cards you see.

You survey the scene, and one by one you name all of the remaining face-down cards. They are turned over to reveal that you are correct.

How it works: The first volunteer can be any audience member—we recommend selecting one known to be skeptical of mathemagic—but the second one is your secret accomplice. Let's call him Larry. When shuffling the deck at the outset, be careful not to disturb the top stock (see page 10), consisting of ten special cards planted there in advance, known only to you and Larry.

The first volunteer deals out and then mixes these cards, thus determining the order in which they are laid in a face-up row. In performance, this randomness should be played up in your comments. The suitable period of reflection we suggest above is to give Larry time to (1) check for a pattern that must occur, as we will soon explain, and (2) memorize the positions of four or five key cards.

After all of the cards are flipped face down again and you turn back to face the audience, those key cards just noted by Larry are the ones that are left face down for you to identify. He turns over the others in the displayed row, either from left to right or from right to left, depending on exactly what he's trying to communicate to you. Note that the communication is entirely mathematical, your back is turned during the crucial stages, and Larry does not speak when he turns over the non-key cards, as you watch.

First let's explain a simpler version that can be performed if you don't wish to pretend—as we've suggested—that you're using ten random cards. Assume that the cards used are the Ace to 10 of any one suit. Surprisingly, it's possible for you to identify the four or five face-down cards you see at the end because Larry can ensure that they are *in numerical order*. This follows from an old result due to mathematicians Paul Erdős and Gyorgy Szekeres.

40% Slippery Slope Principle ▶◀

In any arrangement of ten different numbers, there are always at least four among the ten, not necessarily beside each other, that are in numerical order. Hence, there is always a slippery slope of length four (or more), forming either a rising run or a falling run.

For instance, consider the arrangement $3, 4, 2, 5, 1, 9, 7, 10, 6, 8$ of the first ten counting numbers. The rising runs of interest include $3, 4, 5, 9, 10$ and $3, 4, 5, 6, 8$, whereas there are no falling ones longer than three. In contrast, the arrangement $8, 10, 1, 9, 6, 7, 5, 4, 2, 3$ has no rising run of length four or longer, but it has falling runs of length six.

If we are lucky, as we just saw, slippery slopes of length five or six may be found, but the point is that we are totally guaranteed to find slippery slopes of length four. We defer the proof of this until a little later.

In any case, assume Larry has spotted a slippery slope of length four (or five). Those are the key cards that are left face down at the end for you to identify; you can safely assume that the face-down cards are in numerical order, either rising or falling. That's almost sufficient to enable you to name them once you figure out which values they have. Since the non-key cards are on display, face up, you can see exactly which values are missing, and hence you know which ones comprise the face-down cards.

It remains to be explained how to address the rising or falling issue. We have suggested that Larry turn over the non-key cards after you turn back. He can do this from left to right if the hidden cards are rising, and from right to left if they are falling, so that you know which kind of slippery slope you are dealing with.

The problem with the version just outlined is that the 1–10 order is too obvious, and people may catch on quickly, especially if you were to offer a repeat performance. A better idea is to use the alphabetical order of the value names:

Ace, Eight, Five, Four, Nine, Seven, Six, Ten, Three, Two.

It takes practice to learn to think of these numbers in this unfamiliar order, but the payoff makes it worth the effort. It's puzzling to most audiences, even those who know the general underlying principle.

In practice, we recommend using these cards stacked at the top of the deck at the outset:

$$A\clubsuit, 2\clubsuit, 3\clubsuit, 4\blacklozenge, 5\blacklozenge, 6\blacklozenge, 7\heartsuit, 8\heartsuit, 9\heartsuit, 10\spadesuit.$$

Notice that *the suits are in alphabetical order here.* Also, the Red cards are sandwiched in between the Black ones—an observation that remains true when we alphabetize by value instead, to get

$$A\clubsuit, 8\heartsuit, 5\blacklozenge, 4\blacklozenge, 9\heartsuit, 7\heartsuit, 6\blacklozenge, 10\spadesuit, 3\clubsuit, 2\clubsuit.$$

(Alternatively, you could work with this value order, but throw in a memorable suit order of your own choosing.)

Now, let's suppose the cards displayed by the first volunteer are in this order: $7\heartsuit, 2\clubsuit, 5\blacklozenge, 10\spadesuit, 6\blacklozenge, 8\heartsuit, A\clubsuit, 9\heartsuit, 4\blacklozenge, 3\clubsuit$. Remembering to use alphabetical, not numerical, order, Larry see that there are no rising runs longer than three, but there is a falling one of length five, namely, $2\clubsuit, 10\spadesuit, 6\blacklozenge, 9\heartsuit, 4\blacklozenge$.

These are the cards you will have to identify in due course. Right before the first volunteer turns all ten cards face down, Larry carefully notes their key locations. In due course, he flips face up the *other* five cards, namely, $7\heartsuit, 5\blacklozenge, 8\heartsuit, A\clubsuit, 3\clubsuit$. This is done in your presence, and Larry is careful to turn them over from right to left to tip you off to the falling nature of the key face-down cards. What you see as a result is those five face-up cards (see Figure J.1), scattered in a row of ten cards, the other five being face down.

At a glance you can tell that the missing cards are Two = $2\clubsuit$, Four = $4\blacklozenge$, Six = $6\blacklozenge$, Nine = $9\heartsuit$, and Ten = $10\spadesuit$. Finally, you *reverse alphabetize* these as $2\clubsuit, 10\spadesuit, 6\blacklozenge, 9\heartsuit$, and $4\blacklozenge$. This is the order of the hidden cards, from left to right, as shown in Figure J.2.

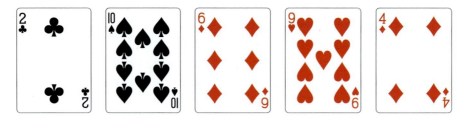

Figure J.2. The face-down cards you can't see.

Name them in any order, and have them exposed to confirm the accuracy of your revelations. (The color separation in both the seen and unseen card sequences here is coincidental, and not significant.)

Source: Original. Published online at AMS.org in "Smells Like Team Spirit" in October 2000 [Mulcahy 00], and at MAA.org in the June 2005 *Card Colm* "A Little Erdős/Szekeres Magic" [Mulcahy 05_06]. It appeared in print in January 2002 as part of "All You Need Is Cards," by the reclusive Brain Epstein, in the Martin Gardner tribute book *Puzzlers' Tribute: A Feast for the Mind* [Epstein 02].

This was inspired by a remark of Martin Gardner [Gardner 87, page 5] that was in turn inspired by a result of Erdős and Szekeres [Aigner and Ziegler 04, page 124], which is discussed in some detail in "Slippery Slopes" below.

Presentational options: Another way that your accomplice can convey whether the four or five face-down cards are rising or falling is to have him collect the row of ten cards before you turn back, and then a moment later, deal them out into a face-down row under your watchful eye. He deals all of the cards left to right if the key cards are rising, and right to left if they are falling, before he flips five or six of them over. With this approach, the non-key cards can always be turned over in left-to-right order.

An alternative handling is to have all ten cards turned face down after Larry has mentally noted which six he needs to have revealed, and then, in your presence, he points to those six, from left to right to indicate that the remaining four are in rising order, or from right to left to indicate that the remaining four are in falling order. Somebody else can flip them over, so that Larry never needs to touch the cards at all.

A further possibility is discussed at the end of "Erdős Numbers" (J♠).

Lovely Assistance

The cat is out of the bag. We have entered a new realm of mathemagic, in which extraordinary things are possible with the help of an assistant, whether the latter is openly acknowledged or not. We suggest concealing the fact that you're getting help, but this condition can be relaxed if you wish, especially for a mathematically savvy audience. If somebody says, "You must be using an assistant!" simply reply, "Mathematical assistance works for me; mathematics is what really makes this possible."

In the worlds of stage and street magic, it's not unusual for miracles to be pulled off with the help of a "plant," "shill," or "stooge." Such assistants (lovely or otherwise) are generally frowned upon by the classier members of the profession, perhaps being viewed as the start of the slippery slope toward magic that's less skill-based.

As we have just seen, we have a different kind of slippery slope in mind, and from this point on, we're happy to embrace assistance of a mathematical nature. We must stress that all assistant-based communication in the remainder of these pages is guaranteed to be 100% mathematical. There are no verbal or physical cues, as there might be in the general arena of assistant-enhanced magic.

Even if you are cheerfully up-front about your use of an assistant, a general audience may suspect that there is some hanky-panky going on between you and that person. To eliminate that possibility, impersonal forms of communication such as phone, email, or instant message can in most cases be considered, or you might enlist the services of an obviously innocent go-between. (Fair disclosure: those options won't work for some of the effects in this chapter, but will for most of the ones in the last two chapters.)

You may choose to dress up the effects a little—for instance, as examples of mind reading. They are likely to stump all but the most alert and sophisticated onlookers.

Slippery Slopes

"Ten Soldiers" (J♣) used the fact, still to be justified, that you can deduce the position of every one in a row of ten known cards once you know the positions of five or six special ones:

> In each list of ten there is bound,
> To be four that do rise; is that sound?
> In a paper with Erdős,
> By Gyorgy Szekeres,[1]
> A counterexample is found.

That tongue-in-cheek limerick—like many of the effects in this chapter—derives from the following remark in Martin Gardner's *Riddles of the Sphinx and Other Mathematical Puzzle Tales* [Gardner 87, page 5]:

> [in any] row of ten soldiers, no two of the same height ...
> no matter what the order, there will always be at least four
> among the ten, not necessarily standing next to each other,
> who will be in ascending or in descending order.

Since it works for a row of ten soldiers, it clearly also works for longer rows. However, as we will see in due course, if we want to find five soldiers in ascending or descending order by height, with 100% certainty, then we'll need at least seventeen soldiers in the original row.

As Gardner points out, the four of ten soldiers remark is a very special case of the following 1935 theorem by two distinguished Hungarian mathematicians.

Erdős and Szekeres Result

In any arrangement of $(k-1)^2+1$ (or more) different numbers, there are always at least k, not necessarily beside each other, that are in numerical order. Hence, there is always either a rising run or a falling run of length k (or more).

As on page 265, we refer to such rising or falling runs of numbers as slippery slopes. We have already seen this principle in the case $k = 4$: it guarantees slippery slopes of that length in any arrangement of the numbers 1–10. It works for any ten different numbers; there is no loss of generality in assuming that we are dealing with 1–10. The result is really about totally ordered sets, so that identical card values pose no problem, provided we have in mind a total ordering on the set of cards being considered.

It is important to note that the stated result is as good as we can hope for, in the sense that the integers from 1 to $(k-1)^2$ can always be arranged so as to avoid having any slippery slope of length k. For instance, the arrangement $4, 6, 1, 9, 3, 5, 2, 8, 7$ of 1–9 (which, coincidentally, has 1935 embedded in it) has no slippery slope of length four.

Taking $k = 5$, we know that every arrangement of 1–17 must contain a slippery slope of length five. Can you find an arrangement of 1–16 that avoids slippery slopes of that length? (In one sense, there are lots of them.) The lists of arrangements of numbers that avoid slippery slopes of desirable lengths is quite short, but that won't deter us from trying to squeeze some more magic out of the situation, in due course.

The arrangements $2, 3, 4, 5, 6, 7, 1$ or $10, 9, 8, 7, 6, 5, 4, 1, 2, 3$, are strongly tilted in one direction or the other, having highly visible long slippery slopes. In one sense, the Erdős and Szekeres result can be thought of as guaranteeing a minimal amount of numerical order in less extreme, more middle-ground, kinds of arrangements.

A proof of the Erdős and Szekeres theorem can be found in *Proofs from The Book* [Aigner and Ziegler 04], and many interesting proofs are available on the Internet. The result is a finite, quantifiable version of one of the cornerstones of real analysis (the fuel that keeps calculus ticking over): *every infinite sequence has an infinite monotone subsequence* [Rudin 76].

Off-Centered Hungarians

Since the arrangement $4, 6, 1, 9, 3, 5, 2, 8, 7$ of 1–9 has no slippery slopes of length four, it certainly has no rising runs of length five. But it has lots of falling runs of length three, such as $6, 3, 2$, or $9, 5, 2$, or $9, 8, 7$.

It turns out that for any arrangement of different numbers, the absence of sufficiently long rising runs is compensated for by the presence of shorter falling runs. There is trade-off between rising and falling slippery slopes of different lengths, and we can quantify it.

Note that in the above example we can write the arrangement length nine as $(5-1) \times (3-1) + 1$. We have stumbled on the asymmetric version of the statement made earlier.

Asymmetric Erdős and Szekeres Result
Any arrangement of $(s-1) \times (t-1) + 1$ different numbers contains either s numbers in rising numerical order, or t numbers in falling numerical order. (These numbers need not be beside each other.) This also holds if we exchange the words rising and falling.

So there is always a slippery slope of length s forming a rising run, or one of length t forming a falling run, or vice versa, in any arrangement of $(s-1) \times (t-1) + 1$ different numbers.

Now, since $13 = (5-1) \times (4-1) + 1$, we know that any arrangement of any Ace–King that fails to have a rising run of length five must make up for that by having at least one falling run of length four. Likewise, arrangements that fail to have a falling run of length four will have at least one rising run of length five. It's okay to fail on both counts—these have both rising and falling slippery slopes of length four—but such examples turn out not to be abundant.

It's also true that $13 = (7-1) \times (3-1) + 1$, and so any arrangement of Ace–King that fails to have a rising run of length seven must have at least one falling run of length three, and vice versa.

Of course, $13 = (13-1) \times (2-1) + 1$ also. This puts into perspective the failure of the close-to-extreme arrangement Ace, King, Queen, Jack, 10, 9, 8, 7, 6, 5, 4, 3, 2 to have a falling run of length thirteen: it's because of the opening (and only) rising run of length two! (Ace = 1 here.)

A Certain Uncertainty

Let's return to the symmetric perspective once more. We know that arrangements of just under $(k-1)^2 + 1$ distinct numbers need not have slippery slopes of length k, but as k gets larger, it seems that most such arrangements do, provided they aren't too far short of their target length.

Taking $k = 4$, there are arrangements of 1–9 avoiding slippery slopes of length four or more. We saw one earlier, but there is a reason it took a little thought to come up with it, as there are only 1,764 such examples out of the $9! = 362,880$ possible arrangements. In the other 362,116 cases, or about 99.5% of the time, there are slippery slopes of length four or more.

Taking $k = 5$, a similar result presumably holds for arrangements of any sixteen cards that don't contain a slippery slope of length five. The same is true if we scale back to thirteen cards. For instance, using an Ace–King of one suit, then Jack, King, 4, 2, 7, 5, Ace, 10, 9, Queen, 3, 6, 8 is one example of an arrangement that avoids a slippery slope of length five, and 10, 3, 6, 5, Queen, Ace, Jack, King, 2, 7, 9, 4, 8 is another. However, if you shuffle such a packet and then lay it out in a face-up row, it seems that there almost always are five (or more) cards that are either rising or falling. Here's why:

> **Probabilistic 38% Slippery Slope Principle** ▶◀
> *In about 98.4% of the 13! (more than 6 billion) possible arrangements of any thirteen distinct numbers, there are at least five, not necessarily adjacent, that are in numerical order, either rising or falling.*

(The name derives from the crude approximation $\frac{5}{13} = 0.38$.)

For an entire deck, considered with respect to any fixed order (see page 13), we have $(8-1)^2 + 1 \le 52 < (9-1)^2 + 1$, and so only slippery slopes of length eight are guaranteed. However, slippery slopes of length eleven or more are present in about 97% of the unimaginably vast total number of possible arrangements (see page 17).

Such matters have been studied at least as far back as 1962, when Stan Ulam did early computer simulations to approximate the true counts. We're very grateful to mathematician Neil Calkin of Clemson University for sharing the above estimates. They were obtained using MATLAB software.

Let's gamble on getting long slippery slopes in the case of a full suit shuffled to the max.

J♥ Slippery Enough

How it looks: *Select two volunteers from the audience. Hand the deck to the first one for shuffling. Now turn away while giving the following directions slowly: "Please fan the deck so that you can see the card faces, and extract the Diamonds as you find them. Display those thirteen cards in a face-up row, and set aside the rest of the deck. Now turn the Diamonds face down."*

Turn around, and ask the second volunteer to silently turn some of the cards face up once more. You may see something like what's depicted in Figure J.3, interspersed with the rest of the Diamonds face down.

You survey the scene, and one by one you name the remaining face-down cards. They are turned over to reveal that you are correct.

Figure J.3. The face-up Diamonds you see.

How it works: As in the last effect, the second volunteer is your secret accomplice, Larry. The first volunteer genuinely shuffles the deck, and thus determines the order in which the Diamonds are laid in a face-up row. Play that up in your comments. Larry then checks for a slippery slope of length five, which is very likely to occur, though not 100% certain, and he also memorizes the positions of those key cards. If no such slippery slope can be found, Larry hands these cards to somebody else, for "one last shuffle, to really mix the cards" before trying again. Chances are, he'll find what he seeks the second time around.

Once the cards are flipped face down again and you turn back to face the audience, it is those key cards that will remain face down for you to identify. Larry turns the others face up as you watch, either from left to right or from right to left, depending on whether he's trying to convey to you that the hidden slippery slope is rising or falling, respectively.

For instance, suppose the first volunteer displays the Diamonds as 6, 8, 3, K, 4, A, 5, J, 9, Q, 2, 7, 10. Then 3, 4, 5, 9, Q is among several (rising) slippery slopes of length five. Those could be the key cards left face down at the end for you to identify; however, as in the previous effect, we recommend thinking in terms of the less obvious alphabetical order: Ace, Eight, Five, Four, Jack, King, Nine, Queen, Seven, Six, Ten, Three, Two. Reexamining 6, 8, 3, K, 4, A, 5, J, 9, Q, 2, 7, 10 in this light, Larry sees that 8, 4, J, 9, Q, 7, 10 is a rising slippery slope of length seven. So when he exposes the other six cards, namely, the 6, 3, K, A, 5, 9, 2, from left to right, you know which ones are missing, and you can name the seven mystery cards as they appear in the face-down row, just by saying their names in alphabetical order.

Source: Original. June 2012.

Presentational options: Mixed suits can be used if you can speed memorize. Have the deck shuffled and then say, "Let's just use one card of each value," going through the faces and tossing out thirteen carefully preplanned and remembered cards as you find them.

The options mentioned at the end of "Ten Soldiers" (J♥) are also worth considering.

♣   ♥   ♠   ♦

The 40% Slippery Slope Principle is the $k = 4$ case of the Erdős and Szekeres result. The smallest interesting case of this theorem actually occurs when $k = 3$, yielding the following little-known fact.

60% Slippery Slope Principle

In any arrangement of five distinct numbers, there are always at least three among the five, not necessarily adjacent, that are in numerical order, either rising of falling.

Compared to the 40% Slippery Slope Principle we've halved the list size while simultaneously increasing the slippery slope conclusion by 50%. Given five known items, we're saying that full knowledge of how they are arranged can be gleaned from less than half of the information seemingly required.

Here's an elementary argument as to why this holds, given five different numbers arranged as A, B, C, D, E:

Case 1: A < B. We claim that if there isn't a rising run of length three, then there is a falling one instead. Assume there is no such rising run. Then each of C, D, and E must be less than B; otherwise we'd have a rising run of length three starting with A, B. If it happens that C > D, then B, C, D is a falling run of length three, whereas if D > E, then B, D, E is a falling run of length three; in either case we are done. One of those subcases must in fact occur; otherwise we'd have both C < D and D < E, and hence a rising run C, D, E of length three, contrary to our assumption.

Case 2: A > B. The argument is the same as in Case 1, reversing the direction of each inequality, and switching the words "rising" and "falling."

For a tongue-in-cheek application of this in a political context, see the main puzzle in Pradeep Mutalik's "Numberplay: Order in the Ranks" in the *New York Times* Wordplay blog in May 2010 [Mutalik 10_05]. The solution is buried at the end of "Numberplay: The Playful Mr. Gardner," which appeared four weeks later following the death of the best friend mathematics ever had [Mutalik 10_06].

For a different kind of application, which also takes advantage of the General Gilbreath Principle, and uses ESP (or Zenner) cards, see "An ESPeriment with Cards" in the February 2007 issue of *Math Horizons* [Mulcahy 07_02a].

♣   ♥   ♠   ♦

We're ready now for our chapter highlight.

J♠ Erdős Numbers

How it looks: *Select two volunteers from the audience. Shuffle the deck and hand it to the first volunteer. Now turn away and give the following directions, in stages: "Deal the top five cards into a pile, and set the rest of the deck aside. Mix up those five cards further and display them in a face-up row. Note how random they are. Turn each of them face down."*

Turn around and ask the second volunteer to turn two of the cards face up once more. With little hesitation, you identify the three face-down cards one by one. They are exposed to reveal that you are correct.

How it works: Once again, the first volunteer can be any audience member, but the second one is your trusty accomplice Larry. You do indeed shuffle the deck at the outset, but you are careful not to disturb the top stock. The top five cards are planted there in advance, known to both you and Larry, and they need to stay there throughout these first shuffles.

The first volunteer deals and then genuinely mixes these cards, hence determining the order in which they are laid in a face-up row. This randomness should be played up in performance. Larry then rapidly checks for a slippery slope of length three, with respect to a pre-agreed-upon order of the five cards in question, and also memorizes the positions of those cards.

After all five cards are flipped face down once more, and you turn back to the audience, it is the three cards just noted by Larry that he leaves face down for you to identify. He turns over the other two in the displayed row, either from left to right or from right to left, depending on what he's trying to communicate to you. You then see something like what's shown in Figure J.4.

Figure J.4. What you see is what you get.

Again, let's start with a simpler version. Suppose that the five cards at the top of the deck are Ace (= 1), $2, 3, 4, 5$ (ignoring suits). By the 60% Slippery Slope Principle, nothing can go wrong. For instance, given $3, 4, 2, 5, 1$, we see that $3, 4, 5$ is rising, and $3, 2, 1$ is one of several falling runs. On the other hand, $4, 1, 5, 3, 2$ has no rising run of length three, but it has two falling rows of that length. There may be longer slippery slopes, but here it's best to turn a blind eye to these. (Can you see why?)

In any case, let's assume Larry has spotted a slippery slope of length three. He leaves those cards face down at the end for you to identify. Since you will see the other two cards face up, you'll know which three are missing, so all that remains is to address the rising/falling issue. Larry exposes the two cards from left to right if the three hidden cards are rising, and from right to left if they are falling.

In practice, we suggest performing the effect with the cards 3♣, 4♠, 5♣, J♥, and Q♦, stacked at the top of the deck at the outset. Furthermore, we consider these cards to have the natural order displayed in Figure J.5.

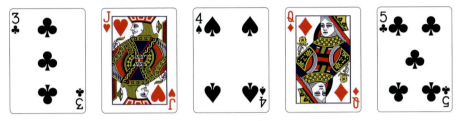

Figure J.5. Erdős is in the cards.

Note the Black Pythagorean triple interspersed with the Red mother and son. Also, the suits cycle in CHaSeD order. To see why we suggest these particular cards, fan them out the "wrong" way so that, in contrast to the fan shown in Figure 12 on page 12, only the pips in the bottom right corners are showing (along with the entire 3♣ at the packet face on the left). With a little imagination, the upside-down 3, J, 4, Q, 5 then on view can be thought of as a distorted version of the word "Erdős"—that word can be considered as a kind of mnemonic.

We work through two examples using these cards. For the first one, let's suppose the cards displayed by Larry are in this order: J♥, 4♠, Q♦, 3♣, and 5♣, as shown in Figure J.6.

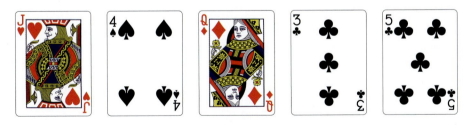

Figure J.6. The reality.

Then both J♥, 4♠, Q♦, and J♥, 4♠, 5♣ form rising runs of length three. Let's suppose that Larry decides to go with the first possibility. The first volunteer turns all five cards face down. After you turn around again, Larry flips the last two cards face up, from left to right, as you watch. You then see the display shown earlier in Figure J.4.

At a glance, you can tell that the 4, Jack, and Queen are missing, and you deduce their order by remembering the word Erdős; therefore, the three face-down cards must be J♥, 4♠, and Q♦.

Here's a second example, assuming that the cards displayed by the first volunteer are in this order: Q♦, 5♣, 4♠, J♥, and 3♣, as shown in Figure J.7.

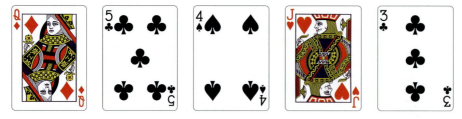

Figure J.7. Another reality.

This time, there are no rising runs of length three, but Q♦, 4♠, J♥ is one of several falling runs of that length. After the first volunteer turns all five cards face down and you turn around, Larry flips the second and fifth cards face up *from right to left* (i.e., flipping over the fifth card before the third one), as you watch. Hence, you see the display in Figure J.8.

Figure J.8. Another what you see is what you get.

At a glance, you can tell that the 4, Jack, and Queen are missing, and you deduce their order by remembering the word Erdős in reverse. Hence, the three face-down cards must be Q♦, 4♠, and J♥, in that order from left to right.

Source: Original. Published online October 2000 at AMS.org as "Smells Like Team Spirit" [Mulcahy 00] and June 2005 at MAA.org as the *Card Colm* "A Little Erdős/Szekeres Magic" [Mulcahy 05_06]. Needless to say, Brain Epstein included it in his truly fab "All You Need Is Cards" [Epstein 02].

Presentational options: The whole issue of left-to-right or right-to-left dealing (in your presence) can be side-stepped if you don't mind a slight departure from our declared policy of purity when it comes to communicating information. (This is in addition to the options mentioned at the end of "Ten Soldiers" (J♥).) For this option, Larry always deals the cards left to right before you turn back, having first flipped some over in his hands, so that a moment later you see only a static display. If a rising run is available, he deals in the usual way, in an ordinary row. If falling runs are all that are available, he also deals left to right, but in a slightly overlapping row. In the last example above, you would hence see a display such as the one in Figure J.9 when you turn back.

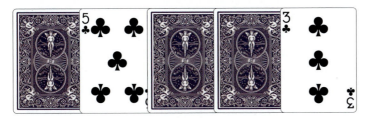

Figure J.9. A possibility for statically communicating a falling run.

Moving on, if a slippery slope of length four (or five) arises using the special card order suggested, Larry can make you appear to be even more talented, but not by turning over one (or no) cards, as that won't tell you what you need to know, namely whether what remains is rising or falling. The first option made at the end of "Ten Soldiers" (J♣) helps if you plan to have all five cards laid out face down as you watch, before some (or maybe none) are turned face up again.

Could you identify all five cards, one by one, in order, if they all remained face down and you only learned what two of their values added up to? Since $3 + 12 = 4 + 11$, it would appear not, but if we trade the J♥ for 9♥, so that the five planted cards are 3♣, 4♠, 5♣, 9♥, and Q♦, then you have a chance. This is because $\{3, 4, 5, 9, 12\}$ has the sum-rich property explored in Chapter 2: knowing the five cards involved and the total of any two of their values, you can indeed figure out what those two cards are.

In practice, Larry would need to peek at the two cards that are not part of the rising or falling subsequence, and this peeking should be done *in one of the two possible orders while you watch*, before he leaves all cards face down. Then, he tells you the total of the two key card values, which is enough for you to figure out what the cards are, as well as which is which. The other three you know by elimination, and their order is also known.

Given that nobody has any idea that you know anything about any of the five cards, your performance should seem quite miraculous.

Less Drama Is More Revealing

Let's take a closer look at the bookkeeping when we randomly mix five different (known) things, denoted by A, B, C, D, E. There are $5! = 120$ possible arrangements of any five items in a row.

In the last effect, the Erdős and Szekeres result was used to enable the identification of three special positions in a randomly arranged row of five known cards. The idea was that if these cards were shown to you face down, and your accomplice flipped over the cards in the other two slots, *in a particular order*, then you'd be in no doubt as to which card is which among the three face-down ones.

How many different experiences can *you* have in this little drama? At first, you see five face-down cards. One of them is flipped over, and it could turn out to be any of the five known cards. So Act 1 could unfold in any of $5 \times 5 = 25$ ways. Next, one of the remaining four cards is flipped over to reveal one of the other four card faces. So Act 2 could unfold in any of $4 \times 4 = 16$ ways. Over all, it seems that this whole drama can unfold in $25 \times 16 = 400$ different ways.

The upshot is that of the 400 possible experiences that appear to be available to you, fewer than a third of them (namely, the 120 corresponding to the possible card orders) could actually transpire in practice. That suggests that we can do better.

Can we drop the part where your accomplice Larry turns over two cards in a particular order while you watch? Yes, we can do without the drama. Instead, you see only a static row of cards, two of which are already face up. There are $\frac{5!}{2!3!} = 10$ options as to which two of five positions contain face-up cards, and $5 \times 4 = 20$ options as to what cards occupy those slots. So, overall you will see one of $10 \times 20 = 200$ static scenes. Certainly, since $200 > 120$, there is hope.

Table J.1 matches each of the 120 possible arrangements to one of the 400 possible displays. Hence, when presented with something like B, C, D, E, A, your accomplice knows which cards—C, E, and A in this case— to display face down. Even better, upon seeing a display such as B, down, D, down, down, you can determine which face-down card is which.

Where did this correspondence come from? The Erdős and Szekeres method we have been exploring does not help at all here. Using that approach, if presented with B, C, D, E, A, your accomplice would probably focus on the leading rising B, C, D, and opt to flip over the E and A in that order after displaying the five cards face down. However, in the current situation, you see only the end result: namely, down, down, down, E, A. Unfortunately, you have no way of distinguishing this from the situation where you walked in at the end of Act 2, your accomplice having started with D, C, B, E, A, and then focused on the falling D, C, B, displaying

See	Show	See	Show	See	Show	See	Show
ABCDE	?B?D?	ACBDE	A??D?	ADBCE	?DB??	AEBCD	A??C?
ABCED	??C?D	ACBED	?C??D	ADBEC	A???C	AEBDC	A?B??
ABDCE	?BD??	ACDBE	???BE	ADCBE	A???E	AECBD	A???D
ABDEC	?B??C	ACDEB	A?D??	ADCEB	A??E?	AECDB	A?C??
ABECD	?BE??	ACEBD	??EB?	ADEBC	A?E??	AEDBC	A??B?
ABEDC	AB???	ACEDB	AC???	ADECB	AD???	AEDCB	AE???
BACDE	??C?E	BCADE	B??D?	BDACE	?DA??	BEACD	B??C?
BACED	??CE?	BCAED	?C?E?	BDAEC	B???C	BEADC	B?A??
BADCE	?AD??	BCDAE	?C??E	BDCAE	B???E	BECAD	B???D
BADEC	?A??C	BCDEA	B?D??	BDCEA	B??E?	BECDA	B?C??
BAECD	??EC?	BCEAD	?CE??	BDEAC	B?E??	BEDAC	B??A?
BAEDC	BA???	BCEDA	BC???	BDECA	BD???	BEDCA	BE???
CABDE	C??D?	CBADE	???DE	CDABE	?D??E	CEABD	?E??D
CABED	??B?D	CBAED	???ED	CDAEB	?D?E?	CEADB	C?A??
CADBE	C?D??	CBDAE	??D?E	CDBAE	C???E	CEBAD	C???D
CADEB	C???B	CBDEA	??DE?	CDBEA	C??E?	CEBDA	C?B??
CAEBD	C??B?	CBEAD	??E?D	CDEAB	C?E??	CEDAB	C??A?
CAEDB	CA???	CBEDA	CB???	CDEBA	CD???	CEDBA	CE???
DABCE	??B?E	DBACE	???CE	DCABE	??A?E	DEABC	?EA??
DABEC	??BE?	DBAEC	???EC	DCAEB	??AE?	DEACB	D?A??
DACBE	?A??E	DBCAE	?B??E	DCBAE	D???E	DEBAC	D???C
DACEB	?A?E?	DBCEA	?B?E?	DCBEA	D??E?	DEBCA	D?B??
DAEBC	?AE??	DBEAC	??E?C	DCEAB	??EA?	DECAB	D??A?
DAECB	DA???	DBECA	DB???	DCEBA	DC???	DECBA	DE???
EABCD	?A?C?	EBACD	???CD	ECABD	??A?D	EDABC	??AB?
EABDC	?AB??	EBADC	?BA??	ECADB	?CA??	EDACB	E?A??
EACBD	?A??D	EBCAD	?B??D	ECBAD	E???D	EDBAC	??BA?
EACDB	?AC??	EBCDA	?BC??	ECBDA	?CB??	EDBCA	E?B??
EADBC	?A?B?	EBDAC	?B?A?	ECDAB	?C?A?	EDCAB	E??A?
EADCB	EA???	EBDCA	EB???	ECDBA	EC???	EDCBA	ED???

Table J.1. A correspondence that makes a great cheat sheet.

the five cards face down before flipping over the E and A in reverse order. You see the same actors on stage in both cases (down, down, down, E, A), and you don't have enough information to unmask the first three.

Clearly a different approach is called for. The mathematics required is significantly more complicated. Recently, mathematician Mark J. Tilford was kind enough to share his insight that such a no-drama version of the effect is indeed possible. Table J.1 (and its companion Table J.2 to follow) is passed on to you with his permission, along with a hint of how he came up with the correspondences in the first place.

In some ways it's a big improvement on "Erdős Numbers" (J♠), as it turns it into an effect that can be performed via phone, email, instant message, or any other impersonal form of communication. There is a price to be paid, however. A dramatic element has been removed, but you and your accomplice now have a lot of new lines to learn. So many, in fact, that you'll both have to resort to reading them all. Ironically, it's with the help of cheat sheets (such as Table J.1) that you can impress the masses with a seemingly "cheat-proof" version of the last effect. It's not so much that this effect *can* be performed via phone, email, or instant message, but we strongly suggest that it *should* be done in such a way.

To smooth the transition to the next chapter, we now switch two key human roles. This time, *you* decide what information is revealed, and it is your lovely assistant—whose existence can be admitted at the outset—who identifies the hidden cards, almost as soon as he is given the requisite hints.

J♦   No Drama Queen

How it looks: *A volunteer is sought from the audience, and given five cards to mix up. When the volunteer is satisfied that they are randomized, have them dealt into a face-up row on the table. After a suitable period of reflection, you flip three of them face down again.*

Ask the volunteer to communicate what is now visible on the table to your lovely assistant, by telephone, email, or instant message. He immediately responds, using the same technology, and correctly identifies the three face-down cards.

How it works: The volunteer genuinely mixes the five cards, hence determining the order in which they are laid in a face-up row. Your job is to decide which two to leave face up, so that when Larry learns the details of the resulting display—the volunteer should communicate something like "The second card is the Jack of Hearts, and the last is the Four of Spades"—then he is able to respond speedily with the names of the three face-down cards.

This time you need help on a large scale. The 120-entry cheat sheet of Table J.1 works for any five cards denoted A, B, C, D, E, which we now assume represent the cards used in the last effect, namely, 3♣, J♥, 4♠, Q♦, and 5♣, in that "Erdős" order.

Suppose the cards are displayed by the first volunteer in this order: J♥, 4♠, Q♦, 3♣, 5♣. That's B, C, D, A, E. Consulting Table J.1, you see that you should leave only C and E face up, so you turn the other three face down.

Everyone present now sees what is depicted in Figure J.10.

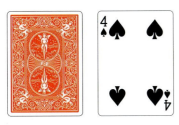

Figure J.10. What you see is what you get.

The volunteer conveys this to Larry, with words such as "the second card is the Four of Spades, and the last is the Five of Clubs." Unknown to all but you, Larry looks up ?C??E in the second cheat sheet (Table J.2), finds BCDAE, and promptly responds that the three face-down cards must be J♥, Q♦, and 3♣, in that order.

Why it works: The mathematics used to derive the correspondence given in the two cheat sheets (Table J.1 and J.2) is more complex than we can explain here. Mathematician Mark J. Tilford writes that he came up with these tables as follows (see [Gould 12] for explanations of the concepts, technical terms, and algorithm mentioned):

   i. Create a bipartite graph in which each left node corresponds to one of the 200 orderings of five distinct cards with two face-up cards, and three face-down, and each right node corresponds to one of the 120 arrangements of all five cards.

   ii. Connect a left node to a right node if they agree on all face-up cards.

   iii. Feed that graph to a maximum bipartite graph matching algorithm.

Of course, not all 200 left nodes are used in his solution; for example, A???B is one of many that are unavoidably left out in the cold.

Source: This is Mark J. Tilford's well-spotted necessary improvement on the last effect, shared November 2012.

Presentational options: We've explained the new, more mysterious workings of this effect assuming that we use the same five Erdős cards as before (on which the previous slippery slope incarnation was based). These, or any other five cards that you and your accomplice can remember, may be used. For once, we've not suggested that you pretend these are random cards, although that's still an option.

Deduce	See	Deduce	See	Deduce	See	Deduce	See
ABEDC	AB???	ACDEB	A?D??	ADBEC	A???C	BEDCA	BE???
ACEDB	AC???	ADEBC	A?E??	AECBD	A???D	BEADC	B?A??
ADECB	AD???	AEDBC	A??B?	ADCBE	A???E	BECDA	B?C??
AEDCB	AE???	AEBCD	A??C?	BAEDC	BA???	BCDEA	B?D??
AEBDC	A?B??	ACBDE	A??D?	BCEDA	BC???	BDEAC	B?E??
AECDB	A?C??	ADCEB	A??E?	BDECA	BD???	BEDAC	B??A?
BEACD	B??C?	CAEDB	CA???	CADBE	C?D??	CADEB	C???B
BCADE	B??D?	CBEDA	CB???	CDEAB	C?E??	CEBAD	C???D
BDCEA	B??E?	CDEBA	CD???	CEDAB	C??A?	CDBAE	C???E
BDAEC	B???C	CEDBA	CE???	CAEBD	C??B?	DAECB	DA???
BECAD	B???D	CEADB	C?A??	CABDE	C??D?	DBECA	DB???
BDCAE	B???E	CEBDA	C?B??	CDBEA	C??E?	DCEBA	DC???
DECBA	DE???	DCBAE	D???E	EDBCA	E?B??	DAEBC	?AE??
DEACB	D?A??	EADCB	EA???	EDCAB	E??A?	EADBC	?A?B?
DEBCA	D?B??	EBDCA	EB???	ECBAD	E???D	EABCD	?A?C?
DECAB	D??A?	ECDBA	EC???	EABDC	?AB??	DACEB	?A?E?
DCBEA	D??E?	EDCBA	ED???	EACDB	?AC??	BADEC	?A??C
DEBAC	D???C	EDACB	E?A??	BADCE	?AD??	EACBD	?A??D
DACBE	?A??E	ABCDE	?B?D?	ECBDA	?CB??	BDACE	?DA??
EBADC	?BA??	DBCEA	?B?E?	BCEAD	?CE??	ADBCE	?DB??
EBCDA	?BC??	ABDEC	?B??C	ECDAB	?C?A?	CDAEB	?D?E?
ABDCE	?BD??	EBCAD	?B??D	BCAED	?C?E?	CDABE	?D??E
ABECD	?BE??	DBCAE	?B??E	ACBED	?C??D	DEABC	?EA??
EBDAC	?B?A?	ECADB	?CA??	BCDAE	?C??E	CEABD	?E??D
EDABC	??AB?	CABED	??B?D	CBDAE	??D?E	ACDBE	???BE
DCAEB	??AE?	DABCE	??B?E	DCEAB	??EA?	EBACD	???CD
ECABD	??A?D	BACED	??CE?	ACEBD	??EB?	DBACE	???CE
DCABE	??A?E	ABCED	??C?D	BAECD	??EC?	CBADE	???DE
EDBAC	??BA?	BACDE	??C?E	DBEAC	??E?C	DBAEC	???EC
DABEC	??BE?	CBDEA	??DE?	CBEAD	??E?D	CBAED	???ED

Table J.2. Your assistant's cheat sheet, looked up by right column.

Parting Thoughts

Don't miss the suggested magic move at the end of this list.

- We've implied that for arrangements of $1, 2, 3, \ldots, 10$, it's not unusual to find slippery slopes of length exceeding four. Can this be quantified? For what proportion of the $10! = 3{,}628{,}800$ possible arrangements is this true? It might be easier to start with the case of arrangements of $1, 2, 3, 4, 5$, and slippery slopes of length four or five (those are not that common there).

- Mathematician Satish Shirali recently sent the following interesting observation:

  > It is known that the integers from 1 to n^2 can always be arranged in a sequence having no monotone subsequence of $n + 1$ terms. Call such a sequence an n-ES sequence. What may not be so well known is that, given an n-ES sequence, one can generate an $(n + 1)$-ES sequence from it by placing the $n + 1$ integers $n^2 + n + 1$ to $(n + 1)^2$ in that order (ascending) at the beginning of the given n-ES sequence and placing the n integers from $n^2 + n$ to $n^2 + 1$ at the end, in that order (descending). In particular, the terms of the given sequence appear as consecutive terms in the new sequence.

  For instance, 2, 1, 4, 3 has no slippery slope of length two, and its extension to 7, 8, 9, 2, 1, 4, 3, 6, 5 has no slippery slope of length three. Can this "putting a bad example on a pedestal" idea be used for a card effect?

- What percentage of all arrangements of 1–9 do not contain a slippery slope of length four?

- What percentage of all arrangements of 1–16 do not contain a slippery slope of length five?

- Repeat for 1–13, by finding the probability of *not* being able to pull off our second effect above successfully if we disallow the extra "emergency" shuffle suggested there.

- Justify the Probabilistic 38% Slippery Slope Principle.

- Repeat the above, but counting rising slippery slopes only. If a sufficiently high proportion of them can be guaranteed, it eliminates the need for your accomplice to have a method of indicating rising or falling. For instance, what proportion of arrangements of 1–13 contain rising runs of length four?

- Can the ideas that Mark J. Tilford used to improve the five-card Erdős and Szekeres effect be adapted to give better results for other packet sizes, such as, replacing our slippery slope approach to packets Ace–10 (certain) or Ace–King (likely)?

- Here's a sketch of a card-handling application that may interest magicians. Let's suppose you desire to have five cards in a specific, but secret, order, perhaps as the staging for a later effect, while giving the impression that they are truly shuffled. Hand them out for mixing, and upon getting them back, flash them face up in a fan, pointing out how jumbled they are. You can now separate this fan into two parts, by pulling out two cards with one hand, leaving three in the other hand that are in the order you want—guaranteed to be doable by the 60% principle—and then reassembe the little packet by reintegrating the two cards among the three in such a way that all five are now lined up as you wish. Yes, this involves some quick on-the-hoof thinking and bold moves, such as pushing a card or two into gaps between the three "good" cards. It may also be necessary to drop the hand holding the two cards briefly, and switch the order of those cards with your fingers. However, in our experience, with practice and a relaxed nonchalance, your audience will be oblivious to the fact that you just pulled a fast one.

Hamming It Down

The effects in this chapter are based on the observation that it is possible to start with a display featuring randomly selected face-up cards, have an audience member implement a color change of one card, and then have a second person—who joins the scene for the first time—detect which card was changed. It should come as no surprise that the second volunteer is your mathematically-trained accomplice, and here we suggest openly admitting this person's role, perhaps presenting her as a mind reader.

Above we said the card display featured randomly selected cards, but we didn't say they were all random, although that is the case for "Poker-Faced" (Q♠). Fair warning: it's one of the most demanding effects in this book to pull off, as might be guessed from the four-page proof that it's even possible.

> *An audience member who likes poker shuffles the deck, and handpicks five cards for a good poker hand. You show them around, one by one, making comments such as, "A pair of Tens and a pair of Kings, not bad!" Spread the cards in a neat face-up row.*
>
> *Next, have the poker fan swap any one of those cards for another from the rest of the deck, with one proviso: there is a color switch. If you encourage a better poker hand, there will likely now be three Tens or three Kings in the case above.*
>
> *Call in your accomplice from the next room. "This is Jeff. He has studied human psychology in his spare time, and he reckons he can figure out which card here is the one you swapped by asking you to read out the card names in any order you wish and listening to your voice modulations." Jeff delivers!*

The trick (so to speak) is to use an application of linear algebra that has been around since the middle of the last century, namely linear error-correcting codes (see [Pless 82]). The good news is that its implementation in card effects is relatively straightforward, and does not require a thorough grasp of the mathematics behind it, although there is no denying that some mental agility is required.

If the theoretical framework we outline seems daunting, fear not—for each effect we also express all the necessary steps in plain language.

A Small Binary Linear Code

Error-correcting codes apply to situations in which messages are converted from alphabetical (or more generally, ASCII) form to numbers in binary form, and these are sent over a communication channel to a recipient, who may well receive a corrupted version containing errors. Just as in the case of product bar codes, which are not always reported or scanned correctly, the basic idea is to tack on so-called parity check digits before the message is sent, thereby building in some (but not too much) redundancy. The hope is that errors might then be detectable and perhaps also correctable.

One simple approach is multiple repetition: if 0 is sent as 0000, but 0100 is received, it's fairly easy to guess what the original message was. Of course, that assumes that three errors didn't occur. This approach is very wasteful, however; we can do much better.

We begin by explaining a binary linear code in which 2-bit messages $M = (a, b)$ are encoded as 5-bit codewords $C = (a, b, a, b, a + b)$. Bit here refers to numbers such as a, b, which are either 0 or 1. The addition mod 2 rules of binary arithmetic apply; hence, $0 + 0 = 0$ and $0 + 1 = 1$, whereas $1 + 1 = 0$ (see page 16). The three new bits appended to the original message M are the *parity check digits*.

This code is simpler—although much less efficient—than those considered to be on the ground floor of the modern edifice of error-correcting codes. However, here we explore it in detail, because of its magical application potential.

There are $2^2 = 4$ possible messages, coded as in Table Q.1. If these 2-bit messages are transmitted via the 5-bit codewords listed there, and possible errors are introduced—meaning that some of the bits get flipped from 0 to 1, or vice versa—then any of $2^5 = 32$ messages could be received, only four of which are as intended, that is, uncorrupted.

message M	codeword C
00	00000
01	01011
10	10101
11	11110

Table Q.1. Simple binary code.

The weight of a codeword is the number of nonzero bits in it (in this case, it just counts the visible 1s). As Table Q.1 reveals, among the (three) nonzero legitimate codewords, the minimum weight is 3. Error-correcting code theory says that since $3 = 2 \times 1 + 1$, *this code detects, and enables the correction of, single errors* [Pless 82].[1]

Of the thirty-two possible received messages, eight have double errors. We focus on the $32 - 4 - 8 = 20$ corrupted messages R arising from the five different possible single-error transmissions for each of the four legitimate codewords C. These are shown—together with the weights of the received messages R—in the first three columns of Table Q.2.

Correct word C	Corrupted word R	Weight of R	P based on R	Where R and P disagree
00000	10000	1	10101	$x, x+y$
00000	01000	1	01011	$y, x+y$
00000	00100	1	00000	x
00000	00010	1	00000	y
00000	00001	1	00000	$x+y$
01011	11011	4	11110	$x, x+y$
01011	00011	2	00000	$y, x+y$
01011	01111	4	01011	x
01011	01001	2	01011	y
01011	01010	4	01011	$x+y$
10101	00101	2	00000	$x, x+y$
10101	11101	4	11110	$y, x+y$
10101	10001	2	10101	x
10101	10111	4	10101	y
10101	10100	2	10101	$x+y$
11110	01110	3	01011	$y, x+y$
11110	10110	3	10101	$x, x+y$
11110	11010	3	11110	x
11110	11100	3	11110	y
11110	11111	5	11110	$x+y$

Table Q.2. Correctly coded messages and their corrupted versions.

We need to find a way to work back from a likely garbled R to the original M via the C that was transmitted. In other words, given a received, corrupted message R, we want to determine the original message M (namely, the first two bits of C).

---

[1]The same theory says that for codes with minimum weight $7 = 2 \times 3 + 1$, triple errors can be found and fixed.

A study of Table Q.2 reveals that if the weight of R is 1, then $C = 00000$, whereas if the weight of R is 3 or 5, then $C = 11110$. The other half of the cases are not so obvious at first glance, but these observations are worth bearing in mind when performing associated card effects.

Consider the fourth column of the table: this gives the "pretend" (consistent) codewords P that are obtained from the first two bits of the erroneous Rs. In other words, the third, fourth, and fifth entries of each P are the parity check digits $(x, y, x + y)$ of the first two bits (x, y) of the corresponding R. Finally, the last column of the table identifies the exact locations in which P and R differ—namely, which of the parity check digits fail to agree. These are listed in terms of their constituent entries $(x, y, x + y)$. Now a clear pattern emerges.

Decoding Algorithm for Binary Code

Given a codeword R, known to contain exactly one error, use its first two bits to generate and append "pretend" parity check digits, thus forming a new self-consistent codeword P. See where R and P differ.

If it's in positions 3 and 5, the first bit is wrong.

If it's in positions 4 and 5, the second bit is wrong.

If it's in position 3 only, the third bit is wrong.

If it's in position 4 only, the fourth bit is wrong.

If it's in position 5 only, the fifth bit is wrong.

Of course, if the weight of R is odd, we can always fall back on our earlier "odd, weighty" observation to decode rapidly.

At last, we are ready to have some fun applying this to card magic. Instead of the obvious division of the cards into two types—even (0) versus odd (1) values—we focus on card colors (0 for Black and 1 for Red). Also, even though we could work with five card positions, it's really no harder to use six, and that seems more fair.

Q♣ A Horse of a Different Color

How it looks: *An audience member is invited to select any three cards from the deck and lay them in a face-up row on the table. You supplement this row with three more face-up cards of your own choosing.*

Say, "Think of these cards as six horses in a stream." Before that sinks in, add, "No doubt you've heard the expression, 'Don't change horses in the middle of a stream.' Actually, that's exactly what I want you to do. Please change any one horse—for a horse of a different color!"

The audience member replaces any one of the cards on the table with a new card from the deck, subject to the provision that the new card must not be the same color as the one it replaces. Your accomplice now enters the room, and soon identifies which card on the table was switched.

How it works: Your three choices are based on the colors of the audience member's first two cards; his third card color (denoted by X below) is ignored. Use the convention that 0 represents Black (B) and 1 represents Red (R), and code (a, b, x) as $(a, b, x, a, b, a + b)$. Hence:

If he picks BBX, you pick BBB.
If he picks BRX, you pick BRR.
If he picks RBX, you pick RBR.
If he picks RRX, you pick RRB.

After the switch is made, your accomplice, let's call her Moe, knows that it's the third card if and only if the other five determine a legal codeword (i.e., are self-consistent).

Otherwise, one of the twenty cases in Table Q.2 from before arises for the five positions (or cards), not counting the third one. The algorithm suggested above can be used to determine which color sequence was in place before the switch, and hence which card is the switched one.

For example, suppose the audience member picks 10♣, Q♦, 2♥, and you—concerned only with the colors of their first two selections—pick J♠, 5♥, 7♦. Before the switch, everybody sees the display in Figure Q.1.

Figure Q.1. Consistently correct.

Let's assume now that the audience member discards the second card, Q♦, and in its place puts 8♣. When Moe enters the room, she sees the display in Figure Q.2.

She ignores the third card, and converts the other five colors to $R = 00011$. She checks then to see whether this is self-consistent—it isn't! So the ignored third card isn't the problem. She has also just worked out $P = 00000$, which differs from R in the b and $a + b$ (i.e., the fourth and fifth) positions. Hence, she knows that the error in R is in the second bit, and so the 8♣ must be the switched card.

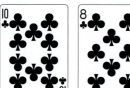

Figure Q.2. Stop the deliberate error!

Source: Original. From late 2002. The idea of applying binary error-correcting codes to two-person card effects was shared by Spelman College mathematician Jeffrey Ehme in October 2002. He suggested using the standard linear binary Hamming code, in which four cards are coded with the help of three parity check cards, as well as the option of an eighth card at the end to facilitate a seemingly fair "you pick four, I pick four" scenario.[2] We simply scaled down those ideas—hence our chapter title.

Published February 2009 by Earnest Hammingway in "Horses in the Stream and Other Short Stories" in the Gardner tribute book *Homage to a Pied Puzzler* [Hammingway 09].

Further exploits with the small binary code used here can be found online in the oddly scheduled November 2010 *Card Colm* "The Three Piles (This Isn't That Trick)" [Mulcahy 10_11].

Presentational options: If the audience member picks three Black cards in the last effect, and you then pick three more, it may come as no surprise to onlookers that Moe is able to spot one Red card among five Black ones! However, this should only happen one time out of eight, on average, and moreover, Moe would have no reason to suspect an initial monochrome arrangement.

Different binary identifiers can be used—perhaps odd/even values—but this creates new difficulties. You don't want to have to explain your new deck division scheme just to force your audience member to do the correct kind of card switch.

(Of course, forcing a replacement card of a specific color raises a different issue: if we could force a desired color every time a selection is made, then we could force any desired color sequence. Then, without any coding theory, the effect would be trivial.)

For a version in which any hint of binary identification is concealed from the audience, see the section "No Face, No Name, No Number" in [Hammingway 09].

---

[2]Spelman College students Andrea Warren and Aminah Perkins gave a presentation by Jeff called "A Trick for Introducing Algebraic Coding Theory" at the Southeastern Section meeting of the Mathematical Association of America at Clemson University in March 2003.

The basic five-card code, $(a, b, a, b, a + b)$, is almost unique up to permutation of the five bits, as a, b, and $a + b$ are the only nontrivial linear combinations[3] of a and b (mod 2). We've been assuming that a and b are freely chosen by the audience member, and hence that the three check digits are forced. If the effect is going to be done several times for the same audience, it may be better to use a different arrangement, or one with different repetitions, such as $(a, b, a + b, b, a + b)$.

Of course, if we do the six-card versions suggested, we can use *any two* of the first three cards as the (a, b) that are encoded (provided we agree on which ones ahead of time, and stick with that choice).

As already remarked, using Black versus Red is just one of many possibilities for binary identifiers for cards, others being even versus odd (where Jack = 11, Queen = 12, and King = 13), or low (say, 2–8) versus high (9–Ace), or maybe composite (4, 6, 8, 9, 10, Queen) versus prime, perhaps lumping Aces in with the 2, 3, 5, 7, Jack, and King (see page 14).

The next effect is a little more ambitious, and uses several of these deck divisions simultaneously.

Q♥ Multiple Personality

How it looks overall: *From a face-up deck you and an audience member pick five cards total and display them in a face-down row. The audience member indicates which card he wishes to switch, and in due course, it is exchanged for any other card of his choosing from the rest of the deck, which is then shuffled and set aside.*

Your accomplice enters, inspects the five card faces, and soon identifies the replaced card. With a little more encouragement and reflection, she rummages through the rest of the deck and soon picks out the card that was originally in that position, to thunderous applause.

How it looks in detail: From a face-up deck, an audience member picks two cards and you pick two, these being displayed in a face-down row. As an apparent afterthought, append the row with one more face-down card. Ask the audience member to slide out any card, to be replaced soon with another card of his choosing. Run through the rest of the face-up deck, seeking your lucky card, which you set face down on the table.

Shuffle the remainder of the deck, and spread it (face up or face down) for the audience member to pick out his replacement card. He switches that card for the one slid out before, and shuffles the dumped card into

---

[3]For us, a *linear combination* of a and b means any $ax + by \mod 2$, where x and y are either 0 or 1.

the rest of the deck. The cards in the row are now flipped over so that all are face up.

Your accomplice Moe enters, passes your lucky card *face down* over the displayed cards, and soon identifies the one that is new. In due course, she picks from the rest of the deck the card that was originally in that position.

How it works: The three cards you choose must simultaneously code three characteristics of the audience member's cards: color (X), status (Y), and parity (Z). To that end, associate a binary triple (X, Y, Z) to each card, where X is 0 if and only if the card in question is Black; Y is 0 if and only if the card value is low, which we take to be 2, 3, 4, 5, 6, or 7 (Aces are considered high here); and Z is 0 if any only if the card value is 2, 4, 6, 8, 10, or Q (Aces are always odd).

For instance, if the audience member picks Q♠ and 4♦, this corresponds to the triples $(0, 1, 0)$ and $(1, 0, 0)$. Computing the binary check digits $(a, b, a + b)$ three times over, where (a, b) is the X, Y, and Z entries of the audience member's cards respectively (namely, $(0, 1)$, $(1, 0)$, and $(0, 0)$), we find that your cards must correspond to the triples, $(0, 1, 0)$, $(1, 0, 0)$, and $(1, 1, 0)$, respectively.

That is to say, your first card must also be Black, high, and even (which gives you five choices), your second card must also be Red, low, and even (which gives you six choices), and your third card must be Red, high, and even (which also gives you six choices). Let's assume you select 8♠, 6♥, and 10♦, which results in the display in Figure Q.3.

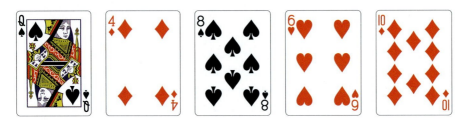

Figure Q.3. Triple coded cards.

As soon as the audience member indicates which card he intends to get rid of, you run through the deck, under the pretext of finding your lucky card, and find the cards you do not want him to pick as the replacement. For instance, if he indicate above that he wishes to dump 6♥, then you must ensure that he does not pick another Red, low, and even card; as that would result in all three error checks—with respect to the different binary codings—coming up blank for Moe. Hence you find and cut to the top and bottom of the deck 2♦, 6♦, 2♥, and 4♥ (the 4♦ is also Red, low, and even, but it's already among the selected cards). Set one of these aside face down as your so-called lucky card, so that there are one or two "to be avoided at all costs" cards at the top and bottom of the deck.

Riffle shuffle, maintaining the key cards at the top and bottom, and have the audience member select a new card from (almost) any part of the deck. The rejected card replaces it, and is then shuffled into the rest.

Unbeknownst to the audience member, the new card in the display will differ *in at least one regard*—color, status, or parity—from the one slid out and now lost in the deck. As a result, when Moe enters the room and does error checking and decoding—three separate times, one each for the various binary characteristics—she will find at least one instance of error, and hence be able to pinpoint the position of the card switch. The use of your lucky card is, of course, totally bogus, but in case the audience is suspicious, make it clear that the face of this card is never seen by Moe.

Better yet, by keeping track of all errors that occur, Moe will know the precise characteristics of the dumped card. Since you obligingly collected all matching cards (*except* the correct one) at the top and bottom of the deck earlier, only one card *within* the deck will fit the bill. This is the one she picks out, in due course, to thunderous applause.

For instance, if the audience member replaces the 6♥ above with the 3♣, then Moe sees the display in Figure Q.4.

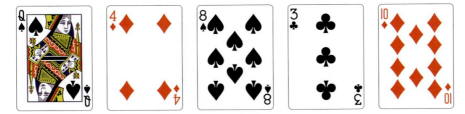

Figure Q.4. What your accomplice sees.

Just by considering color, it's easy for Moe to deduce that the fourth card is the one "out of sequence"—and had that not set off alarm bells, the change in parity in the same position would have done so. There is no problem with the high/low coding, so Moe knows that the dumped card was Red, low, and even. There should only be one such card in the deck, not counting those at the very top and bottom, namely, 6♥, thus allowing her to pull off a seeming miracle.

Source: Published February 2009 by Earnest Hammingway in "Horses in the Stream and Other Short Stories" [Hammingway 09].

Presentational option: This can also be done in a more-fair-seeming six-card incarnation, where the audience member starts by picking any three cards, and you "match" those with three based on two (let's say the first two) of his. The downside is that if the switched card is the third one, Moe will know this but she won't have a clue what was there before the switch. You can either accept a $\frac{5}{6}$ success rate with this effect, on average, or be content to forego the kicker sometimes, provided you have not announced ahead of time exactly what Moe would be able to do.

Given Any Five Cards

Some of our favorite mathemagical entertainments start with the words, "Given any five cards." Our next effect is one of those. It hinges on the following result, whose lengthy proof we give in detail. Perhaps read the effect first, and return here when you wish to learn the nitty-gritty details required for performance.

Five-Card Color Change Principle ⋈
Given any five cards from a deck, it is possible to arrange them in a face-up row in such a way that if one card is exchanged for a new one of a different color, then the position (and hence identity) of the new card can be detected by somebody who sees only the end result.

Specifically, given any five cards, they can be arranged in such a way that they are consistently color coded, in the sense of the binary error-correcting code we have been considering. Recall that for two given cards coded by three extra ones, there are four possible codings, assuming that 0 denotes Black and 1 denotes Red, as shown in Table Q.1, reproduced here as Table Q.3 for convenience. There are thus $2^2 = 4$ possible messages.

message M	codeword C
00	00000
01	01011
10	10101
11	11110

Table Q.3. Simple binary code.

Note that the five-card displays arising from encoding any two cards have no Reds, three Reds, or four Reds. So, if the five cards feature zero, three, or four Reds, we know how to proceed. There are even two ways (in terms of color distribution) to arrange the cards in the three Reds case. For example, if the cards are K♥, K♦, A♥, A♦, A♠, then there are four Reds, and one consistently color-coded way to arrange them is shown in Figure Q.5.

Figure Q.5. A good poker hand.

The real question is, what can we do if the five cards feature one, two, or five Red cards instead? Then it is time for a convention switch.

> ### The Switch (Complement in Five Principle) ▶◀
> *If there are one Red, two Reds, or five Reds among the five given cards, then there must be four, three, or no Black cards, since $5 - \{1, 2, 5\} = \{4, 3, 0\}$. We simply switch the convention: now 0 denotes Red, and 1 denotes Black!*

Fortunately, this can be done in such a way that it's detectable later, by also taking advantage of the values of the cards on view.

Assume throughout that the deck has the total ordering determined by CHaSeD suit order. There are six cases to consider, three easy and three not-so-hard:

Case 0: Zero Red cards. Proceed as usual (but see below).

Case 1: One Red card. Indicate switch and mimic Case 4.

Case 2: Two Red cards. Indicate switch and mimic Case 3.

Case 3: Three Red cards. Proceed as usual (but see below).

Case 4: Four Red cards. Proceed as usual (but see below).

Case 5: Five Red cards. Indicate switch and mimic Case 0.

Before we spell this out in detail, we explain the method of indicating the convention switch. So far we have concerned ourselves only with the card colors, not their values. We can use the values of the Red *and* the Black cards to communicate further information.

In any row of five cards, there will be "subrows" of both Red and Black cards. (If we have all one color, of course, there is no subrow of the other color, and if we have four of one color, the other subrow consists of just one card.)

On the one hand, if we intend the usual convention to be understood (namely, 0 = Black), which is to say Cases 0, 3, or 4 above, we can arrange the cards coded so that any visible subrows are rising, with respect to the total ordering of the deck. On the other hand, if we intend the convention to be switched (namely, $0 = \text{Red}$), as in Cases 1, 2, or 5 above, we can arrange it so that all subrows are falling.

Since the rising or falling considerations apply to just two suits (of the same color) at a time, it's not so difficult for either party involved to deal with the ordering considerations quickly.

The point is that when a card is exchanged later, which might disturb a rising or falling subrow, all is not lost. If the affected subrow is of length four or five—after the exchange—then a single card out of sequence still leaves no doubt as to the original rising or falling status. If, however, after the exchange we have subrows of length two and three, one of three situations arises: one subrow is rising and the other is falling (in which case the one of length three is the "correct" one), both are monotone *in the same sense*, or only one (of length two) is monotone. These last cases will be established in detail below.

It's time to try all of this in practice. This time we present the three "easy" cases first:

Case 0: Zero Red cards. The cards are B_1, B_2, B_3, B_4, B_5, in ascending order with respect to the agreed-upon total ordering of the whole deck. They are arranged in that order: after one exchange, we have either (a rising subrow of) four Blacks, which indicates that the usual convention applies, and it should be decoded to $00000 = BBBBB$ (i.e., it will be obvious that the lone Red card was the exchanged card), as required.

Case 3: Three Red cards. The cards are B_1, B_2, R_1, R_2, R_3, with each subrow in ascending order, and are rearranged as B_1, R_1, B_2, R_2, R_3. Without loss of generality, the cards are 2♣, 2♥, 4♣, 4♥, 6♥. Now consider the possibilities after one exchange. First let's assume a Black card is removed and replaced by a Red card. There are four possible relative values for the Red card: A♥, 3♥, 5♥, or 7♥, and no matter which of the two Black cards it replaces, the new subrow of four Reds is either rising or rising-dominant. If, on the other hand, we assume that a Red card is removed and replaced by a Black one, say A♣, 3♣, or 5♣, then no matter which of the three Reds it replaces, the remaining subrow of two Reds is rising (and the new subrow of three Blacks is either rising or not monotone). In all subcases, after one exchange, it will be clear that the usual convention applies, and that what is seen should be decoded to $01011 = BRBRR$ as required.

Case 4: Four Red cards. The cards are R_1, R_2, R_3, R_4, B, and are arranged in that order. Without loss of generality, the cards are 2♥, 4♥, 6♥, 8♥, and 2♣. Now consider the possibilities after one exchange.

First let's assume the Black card is removed and replaced by a Red card. There are five possible relative values for the Red card, A♥, 3♥, 5♥, 7♥, or 9♥, and the new row of five Reds is again either rising or rising-dominant. Now, assume instead that a Red card is removed and replaced by a Black one, say, A♣ or 3♣. In the latter case, a bogus falling subrow of length two is created, but we still have a rising subrow of length three to guide the way. In the former case, we have two rising subrows. In any case, it will be clear that the usual convention applies, and that what is seen should be decoded to $11110 = RRRRB$, as required.

It only remains to do all of this "reversed" for the other three cases, making sure that the convention switch can be communicated clearly.

Case 1: One Red card. The cards are B_1, B_2, B_3, B_4, R and are arranged as B_4, B_3, B_2, B_1, R. After one exchange, there is a falling-dominant subrow of (at least) three Blacks, and regardless of the two Reds that may be present, it is clear that a switch of convention applies. Hence, what is seen should be decoded as $00001 = BBBBR$, as required.

Case 2: Two Red cards. The cards are R_1, R_2, B_1, B_2, B_3, each subrow in ascending order. These are rearranged as R_2, B_3, R_1, B_2, B_1. Without loss of generality, the cards are 4♥, 6♣, 2♥, 4♣, and 2♣. Consider the possibilities after one exchange. First let's assume a Red card is removed and replaced by a Black card. Without loss of generality it's A♣, 3♣, 5♣, or 7♣, and no matter which of the two Reds it replaces, the new subrow of four Blacks is either falling or falling-dominant. Next, let's assume instead that a Black card is removed and replaced by a Red, say A♥, 3♥, or 5♥. No matter which of the three Blacks it replaces, the remaining subrow of two Blacks is falling (and the new subrow of three Reds is either falling or not monotone). In all subcases, after one exchange, it is clear that a switch in convention applies, and that what is seen should be decoded to $01011 = RBRBB$, as required.

Case 5: Five Red cards. The cards are R_1, R_2, R_3, R_4, R_5, in ascending order with respect to the agreed-upon total ordering of the whole deck. Arrange them in *reverse* order; after one exchange, there will be a falling subrow of four Reds, indicating that the convention is switched. Hence, what is seen should be decoded to $00000 = RRRRR$ (i.e., the lone Black was the exchanged card), as required.

Worked examples are given in our chapter centerpiece effect below.

♣   ♥   ♠   ♦

Q♠   Poker-Faced

How it looks overall: *An audience member who likes poker shuffles the deck, and handpicks five cards for a good poker hand. You show them around, one by one, making comments such as, "A pair of Tens and a pair of Kings, not bad!" Spread the cards in a neat face-up row.*

Next, have the poker fan swap any one of those cards for another from the rest of the deck, with one proviso: there is a color switch. If you encourage a better poker hand, there will likely now be three Tens or three Kings in the case above.

Call in your accomplice from the next room. "This is Jeff. He has studied human psychology in his spare time, and he reckons he can figure out which card here is the one you swapped by asking you to read out the card names in any order you wish and listening to your voice modulations." Jeff delivers!

How it looks in detail: Take the five selected cards and move them around, commenting briefly on the spectator's prospects if a real game of poker were in play. "I'm going to make you an offer you don't usually get in real life," you say as you spread the cards in a neat face-up row. "You may like this hand, or you may not. Frankly, I think it could be better. I'm going to give you the opportunity to replace any one of these cards with a better one! You choose which one you want to get rid of. Take your time, you get to do this just once."

The poker fan points to a card, and you slide it out. "I promised you a better card did I not? You pick it—you can use any card at all from the rest of the deck, provided it's the opposite color of the one you just discarded." Have the poker fan slide the new card into the gap in the row. The discarded card is, er, discarded!

Now call in your accomplice Jeff from the next room. Have the poker fan read aloud the names of all of the cards on display, in any order. Jeff soon points to the card that was switched.

How it works: The basic idea is to arrange the five cards in such a way that regardless of what selection of Black and Red cards we have, they are consistently color coded, in the sense of the error correcting code from earlier. This is possible because of the extensive discussion above.

Let's go through two examples in detail. First, suppose that the initial poker hand picked is K♥, K♦, K♠, J♥, and J♦. There are four Red cards here, so Case 4 from the earlier discussion applies. You must order and display them as: R_1, R_2, R_3, R_4, B, which in this case is J♥, K♥, J♦, K♦, K♠. Figure Q.6 shows the row you deal out and then make distracting comments about.

Figure Q.6. A consistently color-coded poker hand.

Next, the poker fan selects a card to change, perhaps the J♦. Suppose that the K♣ is selected to replace it. When Jeff enters the room, he sees the *RRBRB* display in Figure Q.7.

Figure Q.7. What your accomplice sees.

First he needs to figure out which convention is being used. The dominant rising J♥, K♥, K♦ shows that the usual convention applies, and then Jeff decodes $RRBRB = 11010$ to $11110 = RRRRB$ in the usual way, deducing that the middle card was the one switched.

As a second example, suppose that the cards picked by the poker fan are K♥, K♦, K♠, J♣, and J♠. There are two Red cards here, so Case 2 from earlier applies. First you think of them as R_1, R_2, B_1, B_2, B_3, each subrow in ascending order, namely, as K♥, K♦, J♣, J♠, K♠. You now reorder and display them in a row as R_2, B_3, R_1, B_2, B_1, that is to say, K♦, K♠, K♥, J♠, J♣, as shown in Figure Q.8.

Figure Q.8. Another consistently color-coded poker hand.

Next, the poker fan selects a card to change; for the sake of argument let's suppose it's the J♠. This time, the K♣ may not be selected to replace it, as it's the same color. Unless a poorer-quality poker hand is opted for, the poker fan has little choice but to replace the jettisoned Black Jack with a Red Jack, perhaps the J♥, which of course, does not change the value of the poker hand. When Jeff enters, he sees the *RBRRB* display in Figure Q.9.

Figure Q.9. What your accomplice sees this time.

As before, Jeff first needs to figure out which convention is being used. The subrow of two Blacks is falling, as is the new subrow of three Reds. So Jeff knows that a switch of convention applies, and decodes *RBRRB* = 01001 to 01011 = *RBRBB*, as required, correctly deducing that the fourth card was the one switched.

Source: Original. It dates from around 2004. As mentioned at the end of "A Horse of a Different Color" (Q♣), the idea of applying binary error-correcting codes to two-person card effects was first suggested by Spelman College mathematician Jeffrey Ehme in October 2002. Published by Earnest Hammingway in February 2009 in "Horses in the Stream and Other Short Stories" in the Martin Gardner tribute book *Homage to a Pied Puzzler* [Hammingway 09], and also online as the October 2009 *Card Colm* "Poker-Faced Over the Phone" [Mulcahy 09_10].

A Small Ternary Linear Code

We close this chapter by changing gears a little, and considering a *ternary* linear code in which 2-trit messages (a, b) are encoded as 4-trit codewords $(a, b, a + b, b - a)$. Here, a and b are trits, meaning values 0, 1, or -1, and the rules of balanced ternary arithmetic apply (see page 16), so that $1 + 1 = -1$ and $-1 - 1 = 1$. (Ternary arithmetic usually works with 0, 1, and 2, combined using mod 3 rules (see page 16), but it's convenient here to use the equivalent balanced formulation in which 2 is replaced by -1.) There are $3^2 = 9$ possible messages, coded as in Table Q.4.

If these 2-trit messages are transmitted, via the 4-trit codewords listed in the table, and possible errors are introduced—meaning that some of the trits get changed to other trits—then any of $3^4 = 81$ messages could be received, only nine of which are uncorrupted. The weight of a codeword is

message M		codeword C			
0	0	0	0	0	0
0	1	0	1	1	1
0	-1	0	-1	-1	-1
1	0	1	0	1	-1
1	1	1	1	-1	0
1	-1	1	-1	0	1
-1	0	-1	0	-1	1
-1	1	-1	1	0	-1
-1	-1	-1	-1	1	0

Table Q.4. Ternary coding.

the number of nonzero trits it contains. As Table Q.4 reveals, among the (eight) nonzero legitimate codewords, the weight is always 3. Once again, we can be sure that this ternary code *detects, and enables the correction of, single errors* (see page 287), as we now demonstrate.

Given a received message $R = (s, t, u, v)$, we must determine the original message $C = (a, b, a + b, b - a)$. We assume that precisely one error is present, so we know for sure that exactly three of (s, t, u, v) match up with $(a, b, a + b, b - a)$. (It's worth noting that the trits $(b, a + b, b - a)$ can be thought of as the terms in an infinitely looping arithmetic progression based on $(b, b + a, b + 2a)$, those three terms having sum 0.)

If there were no errors in (s, t, u, v), then any one of the trits could easily be obtained from two of the others, since we'd have $t + u + v = 0$, $s + u = v$, $s + v = t$, and $s + t = u$ (yielding $s = v - u$ also). When there is exactly one error, then exactly one of those equations must fail, and a pleasing pattern emerges.

Once again, we opt to number the four positions from 0 to 3, as the last three play a different role from the ones that precede them.

Decoding Algorithm for Ternary Code

The original message $C = (a, b, a + b, b - a)$ can be determined from the received message $R = (s, t, u, v)$, assuming precisely one error is present, as follows.

The error is in the zeroth trit (i.e., $s \neq a$) iff $t + u + v = 0$.

The error is in the first trit (i.e., $t \neq b$) iff $s + u = v$.

The error is in the second trit (i.e., $u \neq b + a$) iff $s + v = t$.

The error is in the third trit (i.e., $v \neq b + 2a = b - a$) iff $s + t = u$.

("Iff" means "if and only if"—namely, "precisely when"—either both are true or both are false: it's an all or nothing situation.)

There is one final step: identifying in which of the other two possible ternary states the altered trit was before the switch. We leave that for the motivated reader to investigate.

Let's now apply this to card magic.

Q◆ And Now for Something Completely Different

How it looks: *Give out a deck of cards for shuffling. Take it back, and fan it to reveal that the cards are all face down. Comment, "These aren't mixed up very well. Look, they all face the same way!" Split the deck near the middle, and flip over one half, before riffling the two parts together. Perhaps hand the deck out again for additional shuffling. "That's better," you conclude, as you fan the cards again to show that they are well and truly mixed now.*

Invite an audience member to select any two cards from the deck and place them side by side on the table. You rapidly supplement these with two cards of your own choosing, to form a row of four cards.

"Four random cards, some Red, some Black, some face down! And now for something completely different. Please change any one card. For instance, you could just turn one of these cards over, or you could switch a Red card there for a Black one from the deck, or vice versa."

The audience member does as instructed. Your accomplice now enters the room for the first time, and soon identifies which card on the table was switched. Even better, if the switched card is now face down, she can tell whether it was originally Black or Red. Furthermore, if the switched card is face up, she can tell whether it was originally a different color or face down.

How it works: Your choices are, of course, based on those of the audience member. Use the convention that 0 represents a face-down card (D), 1 represents Black (B), and -1 represents Red (R). Since (a, b) is coded as $(a, b, a + b, b - s)$, the following possibilities arise:

If he picks DD, you pick DD.
If he picks DB, you pick BB.
If he picks DR, you pick RR.
If he picks BD, you pick BR.
If he picks BB, you pick RD.
If he picks BR, you pick DB.
If he picks RD, you pick RB.
If he picks RB, you pick DR.
If he picks RR, you pick BD.

After the card switch, let's bring back Moe. She decodes as indicated above. For instance, perhaps the audience member opted for 5♥ followed by 7♠ face down (RD). That corresponds to the 2-trit message $(-1, 0)$, which gets coded as $(-1, 0, -1, 1)$. So, you must append another Red card and then a Black one, both face up (RB). This can be achieved by supplementing the row with K♥ followed by 3♣, so that you get the display shown in Figure Q.10.

Figure Q.10. Consistently correct ternary coded display.

Next, the audience member changes something, perhaps flipping over the last card to yield the display in Figure Q.11.

Figure Q.11. What your accomplice sees.

Upon being confronted with this display, corresponding to $(s, t, u, v) = (-1, 0, -1, 0)$, Moe checks the four conditions from before, one after the other, stopping as soon as one fails. The thought process is as follows.

Is $t + u + v = 0$? No, so the error isn't s. Is $s + u = v$? No, so the error isn't t. Is $s + v = t$? No, so the error isn't u. Hence it must be v that's wrong, and indeed, $s + t = u$, confirming that. So the final face-down card used to be face up. Moe now does what you did earlier: knowing that the first two cards are correct, she notes that $(-1, 0)$ should be coded as $(-1, 0, -1, 1)$. This means that the final face-down card she sees must have been a face-up Black, which she now announces.

Source: Original. This arose from an independent study experience at Spelman College in the spring of 2010. Published online as the December 2010 *Card Colm* "It's Red or Black and Blue All Over" [Mulcahy 10_12].

Parting Thoughts

- In "And Now for Something Completely Different" (Q♦), can we push a little further and also consider the suits and/or values of the card faces? Can the effect be modified, by using the card suits and/or values in addition to their colors and face-up or face-down status, so that your accomplice can do even better in terms of identifying the altered card? Could she ever give more information than above about the nature of the card pre-switch?

- *Wavelets* is a newly developed branch of mathematics that is used to represent and synthesize data, often as an alternative to Fourier methods. The first step in representing data with the simplest Haar wavelet sees pair of numbers (a, b) processed via "averaging and differencing" to yield $(\frac{a+b}{2}, \frac{b-a}{2})$ [Mulcahy 96, Mulcahy 97].[4]

  In the balanced ternary context, since, $2 = -1$, and division by -1 is the same as multiplication by -1, we see that $(\frac{a+b}{2}, \frac{b-a}{2}) = (-a-b, a-b)$. This is very close to the ternary code used in the final effect above, hence partially fulfilling the author's longtime dream of finding an application of wavelets to card magic.

---

[4]Admittedly, $\frac{b-a}{2}$ is really half the difference of a and b.

The Hidden Value of Cards

In this final chapter, we present four more two-person magical effects based entirely on mathematics. As in the first three effects in Chapter J, we make you the star, ably abetted by an assistant who has been trained in advance. As before, we proceed under the assumption that you pretend this person is a volunteer; you may, however, prefer to be up-front that she's in cahoots with you.

Your reputation as a stellar mathemagician should be well and truly cemented once you've nailed our book closer, "Fitch Four Glory" (K♦).

> *Select a volunteer and give her the deck of cards. After you have stepped out of the room, an audience member shuffles the cards and gives the volunteer any four of them. The volunteer then hides one of them and arranges the other three in a row, with some of them face down. You reenter the room, glance at the display, and soon name the hidden fourth card—even in the case where all of the cards are face down!*

At first sight, this seems astonishing. How on earth can you pull this off, given that you have absolutely no control over the identity of any of the cards? Of course, you must choose the so-called "volunteer" wisely; unlike the cards selected, she's far from being random. She's an assistant of yours who has been trained ahead of time; everything you need to know can be gleaned from the cards as displayed, once you two have agreed in advance on some communication conventions.

How can four arbitrary cards allow for such a stunt? The answer, of course, is mathematics, which we'll discuss below in due course, starting with simpler incarnations of the principles involved.

The information you need could be phoned or emailed (or sent via instant message) to you, in a remote location, by another audience member, with you reporting back (by phone, email, or instant message) your unfailingly correct verdict about the hidden card. Absolutely no body language or verbal signaling is used. Indeed, the "Telephone Stud" trick published in 1951, which served as the basis of all of this, was intended to be carried out over the phone (see [Lee 51]). The phone symbol in the margin is used to denote this option. We start with a slight modification of this superb creation of mathematician William Fitch Cheney, Jr, the goal being to make its inner workings less obvious if it's performed several times for the same audience.

There is no setup required for any of the remaining effects in these pages, and as such, most of them stand up well to repetition. Indeed, audiences seem to find them increasingly baffling the more often they are performed. Furthermore, both the original trick and more recent generalizations work as well for large audiences as they do for small ones.[1]

The standard incarnation of the Fitch Cheney Trick sees the five random cards given to a mathemagician, who hides one and displays the rest in a row. Then a volunteer, who is secretly a confederate of the mathemagician, but has been out of the room for the selection and card placements, surveys the display and soon identifies the hidden card based on what can be seen. In what follows, we reverse the roles, thus making you (the mathemagician) the one who appears to have special powers. Being up-front about the fact that you are using a trained accomplice adds a level of intrigue for a mathematically inclined audience.

K♣   Fitch Cheney's Five-Card Twist

How it looks overall: *Select a volunteer and give her the deck of cards. After you have stepped out of the room, an audience member shuffles the cards and gives the volunteer any five of them. The volunteer then hides one of them and arranges the other four in a face-up row. You reenter the room, glance at the display, and soon name the hidden fifth card.*

How it looks in detail: Choose a volunteer from the crowd, and direct her to do as follows after you've left the room. (1) She is to hand out a regular deck to the audience for inspection and shuffling, and then ask for any five cards to be given to her, the rest being set aside. (2) She is to briefly look at the faces of those five cards, and hide one (perhaps under a book or under the card case). (3) She is to place the remaining four

---

[1] It's worth investing in a deck of giant cards for stage performances.

cards on the table from left to right in a face-up row. (4) She is to have somebody else retrieve you.

When you are convinced that these instructions have been understood, leave the room. Upon returning to the scene a few minutes later, you inspect the row of four cards, and soon name the hidden fifth card.

How it works: Before you read any further, we strongly recommend that you spend some time—days, if necessary—thinking about just how this famous trick might work. It will give you great pleasure, and you are sure to work out most, if not all, of it. Yes, it really does work given *any* five cards.

Read on only if you are really ready do to so.

There are several distinct parts to the approach presented here. The so-called "volunteer" is, of course, your assistant. Let's agree that she's Curly.[2] Also, while the five selected cards are indeed random, Curly gets to decide which one to hide and in what order to display the others.

For instance, Curly may be handed 2♣, 10♠, 2♥, J♣, and 8♦, as shown in Figure K.1. What one thing can be said for sure about these, or indeed, any five cards?

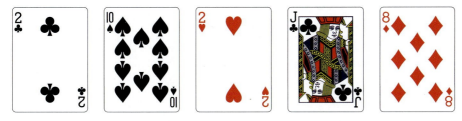

Figure K.1. Given any five cards—what can we say for sure?

Since there are only four possible suits, then by the Pigeonhole Principle (page 23), at least two cards have the same suit.[3] Without loss of generality, we may suppose that there are two (or more) Clubs.[4] Of course, there may be more than two Clubs, but let's focus on two in particular. Curly hides one Club, and then by placing the remaining four cards in some particular order, effectively tells you the value of the hidden Club.

Let's assume you and Curly agree in advance to: (1) use the second of the four positions available on the table for the retained Club—*which determines the suit of the hidden card*—and (2) use the other three for

---

[2]Mine certainly was, back in 1999; Shemp wasn't available.

[3]By the same principle, given any three cards, at least two have the same color, since there are only two colors available.

[4]The argument works just as well if we assume there are at least two Hearts, or at least two cards of any other specific suit.

the placement of the remaining cards. (Later, we suggest an alternative to this "second card is always the suit-giver" convention.)

Now we shift attention to the other three cards: they act as "value-givers." Note that three items can be arranged in $3! = 6$ ways. If the two of you also agree ahead of time on an unambiguous way to link those possible arrangements to the numbers $1, 2, \ldots, 6$, then Curly can communicate any one of six things to you, depending on where she places those three cards in the row of four.

What *can* one say about these other three cards? Not much—for instance, some (or indeed all) of them could be Clubs, there could be other suit matches, and so on. However, *one thing is certain*: they are all distinct, so with respect to some total ordering of the entire deck (see page 13), one of them is low, one is medium, and one is high. As we explain shortly, this allows for a well-defined and easily remembered way to communicate a number between one and six.

But surely six isn't enough! After all, in general the hidden card could be any one of the twelve Clubs not in position 2.

This brings us to the final key idea: Curly must be very careful as to which Club to hide!

Poles Apart Principle

Consider the thirteen possible card values, Ace, 2, 3, ..., 10, Jack, Queen, King, arranged clockwise in a circle (see page 15). Then, given any two of them, they are at most six "hours" apart, so that counting clockwise, one of them lies at most six positions (or "hours") past the other.

The reason this holds is simply because, if it didn't, we'd have two gaps of size six or more between the two card values (hours), which would require the clock to have at least fourteen hours.

The "higher"-valued Club (i.e., the one within six "hours" of the other, going clockwise) is the one that is hidden. If the "lower" Club is then placed in a particular position in the row of four displayed cards—we've already suggested position 2—and the other three cards are positioned according to their low, medium, high ranking, the upshot is that when the time comes, you'll be able to deduce the identity of the hidden card as required.

For example, if Curly has the 2♣ and 8♣ as the suit match of note among her five cards, then she hides the 8♣, which is six hours past the Two. If she has the 2♣ and J♣, as is the case for the five cards in Figure K.1, then she hides the 2♣, which is four hours past the Jack, since $11 + 4 = 2 \mod 13$. (The Jack is more obviously nine past the Two, but that puts it beyond reach with reference to the latter, as the communication of numbers larger than six is not an option here.)

Fitch Cheney Trick Principle

No matter what five cards your accomplice is handed, there must be two of the same suit such that she can hide one of them and communicate its identity to you by displaying the other four cards as follows: (1) one card of the same suit is placed in an agreed-upon position of the display, and (2) a particular arrangement of the remaining three cards tells you what number k between 1 and 6 to add (mod 13) to the value of the suit-giver card, to yield the hidden card.

For the communication of k, you and Curly agree ahead of time on a convention such as the following.

Rank Ordering Convention for Three Cards

Consider the following CHaSeD total ordering of the deck:[5] $A♣, \ldots, K♣, A♥, \ldots, K♥, A♠, \ldots, K♠, A♦, 2♦, \ldots, K♦$. With respect to this, any three cards match up with one of L (low), M (medium), and H (high). Now, we agree to order the six arrangements of $\{L, M, H\}$ by rank: $1 = LMH$, $2 = LHM$, $3 = MLH$, $4 = MHL$, $5 = HLM$, and $6 = HML$, hence yielding a number from one to six for each possible arrangement of the three cards.

Curly can place the cards in question from left to right—in positions one, three, and four, respectively—according to this scheme.

Returning to the situation where Curly has 2♣, 10♠, 2♥, J♣, and 8♦, we've seen that she hides the 2♣, and uses the rest to convey its identity to you by nudging you to add $k = 4$ (mod 13) to the visible J♣ in position 2.

In standard LMH order, the other cards are 2♥, 10♠, 8♦—because the suits are different, we're really ranking according to CHaSeD order, ignoring values, here—so to communicate 4 = MHL, Curly must display them in the order 10♠, 8♦, 2♥, in the appropriate positions. Putting all of this together, Figure K.2 is what you should see when you reenter the room.

Here's how it works from your perspective, starting with only this display. You immediately note what's in position 2, and deduce that the hidden card is within six of J♣, counting clockwise around a thirteen-hour clock. With respect to the deck ordering agreed upon, you note that the other three cards shown, i.e., 10♠, 8♦, 2♥, are in order MHL, which corresponds to 4. So you mentally add $11 + 4 = 15 = 2 \mod 13$, and conclude that the hidden card is the 2♣.

---

[5]Indeed, combining this with an extended version of the first panel in Figure A.8, we obtain fifty-two shades of gray.

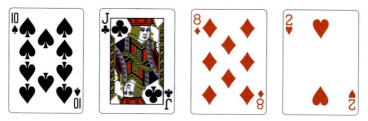

Figure K.2. Fitch Cheney's five-card twist—version 1.

This method is fine for one-time performances, but the invariant use of position 2 for the "suit-giver" is soon spotted by alert audiences if the trick is repeated. We now explain how to sidestep this weakness, and provide two examples of this harder-to-crack version in action.

Variable Suit-Giver Principle ▸◂

Since both you and your assistant get to see the four cards that are displayed in a row, before she arranges them she sums these four values and reduces mod 4, *using that number for the position of the suit-giver card.*

She uses position 4 if she gets 0 as the remainder of the card value sum upon division by 4 (see page 15).[6]

As before, the remaining cards, when placed in the correct order in the other three positions, tell you what k to add to the value of the suit-giver card to determine the value of the hidden card.

In the 2♣, 2♥, 10♠, J♣, and 8♦ example, in which the cards to be displayed are a Two, Ten, Jack, and Eight, Curly mentally computes $2 + 10 + 11 + 8 = 3 \mod 4$, and hence knows to use position 3 for the J♣ (the suit-giver). This time, the other three cards are placed in the correct order (MHL) in positions 1, 2, and 4, to tell you to add 4 to the numerical value of the Jack as before. Figure K.3 is what you should see when you reenter the room.

Figure K.3. Fitch Cheney's five-card twist—version 2 (first example).

---

[6]Another option is to use the actual remainder, considering the four positions as being numbered 0–3.

Your perspective upon seeing this display is first to mentally compute $2 + 11 + 10 + 8 = 3 \mod 4$ and deduce that the missing card is within six of J♣, counting clockwise around a thirteen-hour clock. The rest is as above: the other three cards are in order MHL, which corresponds to 4, and hence the hidden card is 2♣.

Here's a new example. Suppose Curly is handed 2♠, A♦, J♠, 9♠, and 9♦, as shown in Figure K.4.

Figure K.4. Fitch Cheney's five-card twist—version 2 (second example).

There are multiple suit matches here, and Curly should immediately focus on the easiest one to communicate: let's assume she goes with the fact that the J♠ is only two past the 9♠. She hides the Jack and decides where to place the Nine by mentally computing $2 + 1 + 9 + 9 = 1 \mod 4$: so it goes in position 1. Now we must communicate "add two" using the other three cards. In LMH order, with respect to the agreed-upon total ordering of the deck, they are as listed above, namely, 2♠, A♦, 9♦. Now they are placed in the appropriate order (LHM = 2) in positions 2, 3, and 4, to tell you to add two to the value of the leading 9♠. When you reenter the room, you should see the display in Figure K.5.

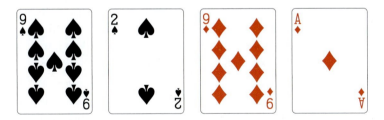

Figure K.5. What you see in the second example of version 2.

Starting with only this display, you mentally compute $9 + 2 + 9 + 1 = 1 \mod 4$, and deduce that the missing card is found by counting past the 9♠. The other cards shown, 2♠, 9♦, A♦, are in order LHM, which corresponds to 2, so the hidden card is J♠.

Source: The suggested method for deciding which card is the suit-giver is original and dates to about 2001. Brain Epstein included it in his

"All You Need Is Cards" contribution to the Martin Gardner tribute book *Puzzlers' Tribute: A Feast for the Mind* [Epstein 02]. It also appeared in the February 2003 *Math Horizons* article "Fitch Cheney's Five Card Trick" [Mulcahy 03] (which was later featured in *The Edge of the Universe—Celebrating Ten Years of Math Horizons* [Mulcahy 06]).

The basic trick is believed to date back to the 1940s. It appeared under the name "Telephone Stud" in the (still available) 1951 book *Math Miracles* by W. Wallace Lee [Lee 51], where it is attributed to William Fitch Cheney, Jr, Chairman, Department of Mathematics, University of Hartford, Connecticut. Martin Gardner mentions it in passing a few times (e.g., see [Gardner 56, page 32], [Gardner 69]), but provides no solution. A much watered-down version, in which all five cards are assumed to be of the same suit and nonmathematical signaling also takes place, appears in Fulves [Fulves 84], with a nod to Cheney.

The reappearance of this effect in mathematical circles in recent decades may be traceable to mathemagician Art Benjamin of Harvey Mudd College, California, who started popularizing it after spotting it in the Lee volume around 1980. Several papers have explored optimal generalizations (e.g., to larger decks), using much more advanced mathematics than what can found in these pages [Holm 03, Kleber 02].

We first learned of the existence of this effect one evening during the Joint Mathematics Meeting in January 1999, in San Antonio, Texas, over a beer with mathematician Paul Zorn of St. Olaf College, Minnesota. It's no exaggeration to say that working out the details a few days later—and in due course coming up with some of the variations that follow—launched our interest in the whole subject of mathematical card tricks, that, in turn, ultimately led to the writing of this book.

Presentational options: Instead of placing the cards in a row, your volunteer can put them in a face-down pile, that you later pick up and inspect. This should allay the suspicions of an audience that thinks there is some subtle signaling going on in the precise placement of the cards on the table.

This could also be presented as a poker effect: have a poker hand selected in the first place, the idea being that by seeing four of the cards you will be able to guess what the fifth one is.

What happens if you have audience members who insist that *they*—rather than your accomplice—get to decide which card is hidden? Have the latter keep cool, by saying, "No problem, it won't make a difference," and then dazzle them all with the following effect as a follow-up! It requires a different, more complicated strategy, and a lot of practice.

K♥   Eigen's Value

How it looks overall: *Select a volunteer and give her the deck of cards. After you have stepped out of the room, an audience member shuffles the cards and gives the volunteer any five of them. The volunteer shows them around for all to see, asking which one should be hidden. That card is then concealed, and she arranges the other four in a face-up row. You reenter the room, glance at the row of four cards, and presently name the hidden fifth card.*

How it looks in detail: The same instructions are given to the so-called volunteer as before, except that this time the audience gets to decide which card is hidden. Then the volunteer places the remaining cards face down in a row.

How it works: The method here requires some serious mental effort. Unlike in the classic Fitch Cheney Trick, Curly must use four cards to communicate which of the remaining $52 - 4 = 48$ possible cards is the hidden one. Since $4! = 24$, it can be narrowed down to only one of two cards by communicating a permutation alone.

First, we need to generalize the $3! = 6$ ways of arranging L, M, H from the rank ordering convention for three cards on page 309. Assume that the entire deck is ordered A♣,...,K♣, A♥,..., K♥, A♦,..., K♦, A♠,..., K♠, and also is thought of as numbered 1–52.

> **Rank Ordering Convention for Four Cards**
> *With respect to the above CHaSeD total ordering of the deck, given any four cards A, B, C, and D, there are $4! = 24$ possible ways to arrange them in a row, varying from ABCD, ABDC, ..., to DCBA, in alphabetical order. Let's agree on the correspondences in Table K.1.*

ABCD	1	BACD	7	CABD	13	DABC	19
ABDC	2	BADC	8	CADB	14	DACB	20
ACBD	3	BCAD	9	CBAD	15	DBAC	21
ACDB	4	BCDA	10	CBDA	16	DBCA	22
ADBC	5	BDAC	11	CDAB	17	DCAB	23
ADCB	6	BDCA	12	CDBA	18	DCBA	24

Table K.1. Correspondences for the twenty-four arrangements of A, B, C, D.

The four cards that Curly displays determine two things: which forty-eight cards remain as possibilities for the hidden fifth card, and a particular permutation from the twenty-four arrangements available. The latter, in turn, results in a number, say 11, being communicated to you. Noting that $24 + 11 = 36$, you can narrow down the identity of the fifth card by determining which of these two possibilities it is: either the 11th or 36th card in the remaining ordered list of forty-eight possible cards. In other words, they are the cards that would usually be in those positions with respect to the agreed-upon total deck ordering, bumped up one for every seen card that occurs before those positions.

Curly must communicate one more bit of information, and numerous possibilities suggest themselves. For instance, we might decree that if the row of four cards is on the table before you inspect them, the first possible card is intended, whereas if Curly lays the cards down after you reenter the room, the second is intended. Another way to handle this twist would be always to have Curly lay the cards out while you watch, either from left to right or right to left, depending on which of the two cards she wishes to communicate.

Let's look at an example in detail. Suppose the cards in question are 2♣, 2♥, 10♠, J♣, and 8♦, as considered earlier in Figure K.1. The audience gets to decide which card is hidden, so five cases arise. The treatments are similar, so we present just the first and last ones in detail, leaving the others for the motivated reader to work through.

Case 1: The audience insists on having the 2♥ hidden. Then think of the four remaining cards in order from lowest to highest with respect to CHaSeD order, as shown in Figure K.6.

Figure K.6. The four cards left in Case 1.

In the full deck, the four depicted cards are in positions 2, 11, 36, and 47, respectively. The target card 2♥ is in position 15. Since two of the cards displayed (we've yet to determine the right order) come before it in the deck, Curly needs to communicate $13 = 15 - 2$.

We see that CABD $= 13$, hence Curly must display the four cards available as in Figure K.7.

How do you decipher this? You see these cards, and mentally put them in CHaSeD order 2♣, J♣, 10♠, 8♦, which determines ABCD, and hence reveals that in Figure K.7 they are in order CABD. Since that corresponds to 13, the hidden card is the thirteenth one of the forty-eight cards in the full deck that are not on display. Since cards 2, 11, 36, and

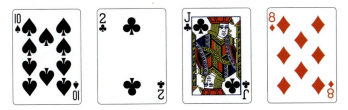

Figure K.7. The four cards displayed in Case 1.

47 are visible, and two of those come before 13, the target card must be
in position $13 + 2 = 15$ in the full deck, i.e., it's 2♥.

Case 5: The audience insists on having the 8♦ hidden. Again, we think
of the four remaining cards in order from lowest to highest with respect
to CHaSeD order, as shown in Figure K.8.

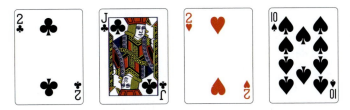

Figure K.8. The four cards left in Case 5.

This time, the depicted cards are in positions 2, 11, 15, and 36, respectively. The target card 8♦ is in position 47, and all four of the cards
displayed come before it in the deck, so Curly needs to communicate
$43 = 47 - 4$. Since $43 = 24 + 19$, she uses the arrangement DABC corresponding to 19 (see Table K.1) in conjunction with whatever was agreed
on to denote "add 24" (see page 314). The cards must be as shown in
Figure K.9.

Figure K.9. The four cards displayed in Case 5.

How do you decipher it this time? Upon seeing this display, you
mentally put the cards in CHaSeD order 2♣, J♣, 2♥, 10♠ to determine
ABCD, which allows you to deduce from Figure K.9 that they are in order
DABC. Since that corresponds to 19, and Curly has also somehow let you

know that you need to add 24, the hidden card is the card in position $19 + 24 = 43$ in the part of the deck consisting of the forty-eight cards that are not on display. Since cards 2, 11, 15, and 36 are visible, and all of those come before card 43, the target card must be in position $43+4 = 47$ in the full deck—it's 8♦.

The other cases are similar.

Source: This is adapted from "Victor Eigen's Trick" as found in Martin Gardner's *The Unexpected Hanging and Other Mathematical Diversions* [Gardner 69, page 152]. There, the assistant locates both of the possible cards in the remainder of the deck, brings them together discreetly, and cuts the deck so as to bring one to the top and the other to the bottom. Upon the naming of the chosen card, the assistant turns over either the top card, or the whole deck to reveal the bottom card. This variable last step—a well-known "out" in magic—makes the trick unsuitable for repeating.

Now we move on to explore the possibilities if some cards are displayed face down. What follows is the oldest[7] trick in the book.

K♠   Ups and Downs

How it looks overall: *Select a volunteer and give her the deck of cards. After you have stepped out of the room, an audience member shuffles the cards and gives the volunteer any five of them. The volunteer then conceals one of them and arranges the other four in a row,* some face up, some face down. *You reenter the room, glance at the row of four cards, and soon name the hidden fifth card.*

How it looks in detail: The same instructions are given to Curly, the so-called volunteer, as in the first effect in this chapter, except that this time stress that some of the cards in the final display can be face down.

How it works: The communication pulled off in "Fitch Cheney's Five-Card Twist" (K♣) seems pretty watertight. This time, surely, there is less to see—fewer card faces visible to code the required information. The impression this creates turns out to be misleading, for the simple reason that $2^3 > 3!$—that is, at this tiny scale, binary bits beat arrangements hands down!

---

[7]Oldest original trick, that is.

Once again, the pigeonhole principle (page 23) guarantees that (at least) two of the five cards are of the same suit. Without loss of generality, Curly has two Clubs. She hides one Club, and communicates its identity to you by placing the remaining four cards in some particular order, where this time some of them are face up.

You and Curly agree in advance to use one designated position, such as the second of the four available on the table for the retained Club, this determining the suit of the hidden card. The other three positions are saved for the placement of the remaining cards, which will tell you exactly which Club is hidden.

Of course, Curly must be very careful as to which Club to hide. As before, she takes advantage of the Poles Apart Principle: the hidden card must be the one that is k (at most 6) "hours" past the other, considering the thirteen card values Ace, $2, 3, \ldots, 10$, J, Q, K arranged in a clockwise circle. The cards in the other three positions thus act as "value-givers."

Unlike before, however, where three card faces would be displayed, here the communication of k is done by using some kind of binary code. Curiously enough, the identities of any face-up cards in the three relevant slots play no role at all.

Ups and Downs Principle ▶◀

In general, your assistant hides one card of a particular suit, and then communicates that card's identity to you, via (1) another card of the same suit in the second position of the display, and (2) a particular face-up/face-down arrangement of the other three cards that tells you what number k to add to the value of the suit-giver card.

Rather than use actual binary representations, let's agree on this convention: UDD, DUD, DDU (only the first, second, or third of the relevant positions is up), and DUU, UDU, UUD (only the first, second, or third such position is down) communicate which of the numbers 1, 2, 3 and 4, 5, 6, respectively, to add to the value of the displayed Club.

For instance, assume that, as in the first example of the first effect in this chapter, Curly is handed 2♣, 2♥, 10♠, J♣, and 8♦. She hides the 2♣, decides to put the J♣ in the second position, and figures out how to tell you to add $k = 4$ to that (to get the 2♣) by displaying the other three cards in DUU form in any one of the many permissible ways. This means that any two of 2♥, 10♠, and 8♦ are displayed face up in the final two positions, in any order, with the remaining card being displayed face down in the first position. Figure K.10 shows one option.

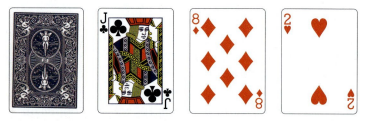

Figure K.10. Ups and Downs: one example.

When you see this row, the card in the second position tells you that the hidden card is a Club. The *DUU* pattern of the cards in the other positions tells you to add 4 to the J♣ value, and you conclude that the hidden card is the 2♣. Note that you totally ignore the suits and values of the cards in the third and fourth positions.

For a second example, assume that Curly is handed A♥, 3♦, J♠, 4♥, and J♦, and opts to take advantage of the Heart suit match. The Four is 3 past the Ace, and 3 is communicated with *DDU*, so Curly hides the 4♥, puts the A♥ in the second position, and displays the other three cards in *DDU* form with any of the three available cards face up in the last position. Figure K.11 shows one of the three options available to her.

Figure K.11. Ups and Downs: a second example.

When you see this, the card in the second position tells you that the hidden card is a Heart. Noting the *DDU* pattern of the cards in the other positions tells you to add 3 to the A♥ value, so that the hidden card must be the 4♥. The suit and value of the fourth card play no role at all.

Source: Original, from early 1999. Published February 2003 in *Math Horizons* [Mulcahy 03] (and again in *The Edge of the Universe—Celebrating Ten Years of Math Horizons* [Mulcahy 06]). Brain Epstein included it earlier in his "All You Need Is Cards" contribution to the Martin Gardner tribute book *Puzzlers' Tribute: A Feast for the Mind* [Epstein 02].

Presentational options: As hinted above, there is a fair bit of latitude as to which card faces are shown in some of the face-up slots; this flexibility can be played up in performance. Curly can solicit audience input as to which card(s) to display, without making it obvious that similar input on the subject of which slots to use is much less welcome.

Since *UUU* is avoided here, if the basic effect in "Fitch Cheney's Five-Card Twist" (K♣) is repeated, the volunteer can throw the audience off the scent by saying, "Should we make it harder this time, and show only *some* of the cards?" Regardless of the answer (and hence, card effect) the audience opts for, as soon as you see the row of cards, you can tell which rules are being followed, and hence identify the hidden card, as desired. If some cards are face down, you should play up the impossibility of the task before you, and the unfairness of it all, before dumbfounding the crowd by correctly concluding the effect.

One more piece of drama can be added. Note that the *DDD* possibility for card placement in the three relevant slots is also avoided, so at least one of these three cards is always face up. Hence, we can use such a face-up card—let's agree on the first such if there are two—to communicate the suit.

When all is said and done, this means that only three cards of the four retained need to be displayed for you, face up or face down; the fourth can be set aside and ignored every time. This fact, too, can be used to spice up the effect in repeat performances. *Before* the five cards are selected, but after you've left the room, Curly can say something like, "We're making it too easy. I'll show one fewer card this time for a real test of mathemagical powers."

A similar trick can be done with fewer cards. This, we believe, is one of our strongest effects.

K◆   Fitch Four Glory

How it looks overall: *Select a volunteer and give her the deck of cards. After you have stepped out of the room, an audience member shuffles the cards and gives the volunteer any four of them. The volunteer then hides one of them and arranges the other three in a row, with some of them face down. You reenter the room, glance at the display, and soon name the hidden fourth card—even in the case where all of the cards are face down!*

How it looks in detail: One last time, the same instructions are given to Curly, the so-called volunteer, except that this time, (1) only four cards are picked at the outset, and (2) it is stressed that some of the three cards in the final display can be face down.

How it works: Since this works even in the case where all three of the cards in the row are face down, such a display can communicate only one possible card. Let's agree that if one of the four randomly selected cards is the A♠, then this is hidden by Curly, and the other three cards are shown face down. (At performance time, try to suppress your excitement at being confronted with three face-down cards, and appear to concentrate suitably before announcing your conclusion.)

How can the pigeonhole principle (see page 23) help Curly in the general case? After all, with only four cards, there may not be any suit matches at all.

Perhaps, (1) the four given cards do include a suit match, and Curly could proceed something like before, or (2) all cards are of different suits, and you and Curly could agree that she always hides the Spade. In the second case, it could be any one of thirteen values, so it would have to be possible for Curly to convey that value to you by using the other three cards—and even if it was possible, how would you know which of the two cases was in force?

Clearly, we need a method that works in all cases from both Curly's perspective and your perspective. Let's redefine the pigeonholes: since we have to worry only about the cases in which the four cards are among the fifty-one in the part of the deck excluding A♠, we break this up into three new "suits" of seventeen cards each.

We suggest the supersuits depicted below, consisting of one of the standard suits, ♣, ♦, ♥, supplemented with four ♠s. Specifically,

$$\text{Supersuit } \alpha = \{A\clubsuit, 2\clubsuit, \ldots, K\clubsuit, 2\spadesuit, 3\spadesuit, 4\spadesuit, 5\spadesuit\},$$
$$\text{Supersuit } \beta = \{A\diamondsuit, 2\diamondsuit, \ldots, K\diamondsuit, 6\spadesuit, 7\spadesuit, 8\spadesuit, 9\spadesuit\},$$
$$\text{Supersuit } \gamma = \{A\heartsuit, 2\heartsuit, \ldots, K\heartsuit, 10\spadesuit, J\spadesuit, Q\spadesuit, K\spadesuit\}.$$

Given four cards selected from the fifty-one indicated, the pigeonhole principle guarantees that (at least) two of the four are from one of the three supersuits. Without loss of generality, assume there are two cards from supersuit α.

Curly hides one of these cards—yes, it matters which one—and by placing the remaining three on the table in some face-up and face-down pattern, she communicates the identity of the hidden card. Consider each of the supersuits laid out in the form of a clockwise circle of seventeen cards (i.e., the seventeen-card analog of the thirteen-hour clock in Figure 13 on page 15).

Extended Poles Apart Principle ▶◀

In any seventeen-hour clock of cards, given any two of them, they are at most eight "hours" apart, so that counting clockwise, one of them lies at most eight past the other.

The basic strategy is to have hidden the "higher" card from supersuit α, which is k past the "lower one," where this time k is between 1 and 8, inclusive. Curly must communicate both the supersuit in question and the value of k to you by using the remaining three cards. In the convention we explain below, at least one card is face up, so once more she can use a face-up card (say, the first such if there are two) to give the supersuit. Note that the hidden and "supersuit-giver" cards here are not always of the same suit in the ordinary sense, and that even if they are, the wrapping around from the King to the Ace of Clubs, Diamonds, or Hearts must take account of four inserted Spades.

The placements UDD, DUD, DDU (one U in the first, second, or third relevant position) and DUU, UDU, UUD (one D in the first, second, or third such position), respectively, can be used by Curly to tell you that k is one of $1, 2, 3, 4, 5,$ or 6.

Curly needs a way to communicate 7 or 8, and fortunately, she has the UUU option at her disposal. If we agree to use one particular U (say, the middle one) to give the supersuit, there are two ways to play the other two: low-high (to convey $k = 7$) or high-low (for $k = 8$) with respect to some total ordering of the deck, such as lining up supersuits α, β, γ in that order.

We now work through some examples in detail that highlight the different kinds of situations that can occur. First, suppose Curly is handed 2♣, 2♥, 10♠, and J♣, as shown in Figure K.12, namely, the first four cards in Figure K.1 from earlier in this chapter.

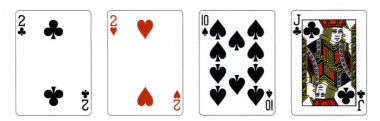

Figure K.12. Given any four cards—what can we say for sure?

There are two Clubs, of course, but this time they belong to supersuit $\alpha = \{A♣, 2♣, \ldots, K♣, 2♠, 3♠, 4♠, 5♠\}$. Considering them as hours in the seventeen-hour clock in question, the 2♣ is eight hours past the J♣. Hence, Curly needs to hide the 2♣, and she knows that the J♣ goes face up in the second position in the row, the other two cards also being face up. With respect to the total ordering of the deck obtained by lining up supersuits α, β, γ, we see that 2♥ is low and 10♠ is high (actually they're both in supersuit γ), and these she arranges in high-low order to communicate "add 8." Upon reentering the room, you see the display in Figure K.13.

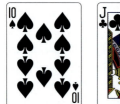

Figure K.13. Fitch Four Glory—one possibility.

You, in turn, decode this as follows. Since all three cards are face up, you know that the second card is the supersuit-giver, and also that you must go 7 or 8 past J♣ in supersuit α to find the hidden card. Noting that 10♠ is after 2♥ in supersuit γ, you know that these two cards are displayed in high-low order, and finally you arrive at 2♣ by counting 8 past J♣ in the seventeen-hour clock of supersuit α.

Here's a second example. Suppose Curly is handed 2♥, 10♠, 8♦, and J♣, as depicted in Figure K.2 on page 310. This has no suit match in the traditional sense, but the first two cards are both in supersuit γ, with the 2♥ being five past the 10♠ in a clock of seventeen cards. Hence, she hides the 2♥ and places 10♠ face up in the first position of the row; the 10♠ is speedily followed by the other two cards in either order, the middle one face down and the third one face up. Upon reentering the room, you see something like Figure K.14.

Figure K.14. Fitch Four Glory—another possibility.

You decode this by noting that since there's at least one face-up card, the hidden card isn't the A♠, and since the first visible card face is a high-valued Spade, then the hidden card is in supersuit γ, and hence is either a Spade or a Heart. Since the displayed cards are UDU, which corresponds to 5, you continue that number of cards past the 10♠ in the seventeen-hour γ clock to conclude that the hidden card is the 2♥.

For our third and final example, suppose Curly is handed 7♠, 2♦, Q♠, and J♥, as shown in Figure K.15.

There are two Spades, but they do not represent a suit match because they belong to different supersuits. In fact, the supersuit match here is 7♠ and 2♦, which are in supersuit $\beta = \{A♦, 2♦, \ldots, K♦, 6♠, 7♠, 8♠,$

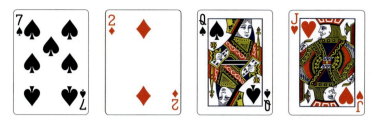

Figure K.15. Given these four cards—what can we say for sure?

9♠}, four hours apart on the seventeen-hour clock in question. Curly needs to hide the 2♦.

Recalling that a count of 4 is communicated by using the DUU option, with the supersuit-giver in the first face-up position, she places the 7♠ in the middle, flanked by the other two cards in either order, as long as the first is face down and the last is face up. Upon reentering the room, you see something like the display in Figure K.16.

Figure K.16. Fitch Four Glory—a third possibility.

You, in turn, decode this as follows. Seeing the DUU configuration, you know to count 4 past what is in the middle position, namely the 7♠. Since that card is in supersuit β, counting four hours around that seventeen-hour clock gets you to the 2♦, which must be the hidden card.

Readers should work through many more examples before trying this effect live; it takes quite a bit of practice to pull it off without fail.

Source: Original. From early 1999. Published February 2003 in *Math Horizons* [Mulcahy 03] (which resurfaced in *The Edge of the Universe—Celebrating Ten Years of Math Horizons* [Mulcahy 06]), and online as the February 2005 *Card Colm* "Fitch Four Glory" [Mulcahy 05_02]. Brain Epstein was the first to get it into print, in his "All You Need Is Cards" offering in the Martin Gardner tribute book *Puzzlers' Tribute: A Feast for the Mind* [Epstein 02].

This also resurfaced as the second puzzle in Pradeep Mutalik's "Monday Puzzle: Magic Phone Calls" in the *New York Times* in December 2009 [Mutalik 09_12a]. A solution, using different supersuits to those suggested above, appeared there three weeks later, including a lovely graphic of a 17-hour clock from talented artist Justin Thyme [Mutalik 09_12b].

In March 1999, following a happy meeting at Rose-Hulman Institute of Technology, operations researcher and scheduling pro Mike Trick of Carnegie Mellon University kindly put together an interactive website that illustrates this, er, trick in action. It uses our supersuit convention, and can be found at http://mat.tepper.cmu.edu/CARD.

Presentational options: Can the Fitch Cheney Trick be scaled down even further, so that given any three cards from a deck, the identity of one of them can be deduced from some sort of display? Essentially, yes, using ideas already mentioned in recent chapters together with one new concept.[8] Details can be found in "Fitch Cheney's Five Card Trick for Three Cards" in *Math Horizons* [Mulcahy 14].

Parting Thoughts

Here are some terrific questions sent by Derek Smith of Lafayette College.

- Can the original Fitch Cheney Trick, in which five cards are randomly chosen and four are lined up face up to point to the fifth, be done if the five cards are drawn from two (identical) decks of fifty-two cards?

- Find a 5-by-5 array of cards, so that reading left to right and top to bottom, we have Fitch Cheney in each row and column. In other words, in each case the first four cards point to the fifth one. Hence, the 4-by-4 upper left-hand subsquare could be revealed with the other nine cards face down, yet "predictable." Can this be done so that we also have Fitch Cheney working on the principal diagonal?

- Can we make *two* such configurations at the same time with a standard deck?

- Can we arrange all fifty-two cards so that every four consecutive cards give the one immediately after those? If not, what's the longest such run that can be devised?

Here are more questions worth pondering.

- It's long been known that the basic Fitch Cheney Trick can be done with a larger deck. Can our approach be modified to work for fifty-three cards, such as a regular deck supplemented with a Joker? How about for a deck of fifty-four distinct cards, perhaps using a different strategy?

---

[8]See the Coda for a hint of what that is.

- What's the largest possible deck that can be accommodated by a "four cards gives the fifth" technique? The answer is widely available on the Internet, and earlier in this chapter we provide some print references, but see if you can come up with the answer unaided. It's surprisingly large.

- How does the Fitch Cheney Trick generalize to a "$n-1$ cards gives the nth" scenario? Rather than thinking of larger values of n, and larger decks, what about the cases $n = 3$ or $n = 4$?

- Show that given all of the Queens and Kings in a deck, no matter which three of them are selected as finalists in the Monarch of the Year competition, you can figure out who the third one is once you are provided with a list of the other two. An incarnation of the ideas at work here, sadly lacking in card content however, can be found online in the April 2010 *Card Colm* "Mathematical Idol 2010" [Mulcahy 10_04], having appeared a year earlier in the book *Mathematical Wizardry for a Gardner* [Mulcahy 09]. Find a simpler method than the one discussed there.

- It is clear that the basic Fitch Cheney Trick cannot be done with a full deck starting with just four cards—that's why we had to resort to turning cards over in "Fitch Four Glory" (having introduced the concept in "Ups and Downs"). After all, given any four cards, there are at most $4! = 24$ things one can do with them: 4 ways to select one to have noted and hidden, and $3! = 6$ ways to arrange the remaining three. Even allowing for the extra bit of information that can surreptitiously be communicated by laying those cards from left to right, or right to left, that allows for only $2 \times 24 = 48$ possibilities. Bearing in mind that three card faces would be face-up—and hence couldn't be contenders for the target card—it seems that at the very most, a deck of size $48 + 3 = 51$ could be accommodated. And that assumes that we could find a strategy that took full advantage of the possibilities just enumerated. Is our final "Fitch Four Glory" offering optimal? Is there any wiggle room at all?

- Can the Fitch Cheney Trick be scaled down even further, so that given any three cards from a deck, the identity of one of them can be deduced from some sort of display involving the other two? Keep reading!

- Hang Chen and Curtis Cooper have another spin on the kinds of effects considered in this chapter in their paper "n-Card Tricks" [Chen and Cooper 09]. It includes nice alternative handlings of both "Eigen's Value" (K♥) and "Fitch Four Glory" (K♦). Stirling work by these authors shows that timing is everything.

Coda

More Advice

Our goals have been simple: (1) to show how mathematical considerations simple and not so simple can lead to entertainment for all when combined with playing cards, and (2) to show how having fun and games with cards leads to interesting questions about mathematics. If there have been moments when a sense of wonder was instilled in readers, we have been successful.

If, in turn, readers have been able to instill a sense of wonder in their audiences, that's even better. A few words of advice are in order, starting with some repeated from "Tips of the Trade" at the start of this volume.

If you are lucky, people may ask, "How did you do that?" The safest answer is always a modest, "Reasonably well, I think." Before saying more, consider who your audience is. If it's the general public, which includes the average friend, relative, or coworker, you may want to leave it at that, resisting all subsequent prompting. If it's somebody with a genuine interest in magic or mathematics, perhaps a student if you are a teacher, you are in a position to turn curiosity into a teachable moment.

As magicians have long known, and our experience confirms, it's a big mistake to launch into an exposé of an effect's inner workings. If you do go down that road, don't be surprised if you later regret it: responses are likely to vary from "Oh, is that all you did?" to "So it's just based on mathematics?". Almost invariably, you've killed the magic, and people think less of you as a result.

As magician Steve Beam has pointed out, when it comes to the value of a secret, there is a parallel with the skillful telling of a good joke: "Nobody wants to hear the same punchline twice." There's another humorous parallel. Card effects, like jokes, are often reverse engineered: the creator starts with a ending that has an irresistible punch, then works backwards to construct an effect that gets one to the desired conclusion in a seemingly logical manner. The real trick is to come up with something that conceals this method of design.

More Mathematics

There is much more mathematics to be explored with a humble deck of cards than we have indicated. For one thing, there are topics ranging from elementary (e.g., magic squares) and intermediate (e.g., Fibonacci bracelets, group operations) to advanced (e.g., Markov chains and statistical moments, such as standard deviation and skewness), which we have not discussed, but have interesting interactions with cards (see [Mulcahy 00, Mulcahy 06_12, Mulcahy 07_02b, Mulcahy 07_04, Mulcahy 07_06, Mulcahy 07_12, Mulcahy 09_12, Mulcahy 12_12b]).

It all started, from the public-access perspective, with Martin Gardner's landmark *Mathematics, Magic and Mystery*. Before its publication, there was very little mathemagic—with or without cards—that general readers could get their hands on. It's still a classic after almost sixty years, crammed with high-caliber material he'd collected from some of the best minds in mid-twentieth-century magic. The year of its publication saw Gardner start a twenty-five year career as a monthly columnist at *Scientific American*. His "Mathematical Games" columns focused mostly on recreational mathematics, and a good number of them involve cards. All 300 of these columns are available on a searchable CD-ROM from the Mathematical Association of America [Gardner 05]. Three *Card Colm*s at MAA.org, published following Gardner's death in 2010, are devoted to some of the highlights of those, in his own words [Mulcahy 10_06, Mulcahy 10_08, Mulcahy 11_10].

Magic Tricks, Card Shuffling and Dynamic Memories [Morris 98] by S. Brent Morris contains everything you ever wanted to know about the mathematics of Faro shuffling, for two and three piles of cards. The same author's short article *A Book of Corollaries: Variations on a Theme* [Morris 04] is also highly recommended. The original master of this kind of collusion of mathematics and magic was Alex Elmsley, whose collected works [Minch 91, Minch 94] are indispensable.

Diaconis and Graham's book *Magical Mathematics* [Diaconis and Graham 11] is the modern bible on mathemagic, containing a lot of fascinating material on shuffles of many types, including Gilbreath and Faro, as well as their connections to Penrose tiles and the Mandelbrot set, along with other deep mathematics with magic applications. Their chapter "Products of Universal Cycles" in the Martin Gardner tribute book *A Lifetime of Puzzles* [Diaconis and Graham 08] also makes for fine reading.

More Card Magic

When it comes to actual cards, we've taken a purist approach and focused on the face-up or face-down status and the visible value in the first case. Hence, we've assumed that any pasteboard shown is in one of fifty-four states: being one of the standard fifty-two card values, a Joker, or a card back (color irrelevant[9]).

Furthermore, displays have been either linear (in a row) or in circles (to accommodate wraparounds). There is no significance to the angles at which cards are shown anywhere in these pages. Only once (in Figure J.9) does it even matter whether or not cards in a row overlap. However, examine the images in Figure 1. Something not considered before in these pages is evident. Can you see what it is?

Figure 1. Orientation issues.

The final card shown is upside down—rotated 180 degrees—compared to the way we illustrated it before. Compare with how it appears in the face of the fan in Figure 10.1 on page 242. On the other hand, the 7♥ is the right way up (as is the Joker). In the cases of J♣ and 10♦, it's impossible to tell, because they look the same either way.

Twenty-two of the fifty-two cards in a normal deck look different when rotated 180 degrees. Those are often called pointers; the second and fifth cards shown in Figure 1 are pointers, as is this particular Joker. In a standard deck, no Royal card or Diamond is a pointer.

New decks have the pointer cards all pointing the same way, but it's perfectly natural in the course of shuffling, dealing, handing out and collecting cards, for the pointers to end up displayed in either of their two possible orientations. With hindsight, this is apparent in some of the photographs in this book. Until now, we've chosen not to draw attention to this subtlety, but with careful planning and card handling, we can take advantage of it for mathemagical purposes.

One interesting question that arises, since almost half of the deck consists of pointers, is how one can use the orientation state in a "given

---

[9]Playing with two decks, one red-backed and one blue-backed, does suggest additional options.

any five (or ten) cards from a shuffled deck" context. Binary codes and the pigeonhole principle spring to mind, if probabilistic considerations are also taken into account.

It's not just the card faces that warrant more attention than we've given them. The card backs have their own secrets to tell. The backs of the Bicycle cards displayed in most of the pages of this volume show a total regularity, such as the two on the left in Figure 2. The same cannot be said of the remaining four images.

Figure 2. Two idealistic—four reality.

Despite the high-quality manufacturing process used to produce Bicycle cards—after all, unlike with many other brands, these allow in-hand Faro shuffling (see page 5)—the standard of printing varies a lot. Some are perfectly aligned and centered, as shown in the first and second images, but as the rest show, the margins may be uneven, and the core blue or red images may tilt one way or the other.

The last four images in Figure 2 depict actual scanned backs, shown in both possible orientations. The blue ones are more off-kilter than the red ones. These imperfect card back images were surreptitiously used a few times earlier, in Chapter K; can you tell which displays there include them?[10]

As a result, many apparently symmetric card backs effectively have *one-way* designs, as do most decks featuring words and/or logos. It takes time to "tune in" to the kinds of differences highlighted above, made more obvious in the much-enlarged images in Figures 3 and 4.

The red images are reasonably straight, but show subtle differences, the top and bottom margins clearly being of different sizes. The blue images, however, show very skewed printing, the left and right margins being clearly distinguishable from each other.

Of course, if the card backs in a particular deck look different when rotated 180 degrees, there's a good chance that the same holds for the card faces. In such a deck, every card is a pointer!

The deviations from perfection vary from deck to deck, but in our experience all of the cards in a given deck show the same idiosyncrasies

---

[10]For the record, the vast majority of the Bicycle backs shown in these pages, including the first two images in Figure 2, are idealized ones modified from those scans.

Figure 3. Red reality close up.

Figure 4. Blue reality close up.

or flaws. Hence, since it's not uncommon to work with the same deck over and over, one can become familiar with its specific irregularities.

There is also a whole shady world of "Marked Cards," as discussed in a chapter of that same name by magician and con man specialist Simon Lovell in his book *Billion Dollar Bunko* [Lovell 03]. There, he remarks, "The prices vary from two dollars a deck to two thousand dollars a deck." Yes, these are specially manufactured decks with very subtle marks, which look exactly like the kinds of decks you can buy in ordinary shops. We strongly suggest sticking with ordinary, legitimate cards!

Because so much card magic involves manipulations of face-down cards, a vast new world of possibilities opens up when the performer can tell the difference between upright and upside down card backs for a specific deck, especially if a little preparation has been done ahead of time. Most audiences will never suspect that this plays any role. A deck that is intentionally one-way, whose card backs make no pretence of rotational symmetry, makes all of that much easier to explore.

Magicians have exploited such ideas at great length and in great depth over time, so don't expect to come up with anything novel in the short term. However, in conjunction with some of the fresh concepts covered in the last thirteen chapters, perhaps there is something new and worthy to be tried out. For instance, given any five cards from such a deck, we might ask if they can be displayed in a face-down row, such as in Figure 5,[11] so as to convey to an accomplice the identity of the last card shown?

Figure 5. What is the last card?

Did we mention that the classic Fitch Cheney Trick can be scaled down to just three cards? All is revealed in "Fitch Cheney's Five Card Trick for Three Cards" [Mulcahy 14].

More or Less?

There seems to be no end to the list of fun things that can be explored with a simple deck of cards. It is indeed a fertile field. Likewise, many intriguing questions are suggested. Back in Chapter 2, on page 52, we asked the following:

> **Conundrum (More or Less?)**
> *Which is more impressive, being able to perform a certain effect when there are just a few cards involved, or when there are more?*

and suggested that readers ponder this conundrum as they explored the effects that followed. The question certainly arose in the last chapter: being able to pull off a variation of the Fitch Cheney Trick for four cards selected from a standard deck seems more impressive that doing it for five cards. Of course, given any seven cards, it's possible to display six of them in a face-up row to convey the identity of the remaining (hidden) one; in fact, given any preexisting row of seven cards, it's possible to turn some (including the last one) face down so as to convey the identity of the seventh card. Many other interesting options suggest themselves. Given any 51 cards, it certainly wouldn't be impressive to be able to display them in some order to convey the identity of the one missing card.[12]

---

[11] Used here with the permission of Princeton University Press.

[12] However, that leads to an interesting question: what is the smallest number n such that given any n cards, they can be displayed in some order so as to convey the

On the other hand, consider "Little Fibs" (4♣): while it might seem harder to have to identify five cards after having that many spectators select one and report on the total of their values, it's actually much easier, because unknown to the audience, it's only necessary for you to subtract that total from 32 to deduce which of the six possible cards is missing!

Many other effects, such as the last two in Chapter 10, are also worth reexamining in this light, and new mathematical questions spring up at almost every turn.

In the meantime, if you're feeling very brave and can think on your feet, and have prepared suitably and practiced a lot, you can wow your audience as follows. Ask for a number between one and ten to be called out. Regardless of the answer, proceed with an appropriate crowd-pleasing effect that uses exactly that many cards, thus giving the illusion that it works no matter what number is selected. Look up *given any* in the Index and master one effect for each number 2–10. If somebody calls for "One," respond with, "I bet you think I'm going to say, 'pick a card, any card.'" That's what ordinary magicians do, but we mathemagicians like to spice it up a little and do something more interesting. Please pick a bigger number."

A Parting Gift

To wrap up what is essentially a collection of parting thoughts, we leave you with a parting gift, from June 2013, courtesy of mathematician Neil Calkin of Clemson University. Stack your favorite 3, 6, Jack, Queen, and King at the top of the deck in any order. Next, shuffle while maintaining that top stock. Find some people who think they've seen it all and give one of them the deck, requesting that the top five cards be dealt out face down and then thoroughly jumbled. Turn away and ask for any two or three to be selected, the rest being shuffled back into the remainder of the deck. Have the selected cards looked at and remembered. Ask for the total of their values, and have those cards also shuffled back into the deck. You promptly announce what each selected card was—or look through the deck and pull them out—even though you were never informed how many cards were involved.

This is a simultaneous generalization of "Subtler Bracelet" (2♥) and "Little Fibs" (4♣), the first of which only allowed the selection of two or three *adjacent cards* from a forced circle, and the second of which only allowed a selection of a known number of cards from a forced collection.

---

identities of the remaining $52 - n$ cards (the latter perhaps being in a specific order, e.g., lined up in a face-down row to the side)?

It works because $\{3, 6, 11, 12, 13\}$ is a dissociated set (see page 103): all of the thirty-one sums of its nonempty subsets are different. A subtle point is that you can also identify or locate all of the selected cards if either one or four are selected, and only the value sum is reported to you, without knowing that one of these more extreme cases holds.[13]

There is another dissociated set of five numbers in the card-friendly range 1–13, but it's very similar to the one above. If we are content to stick with unique sums of just two or three cards, there are two more options, namely $\{1, 2, 4, 7, 13\}$ and $\{1, 2, 7, 10, 13\}$. Using one of these as well as the original $\{3, 6, 11, 12, 13\}$ permits the use of two stocks at the top of the deck, thus facilitating a repeat performance with largely different values.

If that doesn't knock their socks off, what will?

---

[13]The corresponding results are not valid for either "Subtler Bracelet" or "Little Fibs"; in the former case a total of 13 could arise from either $\{6, 7\}$ or $\{1, 3, 4, 9\}$, and in the latter case a total of 13 could arise from either $\{13\}$ or $\{5, 8\}$.

Acknowledgments

Thanks to:

Paul Zorn of St. Olaf College, for inadvertantly setting me down this road, over a beer in Austin, TX, during the Joint Mathematics Meeing in January 1999.

The late Martin Gardner, for suggesting that I write a book on this topic a year or two later—long before that was a good idea—and for being so generous and gracious in correspondence and face-to-face meetings.

Tony Phillips of SUNY Stony Brook and Art Benjamin of Harvey Mudd College for their roles in providing online and print outlets for early creations, coutesy of the AMS and MAA, respectively.

Fernando Gouvea for launching *Card Colm* at MAA.org in October 2004, and the MAA for hosting it bi-monthly ever since.

Ron Gould, Pete Winkler, and the late Thomas M. Rodgers (of G4G) for crucial connectivity help.

Vickie Kearn at Princeton University Press for much encouragement and advice, and classy Klaus Peters formerly of A K Peters for enthusiastically offering me a contract.

All the staff at A K Peters/CRC Press/Taylor & Francis for making this volume a reality, especially editor Sunil Nair for his vision, and project editor Charlotte Byrnes for her technical prowess, great attention to detail, and endless patience.

Matthew Wright, Jim Kinney, and above all Neil Calkin and Scott Anderson, for being sounding boards and trouble shooters on Unix and LaTeX issues, and Kevin Dunn for wise council on matters linguistic.

Neil Calkin of Clemson University for also providing invaluable mathemagical suggestions, as did Derek Smith of Lafayette College and Spelman College colleague Jeff Ehme.

Art Benjamin, Chris Morgan, Max Maven, Bee Bordeaux, Kevin Dunn, Priscilla Smith, Yvonne Prabhu, and Ann & Vicki Powers for eagle-eyed copyediting and proofreading, and above all mathemusician Robert Schneider, for the energy he brought to that task.

Magicians Joe M. Turner and Max Maven for freely sharing their expertise at the drop of a hat.

John H. Conway, Lennart Green, Ron Graham, Persi Diaconis, and Ron Wohl for enthusiastically sharing their mathemagic.

Game changers Glenn Barry, for having introduced me to email, Unix, vi, and more in 1987, and for much help since, and Thomas Hirschmann, for gettting me started with LaTeX at the University of Regensburg in early 1992.

Dan Bascelli (Coordinator for Instructional Technology of Spelman College's Comprehensive Writing Program) for taking all of the photographs.

And Spelman College for fostering a supportive environment in which I could pursue the work required to write this manuscript.

Bibliography

[Adams 82_01] Howard A. Adams. *OICUFESP, Volume 9*. Self-published, January 1982.

[Adams 82_11] Howard A. Adams. *OICUFESP, Volume 10: The Ramasee Prophecy*. Self-published, September 1982.

[Aigner and Ziegler 04] Martin Aigner and Gunter M. Ziegler. *Proofs from The Book*, Third edition. New York: Springer, 2004.

[Bannon 05] John Bannon. *Dear Mr. Fantasy*. Chicago: Bannon Impossibilia, 2005.

[Buckley 46] Arthur Buckley. *Card Control*. Self-published, 1946.

[Burger and Starbird 04] Edward Burger and Michael Starbird. *The Heart of Mathematics: An Invitation to Effective Thinking*, Second edition. New York: John Wiley & Sons, 2004.

[Butler to appear] Steve Butler. "Backwards Addition." In *Proceedings from Gathering 4 Garndner 10*, edited by Scott Hudson. Atlanta: Gathering 4 Gathering, to appear.

[Chen and Cooper 09] Hang Chen and Curtis Cooper. "n-Card Tricks." *College Mathematics Journal* 40:3 (2009), 196–203.

[Curry 01] Paul Curry. *Paul Curry's Worlds Beyond*. Hermetic Press, 2001.

[de Bruijn 87] N. G. de Bruijn. "A Riffle Shuffle Card Trick and Its Relation to Quasicrystal Theory." *Nieuw Archief voor Wiskunde* 4:5 (1987), 285–301.

[Diaconis and Graham 08] Persi Diaconis and Ronald Graham. "Products of Universal Cycles." In *A Lifetime of Puzzles*, edited by Erik D. Demaine, Martin L. Demaine, and Tom Rodgers, pp. 35–55. Wellesley, MA: A K Peters, 2008.

[Diaconis and Graham 11] Persi Diaconis and Ronald Graham. *Magical Mathematics*. Princeton, NJ: Princeton University Press, October 2011.

[Duck 57] J. Russell Duck. "Rusduck 'Stay-Stack' System." *The Cardiste* 1 (February 1957), 12–16.

[Duffie and Robertson 03a] Peter Duffie and Robin Robertson. *Card Conspiracy Vol. 1*. Self-published, 2003.

[Duffie and Robertson 03b] Peter Duffie and Robin Robertson. *Card Conspiracy Vol. 2*. Self-published, 2003.

[Epstein 02] Brain Epstein. "All You Need Is Cards." In *Puzzlers' Tribute: A Feast for the Mind*, edited by David Wolfe and Tom Rodgers, pp. 179–190. Wellesley, MA: A K Peters, January 2002.

[Fulves 68] Karl Fulves. *Riffle Shuffle Set-Ups*, revised third printing (1976). Teaneck, NJ: Self-published, 1968.

[Fulves 79] Karl Fulves. *Self-Working Mental Magic*. New York: Dover, 1979.

[Fulves 75] Karl Fulves. "Lucky 13." *The Pallbearers Review* 10:10 (August 1975), 1035, 1038–1039.

[Fulves 81] Karl Fulves. *Bob Hummer's Collected Secrets*, Second edition. Teaneck, NJ: Self-published, 1981.

[Fulves 84] Karl Fulves. *More Self-Working Card Tricks*. New York: Dover, 1984.

[Fulves 92] Karl Fulves. *Charles Jordan's Best Card Tricks*. New York: Dover, 1992.

[Fulves 01] Karl Fulves. *My Best Self-Working Card Tricks*. New York: Dover, 2001.

[Gardner 56] Martin Gardner. *Mathematics, Magic and Mystery*. New York: Dover, 1956.

[Gardner 66] Martin Gardner. *New Mathematical Diversions from Scientific American*. New York: Simon and Schuster, 1966.

[Gardner 69] Martin Gardner. *The Unexpected Hanging and Other Mathematical Diversions*. New York: Simon and Schuster, 1969.

[Gardner 77] Martin Gardner. *Mathematical Magic Show*. Alfred A. Knopf, 1977.

[Gardner 83] Martin Gardner. *Wheels, Life and Other Mathematical Amusements*. W.H. Freeman & Co, 1983.

[Gardner 87] Martin Gardner. *Riddles of the Sphinx and Other Mathematical Puzzle Tales*. Washington, DC: Mathematical Association of America, 1987.

[Gardner 93] Martin Gardner. *Martin Gardner Presents*. Washington, DC: Kaufman and Greenberg, 1993.

[Gardner 05] Martin Gardner. "Mathematical Games." CD-ROM from the Mathematical Association of America, Washington, DC, 2005.

[Gardner 09] Martin Gardner. "An Amazing Mathematical Trick with Cards." *Normat* 57:1 (2009), 32–33.

[Gilbreath 58] Norman Gilbreath. "Magnetic Colors." *The Linking Ring* 38:5 (July 1958), 60.

[Gilbreath 66] Norman Gilbreath. "One Man Parade." *The Linking Ring* 46:6 (June 1966), 69–88.

[Gilbreath 13] Norman Gilbreath. *Beyond Imagination*. Los Angeles: H & R, 2013.

[Goldstein 02] Phil Goldstein. *Redivider*. Seattle: Hermetic Press, 2002.

[Gould 12] Ronald J. Gould. *Graph Theory*, Revised edition. New York: Dover, 2012.

[Green 84] Lennart Green. *Röda och Svarta Kort*. Gothenburg, Sweden: Self-published, 1984.

[Green 89] Lennart Green. *1–2 Separation*, Lecture notes for Ron MacMillan International Magic Convention, London, UK. Gothenburg, Sweden: Self-published, 1989.

[Green 95] Lennart Green. *The Green Angle Separation*. Gothenburg, Sweden: Self-published, 1995.

[Green 00] Lennart Green. "Green Magic—Volume Two." DVD from A-1 MagicalMedia, Rancho Cordova, California, 2000.

[Hammingway 09] Earnest Hammingway. "Horses in the Stream and Other Short Stories." In *Homage to a Pied Puzzler*, edited by Ed. Pegg Jr, Alan H. Schoen, and Tom Rodgers, pp. 233–247. Wellesley, MA: A K Peters, February 2009.

[Holm 03] Shai Simonson & Tara Holm. "Using a Card Trick to Teach Discrete Mathematics." *PRIMUS* XIII:3 (September 2003), 248–269.

[Hostler 12] John Hostler. *The Triskadequadra Principle*. St. Louis, MO: Self-published, December 2012.

[Johnson 09] Robert Johnson. "Universal Cycles for Permutations." *Discrete Mathematics* 309:17 (2009), 5264–5270399.

[Kleber 02] Michael Kleber. "The Best Card Trick." *Mathematical Intelligencer* 24:1 (Winter 2002), 9–11.

[Lee 51] W. Wallace Lee. *Math Miracles*, 1976th edition. Calgary: Micky Hades International, 1951.

[Lovell 03] Simon Lovell. *Billion Dollar Bunko*. Tahoma, CA: L & L Publishing, 2003.

[Maven 06] Max Maven. "Parade." *The Linking Ring* 86:11 (November 2006), 97–113.

[Mian and Chowla 44] Abdul Majid Mian and S. Chowla. "On the B_2 sequences of Sidon." In *Proc. Nat. Acad. Sci. India. Sect. A, number MR 7,243a in 14*, pp. 3–4, 1944.

[Miller 10] Werner Miller. *Enigmaths*. Self-published, 2010.

[Minch 91] Stephen Minch. *The Collected Works of Alex Elmsley, Volume I.* Tahoma, CA: L & L Publishing, 1991.

[Minch 94] Stephen Minch. *The Collected Works of Alex Elmsley, Volume II.* Tahoma, CA: L & L Publishing, 1994.

[Morris 98] S. Brent Morris. *Magic Tricks, Card Shuffling and Dynamic Memories.* Washington, DC: Mathematical Association of America, 1998.

[Morris 04] S. Brent Morris. "A Book of Corollaries: Variations on a Theme." *Math Horizons* 11 (February 2004), 17–20.

[Mulcahy 96] Colm Mulcahy. "Plotting and Scheming with Wavelets." *Mathematics Magazine* 69:5 (December 1996), 323–343.

[Mulcahy 97] Colm Mulcahy. "Image Compression Using the Haar Wavelet Transform." *Spelman College Science & Mathematics J.* 1:1 (April 1997), 22–31.

[Mulcahy 00] Colm Mulcahy. "Mathematical Card Tricks." http://www.ams.org/featurecolumn/archive/mulcahy1.html, October 2000.

[Mulcahy 03] Colm Mulcahy. "Fitch Cheney's Five Card Trick." *Math Horizons* 10 (February 2003), 11–13.

[Mulcahy 04_10] Colm Mulcahy. "Low Down Triple Dealing." *Card Colm*, http://www.maa.org/columns/colm/cardcolm_10_04.html, October 2004.

[Mulcahy 04_11] Colm Mulcahy. "Low Down Triple Dealing." *MAA Focus* 24:8 (November 2004), 8–9.

[Mulcahy 05_02] Colm Mulcahy. "Fitch Four Glory." *Card Colm*, http://www.maa.org/columns/colm/cardcolm200502.html, February 2005.

[Mulcahy 05_06] Colm Mulcahy. "A Little Erdös/Szekeres Magic." *Card Colm*, http://www.maa.org/columns/colm/cardcolm200506.html, June 2005.

[Mulcahy 05_08] Colm Mulcahy. "The First Norman Invasion." *Card Colm*, http://www.maa.org/columns/colm/cardcolm200508.html, August 2005.

[Mulcahy 06] Colm Mulcahy. "Fitch Cheney's Five Card Trick and Generalizations." In *The Edge of the Universe—Celebrating Ten Years of Math Horizons*, edited by Deanna Haunnsperger and Stephen Kennedy, pp. 273–276. Weshington, DC: Mathematical Association of America, 2006.

[Mulcahy 06_02] Colm Mulcahy. "Many Fold Synergies." *Card Colm*, http://www.maa.org/columns/colm/cardcolm200603.html, February 2006.

[Mulcahy 06_06] Colm Mulcahy. "Better Poker Hands Guaranteed." *Card Colm*, http://www.maa.org/columns/colm/cardcolm200606.html, June 2006.

[Mulcahy 06_08] Colm Mulcahy. "The Second Norman Invasion." *Card Colm*, http://www.maa.org/columns/colm/cardcolm200608.html, August 2006.

[Mulcahy 06_12] Colm Mulcahy. "Quantitative Reasoning in Small Groups." *Card Colm*, http://www.maa.org/columns/colm/cardcolm200610.html, December 2006.

[Mulcahy 07_02a] Colm Mulcahy. "An ESPeriment with Cards." *Math Horizons* (February 2007), 10–12.

[Mulcahy 07_02b] Colm Mulcahy. "Quasi-Masked Forcing Kind of Magic Squares." *Card Colm*, http://www.maa.org/columns/colm/cardcolm200702.html, February 2007.

[Mulcahy 07_04] Colm Mulcahy. "Magic Circles of Eight." *Card Colm*, http://www.maa.org/columns/colm/cardcolm200704.html, April 2007.

[Mulcahy 07_06] Colm Mulcahy. "Gibonacci Bracelets." *Card Colm*, http://www.maa.org/columns/colm/cardcolm200707.html, June 2007.

[Mulcahy 07_08] Colm Mulcahy. "Sixy Alpha Omegas." *Card Colm*, http://www.maa.org/columns/colm/cardcolm200708.html, August 2007.

[Mulcahy 07_10] Colm Mulcahy. "A Magic Timepiece Influenced by Martin Gardner." *Card Colm*, http://www.maa.org/columns/colm/cardcolm200710.html, October 2007.

[Mulcahy 07_12] Colm Mulcahy. "Plurality Events, Standard Deviations and Skewed Perspectives." *Card Colm*, http://www.maa.org/columns/colm/cardcolm200710.html, December 2007.

[Mulcahy 08] Colm Mulcahy. "Low-Down Triple Dealing." In *A Lifetime of Puzzles*, edited by Erik D. Demaine, Martin L. Demaine, and Tom Rodgers, pp. 25–34. Wellesley, MA: A K Peters, 2008.

[Mulcahy 08_02] Colm Mulcahy. "Additional Certainties." *Card Colm*, http://www.maa.org/columns/colm/cardcolm200803.html, February 2008.

[Mulcahy 08_06] Colm Mulcahy. "Sum-Rich Circulants." *Card Colm*, http://www.maa.org/columns/colm/cardcolm200806.html, June 2008.

[Mulcahy 08_08] Colm Mulcahy. "(A) Pi Evolved Set—Harmonic Split Drill." *Card Colm*, http://www.maa.org/columns/colm/cardcolm200808.html, August 2008.

[Mulcahy 08_10] Colm Mulcahy. "Monge Shuffle Cliques." *Card Colm*, http://www.maa.org/columns/colm/cardcolm200810.html, October 2008.

[Mulcahy 08_12] Colm Mulcahy. "What's Black and Red and Red All Over?" *Card Colm*, http://www.maa.org/columns/colm/cardcolm200812.html, December 2008.

[Mulcahy 09] Colm Mulcahy. "Mathematical Idol." In *Mathematical Wizardry for a Gardner*, edited by Alan H. Schoen Ed. Pegg Jr and Tom Rodgers, pp. 155–164. Wellesley, MA: A K Peters, 2009.

[Mulcahy 09_06] Colm Mulcahy. "Two Summer Difference Certainties." *Card Colm*, http://www.maa.org/columns/colm/cardcolm200906.html, June 2009.

[Mulcahy 09_08] Colm Mulcahy. "The Bligreath Principle." *Card Colm*, http://www.maa.org/columns/colm/cardcolm200908.html, August 2009.

[Mulcahy 09_10] Colm Mulcahy. "Poker-Faced Over the Phone." *Card Colm*, http://www.maa.org/columns/colm/cardcolm200910.html, October 2009.

[Mulcahy 09_12] Colm Mulcahy. "The Boldgach Conjecture." *Card Colm*, http://www.maa.org/columns/colm/cardcolm200910.html, December 2009.

[Mulcahy 10_02] Colm Mulcahy. "Tighter Ascertainments: Matching Interest Rates." *Card Colm*, http://www.maa.org/columns/colm/cardcolm201002.html, February 2010.

[Mulcahy 10_04] Colm Mulcahy. "Mathematical Idol 2010." *Card Colm*, http://www.maa.org/columns/colm/cardcolm201004.html, April 2010.

[Mulcahy 10_06] Colm Mulcahy. "In His Own Words: Mathematical Card Tricks From Martin Gardner (1914-2010)." *Card Colm*, http://www.maa.org/columns/colm/cardcolm201006.html, June 2010.

[Mulcahy 10_08] Colm Mulcahy. "Also In His Own Words: More Mathemagical Games (and Tricks) With Cards From Martin Gardner." *Card Colm*, http://www.maa.org/columns/colm/cardcolm201008.html, August 2010.

[Mulcahy 10_11] Colm Mulcahy. "The Three Piles (This Isn't That Trick)." *Card Colm*, http://www.maa.org/columns/colm/cardcolm201011.html, November 2010.

[Mulcahy 10_12] Colm Mulcahy. "It's Red or Black and Blue All Over." *Card Colm*, http://www.maa.org/columns/colm/cardcolm201010.html, December 2010.

[Mulcahy 11] Colm Mulcahy. "Mathemagical Miracles." In *Expeditions in Mathematics*, edited by Tatiana Shubin, David F. Hayes, and Gerald L. Alexanderson, pp. 25–30. Weshington, DC: Mathematical Association of America, 2011.

[Mulcahy 11_02] Colm Mulcahy. "Low-Down Double Dealing with the Big Boys." *Card Colm*, http://www.maa.org/columns/colm/cardcolm201102.html, February 2011.

[Mulcahy 11_04] Colm Mulcahy. "Twisting the Knight Away (No Big Deal)." *Card Colm*, http://www.maa.org/columns/colm/cardcolm201104.html, April 2011.

[Mulcahy 11_10] Colm Mulcahy. "A Third Selection of Mathemagical Amusements with Cards in Martin Gardner's Own Words." *Card Colm*, http://cardcolm-maa.blogspot.com/2011/10/third-selection-of-mathemagical.html, October 2011.

[Mulcahy 11_12] Colm Mulcahy. "Magical Mathematics: Recurring Cycles of Ideas of Cycles." *Card Colm*, http://cardcolm-maa.blogspot.com/2011/12/magical-mathematics-recurring-cycles-of.html, December 2011.

[Mulcahy 12_02] Colm Mulcahy. "Amazon Arrays (Large Action)." *Card Colm*, http://cardcolm-maa.blogspot.com/2012/02/amazon-arrays-large-action.html, February 2012.

[Mulcahy 12_05] Colm Mulcahy. "Like Everybody Else, You Too Can Be Unique. Just Keep Shuffling." *The Aperiodical*, http://aperiodical.com/2012/05/ like-everybody-else-you-too-can-be-unique-just-keep-shuffling, May 2012.

[Mulcahy 12_06] Colm Mulcahy. "Something's Old, Something's True, Something's Borrowed, Something's New." *Card Colm*, http://www.maa.org/ columns/colm/cardcolm201206.html, June 2012.

[Mulcahy 12_08] Colm Mulcahy. "Gilbreeath Shuufling." *Card Colm*, http:// www.maa.org/columns/colm/cardcolm201208.html, August 2012.

[Mulcahy 12_12a] Colm Mulcahy. "Flipping Miracles (or Bar Bets to Amaze Your Friends)." *The Huffington Post*, http://www.huffingtonpost.com/ colm-mulcahy/flipping-miracles-or-bar-bets-to-amaze-your-friends_b_ 2248388.html, December, 2012.

[Mulcahy 12_12b] Colm Mulcahy. "Predictability Outranks Luck." *Card Colm*, http://www.maa.org/columns/colm/cardcolm201212.html, December 2012.

[Mulcahy 13_02a] Colm Mulcahy. "Flushed with Embarrassment." *Card Colm*, http://cardcolm-maa.blogspot.com/2013/02, February 2013.

[Mulcahy 13_02b] Colm Mulcahy. "In My Heart of Hearts: Valentine's Day Special." *The Huffington Post*, http://www.huffingtonpost.com/colm-mulcahy/ my-heart-of-hearts_b_2641875.html, February 2013.

[Mulcahy 13_04] Colm Mulcahy. "Never Forget a Face (Double-Dealing with a Difference)." *Card Colm*, http://cardcolm-maa.blogspot.com/2013/04, April 2013.

[Mulcahy 14] Colm Mulcahy. "Fitch Cheney's Five-Card Trick for Three Cards." *Math Horizons* (2014), to appear.

[Murray to appear] Stewart Murray. *Palindromic Magic Revealed—Theory Techniques Tricks*. Self-published: KR Professional Magic, to appear.

[Mutalik 09_12a] Pradeep Mutalik. "Monday Puzzle: Magic Phone Calls." *New York Times TierneyLab*, http://tierneylab.blogs.nytimes.com/2009/12/07/ monday-puzzle-magic-phone-calls, December 7, 2009.

[Mutalik 09_12b] Pradeep Mutalik. "The Wizard's Clock." *New York Times TierneyLab*, http://tierneylab.blogs.nytimes.com/2009/12/24/the-wizards -clock, December 24, 2009.

[Mutalik 10_05] Pradeep Mutalik. "Numberplay: Order in the Ranks." *New York Times Wordplay*, http://wordplay.blogs.nytimes.com/2010/05/10/ numberplay-order-in-the-ranks, May 10, 2010.

[Mutalik 10_06] Pradeep Mutalik. "Numberplay: The Playful Mr. Gardner." *New York Times Wordplay*, http://wordplay.blogs.nytimes.com/2010/06/ 07/numberplay-the-playful-mr-gardner, June 7, 2010.

[Perfect 12] Christian Perfect. "David's de Bruijn sequence card trick." http://aperiodical.com/2012/08/davids-de-bruijn-sequence-card-trick, August 2012.

[Pless 82] Vera Pless. *Introduction to the Theory of Error-Correcting Codes.* New York: John Wiley & Sons, 1982.

[Rosen 00] Kenneth H. Rosen. *Elementary Number Theory and its Applications, 4th Ed.* Upper Saddle River, NJ: Pearson, 2000.

[Rudin 76] Walter Rudin. *Priciples of Mathematical Analysis.* New York: McGraw-Hill, 1976.

[Scarne 50] John Scarne. *Scarne on Card Tricks.* New York: Crown, 1950.

[Simon 64] William Simon. *Mathematical Magic.* New York: Charles Scribner's & Sons, 1964.

[Sirén 08] Martii Sirén. "Matalan tason Kolmoisjako." In *Jokeri–Taikuutta taikureilley, 3*, pp. 24–25. Kouvola Finland: Markku Purho Ky, March 2008.

[Steinmeyer 93] Jim Steinmeyer. "The Nine Card Problem." *MAGIC* (May 1993), 56–57.

[Steinmeyer 02] Jim Steinmeyer. *Impuzzibilities: Strangely Self-Working Conjuring.* Hahne, 2002.

[Winkler 04] Peter Winkler. *Mathematical Puzzles: A Connoisseur's Collection.* Wellesley, MA: A K Peters, 2004.

Index